Archimedes

Volume 8

Archimedes

NEW STUDIES IN THE HISTORY AND PHILOSOPHY OF SCIENCE AND TECHNOLOGY

VOLUME 8

Archimedes has three fundamental goals; to further the integration of the histories of science and technology with one another: to investigate the technical, social and practical histories of specific developments in science and technology; and finally, where possible and desirable, to bring the histories of science and technology into closer contact with the philosophy of science. To these ends, each volume will have its own theme and title and will be planned by one or more members of the Advisory Board in consultation with the editor. Although the volumes have specific themes, the series itself will not be limited to one or even to a few particular areas. Its subjects include any of the sciences, ranging from biology through physics, all aspects of technology, broadly construed, as well as historically-engaged philosophy of science or technology. Taken as a whole, *Archimedes* will be of interest to historians, philosophers, and scientists, as well as to those in business and industry who seek to understand how science and industry have come to be so strongly linked.

Archimedes

Volume 8
New Studies in the History and Philosophy of Science and Technology

The Alfonsine Tables of Toledo

by

JOSÉ CHABÁS

University Pompeu Fabra, Barcelona, Spain

and

BERNARD R. GOLDSTEIN

University of Pittsburgh, U.S.A.

KLUWER ACADEMIC PUBLISHERS
DORDRECHT / BOSTON / LONDON

A C.I.P. Catalogue record for this book is available from the Library of Congress.

ISBN 1-4020-1572-0

Published by Kluwer Academic Publishers,
P.O. Box 17, 3300 AA Dordrecht, The Netherlands.

Sold and distributed in North, Central and South America
by Kluwer Academic Publishers,
101 Philip Drive, Norwell, MA 02061, U.S.A.

In all other countries, sold and distributed
by Kluwer Academic Publishers,
P.O. Box 322, 3300 AH Dordrecht, The Netherlands.

Printed on acid-free paper

Printed in the Netherlands.

To John D. North, a valued colleague and friend

CONTENTS

LIST OF FIGURES

PREFACE

This book is the result of a close collaboration between authors living far away from each other—even on different continents, Europe (Barcelona) and North America (Pittsburgh)—who have exchanged e-mail messages on an almost daily basis for more than a decade, with occasional transatlantic journeys for face-to-face encounters. The event which first led us to consider research on the Alfonsine Tables was a visit to the Royal Library of Belgium in Brussels to look at a copy of the *Tabulae Resolutae* edited by Johann Schöner (2nd ed., 1551), a well known version of the Alfonsine Tables. Although it was not part of our original reason for consulting this published text, we found in it a table for planetary velocities which had never been described in the secondary literature. We soon discovered that this table was preserved in many manuscripts and the explanation for its neglect seems to be that it was not included in the *editio princeps* of the Alfonsine Tables (1483). Nevertheless, we noticed that this table appears in subsequent editions of the "Alfonsine Tables" printed in the 16th century. We then clearly understood that behind the term Alfonsine Tables there was a much more complex reality than what was revealed between the covers of a single book. By now we have located copies of this table in Arabic and Hebrew manuscripts in addition to a large number of Latin manuscripts and printed editions. This discovery led us to wonder what else might have been ignored in the manuscript tradition surrounding the Alfonsine Tables, and that pursuit has kept us quite busy for a long time.

Our previous research prepared us for this project in many ways, for both of us have devoted much effort to uncovering and elucidating astronomical material produced in the Iberian peninsula in the Middle Ages in Arabic, Castilian, Catalan, Hebrew, Latin, and Portuguese. These traditions are extremely rich and varied in content, but our focus has been on astronomical tables whose analysis requires a certain amount of technical facility as well as familiarity with similar tabular material going back to antiquity and continuing into the 16th century.

The canons to the Alfonsine Tables of Toledo, a 13th-century Castilian text preserved in a unique manuscript in Madrid, is of interest to scholars of medieval Castile and medieval Castilian as well as to historians of astronomy. In fact, the patronage of science by Alfonso X, king of Castile and León (reigned: 1252–1284), known as the Learned (*el Sabio*), was unusual (although not entirely unprecedented) in his time, and his active sponsorship of various scientific disciplines played a significant role in his political strategies both within his realm and in external affairs. Further, the scientific literature written in Castilian or translated into it during his reign is the starting point of a specialized language for astronomy in Castilian. The impact of the Alfonsine Tables on the development of European astronomy from the 13th to the middle of the 16th century was enormous, and it is the reason why Alfonso X is best remembered. Since almost all astronomy written in Latin (other than elementary student texts) until the mid-16th century derives from these tables and associated material, it is not possible to provide a comprehensive study in a single volume. Hence, we have concentrated on the period up to the mid-14th century by which time a large corpus of texts had already been generated, and the displacement of the most widely consulted previous set of tables, the Toledan Tables, was well advanced.

An intriguing problem for historians is how to determine the impact of the Alfonsine Tables of Toledo in the absence of the original Castilian version, for only the canons are extant. Some have contended that the Alfonsine Tables which were diffused from Paris beginning in the 1320s had nothing to do with Castile. However, we argue that this position ignores too much of the evidence for continuity and the complexity of the historical data; indeed, we have taken this opportunity to explore the relationship between Castile and France in the relevant time period, from the 1270s to the 1320s as well as to examine the traces of Castilian astronomical traditions in the Parisian setting of the 1320s.

The reader will notice that we have not included a complete translation into English of these canons; in fact, only the introduction and chapter 27 have previously been translated. The text is written in a difficult style, and our view is that a proper translation is a job for a trained philologist who has the necessary linguistic as well as scientific background to do it justice. We have approached this text as historians of science, not as philologists, attempting to make both the text and its contents accessible to a wider public than has previously been the case. All chapters are the result of the collaboration of the two authors, each with his particular skills and background, bringing to the project his specific approach, and both with the goal of producing a serious and comprehensive scholarly work.

Many libraries in different countries have cooperated in assisting us to locate and consult manuscripts and printed editions. In particular, we would like to thank the Biblioteca Nacional and the Real Academia de la Historia, both in Madrid, for permission to reproduce pages from manuscripts in their collections. We are most grateful to a number of scholars: Alan C. Bowen, José Luis Mancha, Fritz S. Pedersen, Beatriz Porres, and Christopher Schabel for their suggestions concerning Latin texts; Paul Kunitzsch for a discussion of references to star lists; and Ruth Glasner for advice on Hebrew texts. But our principal intellectual debt is to John D. North who helped us in many ways over the years, and whose works have served as an inspiration and a model.

Barcelona, Pittsburgh
February 2003

CHAPTER 1

INTRODUCTION

This is a book on astronomy as it was practiced in the Middle Ages. At its core is an edition of a Castilian text, the canons to the Alfonsine Tables, composed in about 1272 under the patronage of Alfonso X, king of Castile and León (1221 [Toledo] – 1284 [Seville]; reigned: 1252–1284) by two scholars, Isaac ben Sid and Judah ben Moses ha-Cohen. This text, presented in chapter 2, survives in a unique manuscript and has been edited only once, almost a century and a half ago, by Rico Sinobas. As we shall demonstrate in chapter 2, this 19th-century edition does not meet the standards of modern scholarship in a number of ways and a new transcription is clearly required. The Castilian canons to the Alfonsine Tables belong to a long astronomical tradition that began in antiquity and flourished in the Islamic world, and it certainly reflects the intellectual context of Alfonso's time as well. We will clarify the terminology in it in chapter 3, and indicate its relationship with the antecedent tradition in chapter 4 which contains an astronomical commentary. In chapter 5 we provide information on the intellectual context at Alfonso's court, and in chapter 6 we describe the impact of the Castilian Alfonsine Tables on European astronomy, restricting our attention for the most part to the time up to the middle of the 14th century.

It is not surprising that the most widely used tables in the late Middle Ages originated in Castile because Christians in the 13th century had easiest access there to the Arabic scientific material that had reached its highest scientific level in Muslim Spain or al-Andalus in the 11th century. The historical circumstance was that Alfonso's father, Ferdinand III (reigned: 1217–1252) had conquered most of what remained of al-Andalus, including the major centers of Córdoba and Seville. In order to take advantage of this opportunity, Alfonso depended to a significant degree on Jewish intellectuals whose knowledge of Arabic allowed them to read scientific texts in that language and to translate astronomical works into Castilian. To be sure, Alfonso's entourage included many scholars, some of whom were Christian and some Jewish. But Jews were responsible for the bulk of the translations as well as for all the original scientific works produced at his court, where "original" refers to new compositions, including those that hardly contain any innovations (see chapter 5). Nevertheless, in their works we see no specific appeal to any Jewish traditions with the possible exception of the criteria for the visibility of the lunar crescent (see our comments on chapter 36 of the Castilian canons in chapter 4, below).

Scientific activity in 13th-century Castile may be characterized as follows.

1. Concentration on astronomy and related sciences.

At the royal court, works were produced in a wide range of domains, all related to astronomy: instruments, clocks, canons and tables, cosmography, star catalogue, and astrology. It therefore seems reasonable to assume that, in the field of astronomy broadly defined, there was a general plan to cover all its sub-disciplines which became more and more specific over time.

2. Royal patronage.

In this respect, Alfonso followed an established tradition that was practiced by other monarchs shortly before his reign whether in Castile, as was the case with his father King Ferdinand III, or elsewhere, as was the case with Frederick II of Sicily (1197–1250). Alfonso does not appear to have been a patron far removed from his collaborators, for he played an active role in some of the activities of the scholars in his service.

3. Collective effort.

Alfonso gathered around him a group of learned men who engaged in many collaborative efforts. Some of them worked for him during most of his reign, whereas others were only involved in a few projects, but they all mastered more than one language, either because they had come to Castile from foreign countries or had cultural backgrounds other than Christian (Jews, and Arabs whether recently converted to Christianity or not) and had varying levels of scientific knowledge.

4. Independence from the university or the church hierarchy.

The astronomers worked for a secular court led by the king and they had no known connection to such educational institutions as the recently founded University of Salamanca, established *ca.* 1228 by King Ferdinand III of Castile. This situation is rather different from that of other scientists working shortly after the time of Alfonso's astronomers, e.g., the Parisian astronomers in the early 14th century were, for the most part, associated with the University.

5. Science in the vernacular.

Castilian was adopted as the predominant language for science, making the vernacular a new vehicle for scientific communication. The Alfonsine collaborators who wrote on astronomy and related disciplines in Castilian, a language that had never been used previously for that purpose, exerted much effort to create new terminology and new ways to express scientific ideas and, in general, to build a new specialized language in Castilian.

Alfonso X and the scholars working for him appropriated the Islamic astronomical tradition of al-Andalus by means of their translations and original works (for the case of the appropriation of Greek science by the Muslims in the 9th century, see Sabra 1987). It was not merely an attempt to preserve antecedent knowledge, but to extend it and to make it part of the intellectual life of Castile. Nevertheless, it is fair to say that Alfonso was not completely successful, for the scientific terminology introduced by his astronomers was only partially accepted by later generations of scholars writing in Castilian; in fact, new terminology was introduced in the late 15th century and the beginning of the 16th century from which time a continuous tradition has developed (for examples of the differences between modern usage and 13th-century terminology, see chapter 3). In this context, it is important to distinguish translations that maintain a dependence on the source culture as distinct from original works which assert cultural independence, a process of assimilation or naturalization of knowledge in the "target" culture. This process implies an ability to manipulate the previous tradition in one's own terms without necessarily departing from it. The most significant original text of this kind produced at Alfonso's court was the Castilian Alfonsine Tables of which only the canons survive. And it was this achievement that provided King Alfonso with his most enduring claim to fame.

In the prologue to these canons Isaac ben Sid and Judah ben Moses ha-Cohen tell us that they corrected errors they found in the work of the foremost Muslim astronomer in al-Andalus, Azarquiel (d. *ca.* 1100). It is also claimed that the improvements were based on new observations made with carefully constructed instruments, but no details survive except for four eclipses that are reported in Isaac Israeli's *Yesod ʿOlam* (*ca.* 1310). The only other astronomer before the time of Alfonso mentioned in these canons is Ptolemy of Alexandria (2nd century) who, in his *Almagest*, established the basis for mathematical astronomy as practiced until the 17th century. While Ptolemy wrote in Greek, his works were translated into a variety of languages and then widely diffused in different cultural settings. Of particular interest are the translations of the *Almagest* into Arabic because of their subsequent impact in al-Andalus. Astronomical activity in the 8th and 9th centuries in Baghdad, in part under the patronage of the Abbasid caliphs, involved new observations, new parameters, new mathematical techniques, as well as translations from both Greek and Sanskrit. Characteristic of this period is the confrontation and harmonization of Greek and Indian traditions, and the assimilation of scientific practice into Islamic culture. With the Muslim conquest and settlement of the Iberian peninsula, this knowledge was brought from eastern Islamic lands and it was subsequently enriched by scholars in al-Andalus who extended the antecedent astronomical tradition.

In the canons by Isaac ben Sid and Judah ben Moses ha-Cohen we find many traces of this Andalusian tradition in which Azarquiel was the key figure. To be sure, some aspects of it that persisted both in al-Andalus and in the Maghrib are not fully understood because of the limited number of Arabic texts that survive and that have been studied, but the main outlines of it are well under control. Generally speaking, in medieval astronomy there was a twofold tradition: on the one hand, the Greek/Ptolemaic tradition, represented by Ptolemy's *Almagest* and *Handy Tables* and exemplified by al-Battānī and, on the other, the Hindu tradition exemplified by al-Khwārizmī. Both traditions had roots in the eastern Islamic world of the 9th century, and they were maintained in al-Andalus from the 10th century onward (see Ṣāʿid al-Andalusī, trans. Salem and Kumar 1991, pp. 53, 64; Goldstein 1967), although they were not always distinct (e.g., Ibn al-Kammād included two tables for lunar latitude, one with a Ptolemaic parameter and one with a Hindu parameter: Chabás and Goldstein 1994, pp. 20–22). The Andalusian tradition had previously affected the development of Latin astronomy due primarily to the influence of texts translated from Arabic into Latin in the 12th century in Toledo and elsewhere, including the tables of al-Khwārizmī as revised by Maslama al-Majrīṭī (d. 1007). Moreover, the Toledan Tables, a compendium of Arabic astronomical tables compiled in al-Andalus by Ṣāᶜid al-Andalusī and his circle (which included Azarquiel), were also translated into Latin in the 12th century and they were the most widely diffused tables before the Parisian version of the Alfonsine Tables. These Toledan Tables in Latin (of which the original Arabic does not survive) use the Muslim calendar, and, according to F. S. Pedersen (2002, pp. 759–760), most of these tables derive from an archetype in Arabic that probably dates from about 1110. The earliest witness to the Latin version comes from a text by Raymond of Marseille that is dated 1141, and various adaptations to the Christian calendar were compiled in the 12th and 13th centuries, notably: the Tables of Marseille, the Tables of Novara, and the Tables of Toulouse (Benjamin and Toomer 1971, p. 15; Poulle 1994; F. S. Pedersen 1998; and F. S. Pedersen 2002, pp. 760–761).

Thus, Alfonso's astronomers took advantage of the best astronomical traditions of their time. They also followed the traditional structure of Arabic sets of tables (included in handbooks called zijes: from the Arabic *zīj*, plural *zījāt*), as can be presumed from the canons explaining their use, to the point that the Alfonsine Tables of Toledo could properly be called the Alfonsine Zij. The contents of these tables, together with the chapters of the canons associated with each of them, are listed below:

Prologue (chapter 0)
Chronology (chapters 1–13, 54)
Trigonometry and the daily rotation (chapters 20, 37–47, 51, 53)
Mean motions (chapters 14–15, 19)
Equations (chapters 16–18)
Latitude (chapters 21–22)
Planetary visibility and retrogradation (chapters 23–25)
Planetary and lunar velocities (chapters 27)
Syzygies and eclipses (chapters 30–35)
Visibility of the lunar crescent (chapter 36)
Domification and astrology (chapters 26, 28,48, 52)

Equation of time (chapter 29)
Trepidation (chapters 49–50)

However, the canons to the Alfonsine Tables are not a mere rendering of either of the two main currents in medieval astronomy, and they differ in several respects from both of them. For example, in chapter 17 of these canons, the lunar equation is given by a rule that is found in Andalusian sources but not in the zijes of al-Battānī or of al-Khwārizmī. Similarly, chapter 27 describes the use of tables for planetary velocity that also belong to a specifically Andalusian tradition which has not been found in texts from the eastern Islamic world, and this chapter was paraphrased in Latin in a work dated 1322 by John of Lignères (who did not indicate his source). Following Azarquiel, an Andalusian tradition ascribed proper motions to the solar and planetary apogees; these proper motions are not mentioned in the canons to the Castilian Alfonsine Tables and they are not present in the Parisian Alfonsine Tables, but they do appear in the tables of John Vimond (Paris, *ca.* 1320): see chapter 6. Finally, models for the trepidation of the equinoxes are not found in either al-Khwārizmī's zij or of al-Battānī's zij, but they are the legacy of a text, *On the motion of the eighth sphere,* ascribed in the Middle Ages to Thābit Ibn Qurra (9th century), and tables based on Pseudo-Thābit's model already appear in the Toledan Tables (see Toomer 1968, p. 118). But the model for trepidation in the Parisian Alfonsine Tables, and the variant of it that has been identified in the tables of John Vimond, is not extant in any known text from the Iberian peninsula prior to 1300.

A version of the Alfonsine Tables in Latin produced in Paris in the 1320s was at the heart of the most important corpus of tables in the late Middle Ages in Europe. Indeed, other such tables that survived from an earlier epoch (e.g., the Toledan Tables) were largely displaced by this material. Thus the study of the Alfonsine corpus (as we shall call the totality of the astronomical works that ultimately derive from Alfonso's court) is almost coextensive with the study of late medieval astronomy in Latin (excluding elementary texts intended for use by university students). The Parisian Alfonsine Tables in one form or another continued to be printed and consulted by astronomers until the end of the 16th century and, in some cases, even later. They were gradually replaced by Erasmus Reinhold's Prutenic Tables, based on Copernican models, that were first printed in 1551. The duration and dominance of the Alfonsine Tables in Latin Europe may be contrasted with the diversity of comparable material in the medieval Islamic World where a vast number of astronomical tables were produced (see King *et al.* 2001). Moreover, in Aragon, for the most part, astronomical activity in the late Middle Ages did not depend on the Alfonsine corpus, and this is largely true even for Castile until the middle of the 15th century.

An issue that has puzzled historians is the way that the Alfonsine Tables whose origin was in Toledo came to be known in Paris, the main point of diffusion for the rest of Europe. We think it relevant to consider the complex links between the kings of Castile and France throughout this period (see Daumet 1913). Alfonso X had an active international policy, especially with England, France, Italy, and the Roman Curia because of his pretensions to become the Holy Roman Emperor (O'Callaghan 1990, pp. 25–26). However, the relationship with France was privileged, although with varying degrees of friendship. Both the Castilian and the French royal dynasties had strong family ties that began long before this period as the result of various marriage alliances, and there was a continuous interchange, mainly political, between Castile and France through frequent embassies visiting their respective courts. However, two problems led to conflict between Alfonso and the king of France, Philip III (reigned: 1270–1285), son of Louis IX, known as Saint Louis: the dispute over Navarre, and the succession to the throne of Castile (Ballesteros 1984, pp. 795ff.). The first-born son of Alfonso X and heir of the Castilian Crown, Fernando de la Cerda, died in 1275 and left two young sons from his marriage with Blanche of France, sister of Philip III. A bitter dynastic dispute developed between the followers of Alfonso's son, Sancho, and those of Alfonso's grandsons. The king of France was in favor of his nephew, Alfonso's eldest grandson. There were various attempts to overcome this situation. Of special interest to us is the embassy sent to Paris in 1280 in order to arrange a meeting between kings Alfonso X and Philip III to settle their differences (Ballesteros 1984, p. 924). Among the Castilian personalities who traveled to Paris was Petrus de Regio, usually called "protonotarius aulae imperialis", one of the scientific collaborators of King Alfonso and translator of the *Libro conplido en los iudizios de las estrellas* (see chapter 5). This exchange of embassies between the Castilian and French courts continued with great pomp and the giving of presents after Alfonso's death, under the reigns of Sancho IV of Castile (reigned: 1284–1295) and Philip IV of France (reigned: 1285–1314). In July 1288 their representatives drafted the Treaty of Lyon, according to which Blanche was given back her dowry and the properties acquired by her husband. Shortly thereafter, in April 1290, the two kings arrived in Bayonne and ratified the accord of Lyon, adding one detail, namely the number of horses and mules that the king of France could buy in

Castile (Daumet 1913, p. 109). This permanent channel of communication between the courts may have been a way for the Castilian Alfonsine Tables to reach Paris.

As a matter of fact, Ballesteros points out that, according to William of Nangis, chronicler of Saint Louis (d. 1270) and author of a chronicle from 1113 to 1300, knowledge of the Alfonsine Tables was considered to be "one of three great advances in the [intellectual] development of France" (Ballesteros 1984, p. 247). Ballesteros does not give a precise reference to William of Nangis (the date of his death is disputed: 1300 or 1303) and only mentions the chronicle of John of Joinville (d. 1317), that depended upon William of Nangis. We have not been able to locate the exact reference in the text by William of Nangis (Géraud 1843) but, if this story is to be believed, we have an official chronicler at the French court who, prior to 1300, was aware of the Alfonsine Tables and their importance, well before any intervention by the Parisian astronomers.

The scholarship on the astronomy produced by Alfonso's collaborators and its subsequent legacy has been growing apace in the last decades leading to the proliferation of concepts and terms in this field. In particular, we find that the term "Alfonsine Tables", *tout court*, has been given different meanings, thus creating some confusion, and that, by itself, it no longer represents a unique and comprehensible concept. This is the reason why, as has probably already been noticed, we have distinguished between "Castilian Alfonsine Tables" (or "Alfonsine Tables of Toledo"), and "Latin Alfonsine Tables" (or "Parisian Alfonsine Tables"). We also restrict these terms to the specific texts for which they are appropriate, in contrast to the more general concept of the Alfonsine corpus. We have tried to maintain this precise terminology consistently throughout the book, only using "Alfonsine Tables" when the context precludes any confusion.

CHAPTER 2

TEXT

2.1 Need for a new edition

The text we present here has only been edited once, about 140 years ago, and that edition is deficient in a number of ways.

Between 1863 and 1867 Manuel Rico Sinobas edited, under the general title of *Libros del Saber de Astronomía*, all the works related to astronomy produced by the astronomers under the patronage of Alfonso X. Volume 4 contains the book that Rico called *Las Taulas del Rey Don Alfonso*, but herein we will refer to it as the *Libro de las tablas alfonsies*, following the designation given in the text itself. This book consisted of a set of astronomical tables computed for the city of Toledo and a text explaining their use in 54 chapters (or, canons). Unfortunately, only the text has been preserved in a unique manuscript in Castilian: Madrid, Biblioteca Nacional, MS 3306. On the original tables, only conjectures can be offered based on the numerous indications found in the text.

When Rico published the text of this book in 1866, he added some tables after the Castilian canons under the title *Fragmentos numéricos de las taulas alfonsíes*. It has long been known that these tables have nothing to do with the Alfonsine canons; in fact, they are a Portuguese version of a perpetual almanac beginning in 1307. The worst part of this story is that, according to Millás (1943–1950, p. 407), "Rico Sinobas did not act at all in good faith" (*Rico Sinobas no obró con toda la buena fe deseada*), for Rico translated the headings of the tables he reproduced from Portuguese into some sort of medieval Castilian without indicating what he had done.

In Rico's edition, the Alfonsine text is preceded by a table of contents listing the titles of the chapters. It is readily seen that these titles do not coincide with those found at the beginning of each chapter. The explanation is straightforward: no such table of contents exists in the manuscript. In fact, Rico did not transcribe the text that is preserved in a unique manuscript; rather, he presented a text in which he introduced what he took to be more archaic word forms than those that appear in the manuscript. Here follow some examples of systematic changes introduced by Rico in his edition.

MS 3306	Rico's edition
angulo	angol
arco	archo
capitulo	capítolo
cienҫia	sciencia
como	cuemo
haser	fazer
hasta	fata
hora	ora
menester	mester
no	non
ser	seer
si	sipse
tabla	taula
y	e, et

Rico also introduced arbitrary changes in his edition, again without any indication of what he had done. (In our transcription, all sentences are indicated by the chapter number and, within it, the sentence number, separated by a colon.) So, Rico has "exiemplo" instead of "verbigraçia" (8:3), "ouramos" instead of "labramos" (23:15), "et estonce será ascondido" instead of "pues es abscondida" (25:15), and "cuemo sobredicho es" instead of "como se adelanto" (27:17).

The intervention of the transcriber is not only present in some words, but also in phraseology and in presentation. The following examples show the considerable differences between our transcription and that given by Rico as regards words, numbers, punctuation, etc.

Our transcription (0:8):	Rico's edition (vol. 4, p. 111):
(...) e los preçiava. mandoles haser los ynstrumentos que dixo Ptholomeo en su libro del Almagesto. sigud son las armillas y otros ynstrumentos.	(...) et los preciaba. mandó fazer los estrumentes que dixo Ptolomeo en su libro dell Almaíeste. segund son las armillas. et otros estrumentes.

Our transcription (1:20–21):	Rico's edition (vol. 4, p. 120):
E la longura desta çibdad sobredicha del çerco oçidental del orisonte de Arin. donde se parezçen ambos polos es veinte y ocho grados. E del cerco de orizonte deste lugar sobredicho de Arin es çiento y çinquenta y dos grados.	Et la longura desta cibdat sobredicha del çerco occidental dell orizon de (*aryn*). donde aparescen amos polos. es XXVIII grados. Et del çerco dell orizon deste logar sobredicho de (*aryn*) es C et LII.

These examples speak for themselves. But there is more, for we have found some 'slips' which are relevant for understanding the text, e.g., "(...) sacando la maior de la menor" (Rico, vol. 4, p. 132), instead of "(...) sacando la menor de la mayor" (14:4); and the reading "Mares" (Mars), instead of "Mercurio" (24:2). There are also cases where an entire line has been omitted, e.g., in (1:15) the underlined phrase is missing in Rico's edition: "Pues el comienço destas tablas es deste dia sobredicho et la fin de cada dia es en su mesmo dia e assi son todos los dias e lo que se sigue de los meses y de los años." In other instances, Rico has added a line not in the manuscript, as occurs in chapter 2 between sentences 14 and 15, where Rico added: "El noueno mes es september. et es de XXX dias." Nothing like this is found in the manuscript.

In his edition, Rico added a general title: "El libro astronómico dicho de las Tablas Alfonsíes"; a subtitle: "Prólogo" (vol. 4, p. 111); a table of contents (vol. 4, pp. 112–118); and a colophon: "Aquí concluye el libro et se comienzan las taulas" (vol. 4, p. 183); none of these features appears in the manuscript.

After this list of deficiencies, it is easy to agree with Millás's assessment of Rico's skills in transcription (Millás [1956]/1960, p. 223): "(...) Mr. Rico Sinobas was quite in the dark concerning paleography" (*el señor Rico Sinobas estaba bastante ayuno en paleografía*), as well as the harsh criticism made by Gerold Hilty (1954, p. lvi) on the work by Rico concerning his transcription of another Alfonsine book, the *Libro conplido en los iudizios de las estrellas*:

> Rico y Sinobas incluye el prólogo en el tomo III de su edición (*Los Libros del Saber de Astronomía*). Afortunadamente no publicó toda la obra. (...). La justificación de tan severo juicio nos la suministra el cotejo de la transcripción de Rico y Sinobas con el manuscrito: (...) En poco más de una columna –y el manuscrito tiene más de 900– la transcripción contiene 35 inexactitudes. (...) Parece mentira que Rico y Sinobas se haya servido del mismo manuscrito que nosotros. Pero menciona *expressis verbis* el códice de la Biblioteca Nacional, y todos los errores se explican por descuido o mala lectura.

However, despite the unreliability of Rico's transcription, we are grateful to him for having published a version of the manuscript at a time when it was in much better condition that it is today. Moreover, as can be seen from the microfilm made in the 1960s, the manuscript has suffered considerable damage in the last decades. This damage to a text written on very thin paper makes it especially difficult to read some passages as is the case for folios 68 and 69 where the black ink has eaten away the paper, leaving holes where once there was text.

2.2 The manuscript

Madrid, Biblioteca Nacional, MS 3306 (*olim* L.97) is a compendium of several treatises including the unique extant copy of the *Libro de las tablas alfonsies* (ff. 34v–73v). Because of its importance for this study, there follows a description of its contents.

Of greatest interest is the first item of the manuscript (ff. 1r–34v), for it allows us to date the copy of the *Libro de las tablas alfonsies*. This treatise deals with standard subjects of Arabic astrology, whence the general title "Astrología de los árabes" stamped on the spine of the volume. It is divided into 20 chapters (note that they are numbered 1 to 19, with two chapters numbered "fourth"), but the beginning of the first chapter is missing, for the first page have been ripped out, and current folio 1r contains the end of chapter 1 (20 lines), chapter 2, and the beginning of chapter 3. The following folios bear a double numbering. A striking characteristic of this treatise is that it is partly in Castilian and partly in Latin: although most of the chapters are in Castilian, some are in Latin (chapter 3, despite its Castilian title; the first chapter numbered 4; and chapters 14–15). There is even a chapter (13) whose first part is in Latin and the rest in Castilian; the reverse happens in chapter 18. Note also that some titles are given in Castilian, and others in Latin:

Cap°.2 para hallar el almuten del nazçimiento
Cap°.3 otra manera para lo mesmo
Capi°.4 modus inuentionis hilec
Capi°.4 alia opinio de hilec
Capi°.v de alcocoden
Capi°.vi de la complexion del hombre
Capi°.7 de phisonomia façiei
Capi°. 8 dela delgadez o gordura del cuerpo umano
Capi°.9 de magisterio hominis
Capi°.x de honore hominis
Capit°.11 de infirmitatibus
Capi°.12 de annuis [?] profectionibus signorum
Cap°.13 ad iudicandum de infirmitatibus hominem
Capi°.14 de domibus
Capi°.15 de naturali hominum amititia et inimititia
Capi°.16 de aquellos que a las veses son señores a las veses siervos ymperantes y obedientes
Capi°.17 de las significaçiones delos señores de la genitura
Capi°.18 significaçiones delas 12 casas del çielo
Capi°.19 diferençias de los signos

After the last chapter there are notes on the *gradus fortunati*, the *gradus putealis*, the *gradus cazemedus*, the *gradus masculinei et feminei*, and the *gradus tenebrosi et luminosi*. This information is the same as that found in tabular form in standard astrological treatises of Arabic origin such as ᶜAbd al-ᶜAzīz al-Qabīṣī's *Liber introductorius* (see *Alchabitius cum commento, nouiter impresso*, Venice, 1512, ff. a5v–a6r).

The treatise ends with a short text in Castilian whose title is: "Regla para hallar verdaderamente y preçissa verificado el grado aspectus de qualquier naçimiento umano" (f. 33r), and it ends: "hora y media antes de mediodia. Laudatus Christus" (f. 34v).

Several modern scholars (e. g., Zarco, and Cárdenas) have claimed that this treatise is a translation of Ptolemy's *Tetrabiblos*, but a comparison of the two texts shows that they are mistaken.

Ptolemy is frequently cited in the text (ff. 3r, 4r, 6r, 13v), and an astrological treatise attributed to him, the *Centiloquium* (or, *Hundred Sentences*), is explicitly mentioned: in particular, Sentences 52 (f. 5v), 64 (f. 8v), 47, 48, and 49 (f. 13v), and 51 (f. 33v). Book II of Aristotle's *De generatione et corruptione* is also cited (f. 12r), and on the same folio we find a reference to "el almanak o tablas alfonsis". On f. 11r there is a reference to the well known 15th-century astronomer, Regiomontanus, and to a comment by him on Ptolemy's *Tetrabiblos*, here called the *Quadripartitum*. This gives one indication of the date when the treatise was composed, but there are others as well. On f. 8r we read: "en aqueste nuestro tiempo año de 1515", and ff. 8v and 34r mention several dates in the 15th and 16th centuries, leaving no doubt that the treatise was written in the 16th century. Since the *Libro de las tablas alfonsies* is written in the same hand as this astrological treatise, we conclude that the unique text of the canons of the Castilian Alfonsine Tables was copied in the early 16th century from an older manuscript.

This conclusion is corroborated by a new piece of evidence: Madrid, Real Academia de la Historia, MS 9/5707, is a 16th-century manuscript containing the *Libro de las estrellas de la ochaua espera* (see chapter 5, below); it is in two different hands (ff. 1r–7r, 8v; and ff. 9r–105r) of which the first hand corresponds exactly to that of Madrid, Biblioteca Nacional, MS 3306 (see Fig. 2.1 and Fig. 2.2). Unfortunately, we do not know the name of the copyist of Madrid, Real Academia de la Historia, MS 9/5707 but, on f. 95v, there is an owner's mark: "Soy de Don Francisco Tribiño y Calderón", and on f. 103v we read: "De la mano y pluma de D. Francisco Tribiño Calderón". The handwriting of both notes differs from those of the text.

In addition to the astrological treatise mentioned above and the *Libro de las tablas alfonsíes*, Madrid, Biblioteca Nacional, MS 3306, contains three other items. First, there is a Castilian translation of the 1483 edition of John of Saxony's canons to the Alfonsine Tables (ff. 74r–87v) by an otherwise unknown "Bachiller Francisco de Morales, clérigo presbítero" (Martínez Gázquez 1989). Next comes an incomplete treatise on the astrolabe (ff. 88r–94v); and, finally, a copy of the *Libros del Saber* (ff. 98r–302v):

ff. 98r–117r: *Libro de la facion del espera (alcora)*
ff. 118r–150r: *Del astrolabio redondo*
ff. 150v–171r: *Del astrolabio llano*
ff. 171v–204r: *De la lamina universal*
ff. 204v–235v: *Libro de la açafea*

ff. 236r–239v: *Libro del atacir*
ff. 240r–246v: *Libro del quadrante*
ff. 247r–253r: *Libro de las armellas*
ff. 253v–268v: *Libro de las laminas de los siete planetas*
ff. 269r–275r: *Libro de la piedra de la sombra*
ff. 275v–289r: *Libro del relogio de agua*
ff. 289v–293v: *Libro del relogio de argen biuo*
ff. 294r–299r: *Libro del relogio de la candela*
ff. 299v–302v: *Libro en que se fabla de como deven hazer el palacio de las horas.*

Previous scholars did not notice the evidence that has permitted us to date the manuscript to the early 16th century. According to the *Inventario General de Manuscritos de la Biblioteca Nacional* (1984, 10:63–64), the manuscript was written in the 15th century, except for ff. 74r–87v (16th century). Craddock (1974, p. 374, n. 26) considered the manuscript to have been written at the end of 14th century or the beginning of 15th century and Cárdenas (1974, p. liii) added that ff. 98r–302v are in a late 15th-century hand.

2.3 The text

The *Libro de las tablas alfonsies* is not a translation from Arabic as has been suggested by Rosen (1976, p. 163); rather, it is an original treatise that drew on the previous astronomical literature. The two authors, Isaac ben Sid and Judah ben Moses ha-Cohen, were the most active collaborators of Alfonso X. As we shall see in chapter 5, Judah is frequently mentioned in the Alfonsine texts as a translator and a reviser of astronomical treatises, whereas Isaac, probably much younger than Judah, was the most prolific and competent astronomer in the service of the king.

The text is divided into a prologue and 54 chapters, with almost 1,000 sentences or short paragraphs, according to our editorial division. The chapters vary considerably in length; some are very brief, such as chapter 19 on the lunar node, and chapter 42 on a trigonometrical problem (both of which are only 2 sentences long), while others are very long, such as chapter 32 on parallax (with 74 sentences), and chapter 27 on the unequal motion of the planets (with 60 sentences). However, the length of a chapter is not necessarily proportional to its difficulty: for instance, chapter 37, on the construction of a sine table, is long and descriptive (59 sentences), but offers nothing profound or unusual, whereas chapter 22, on planetary

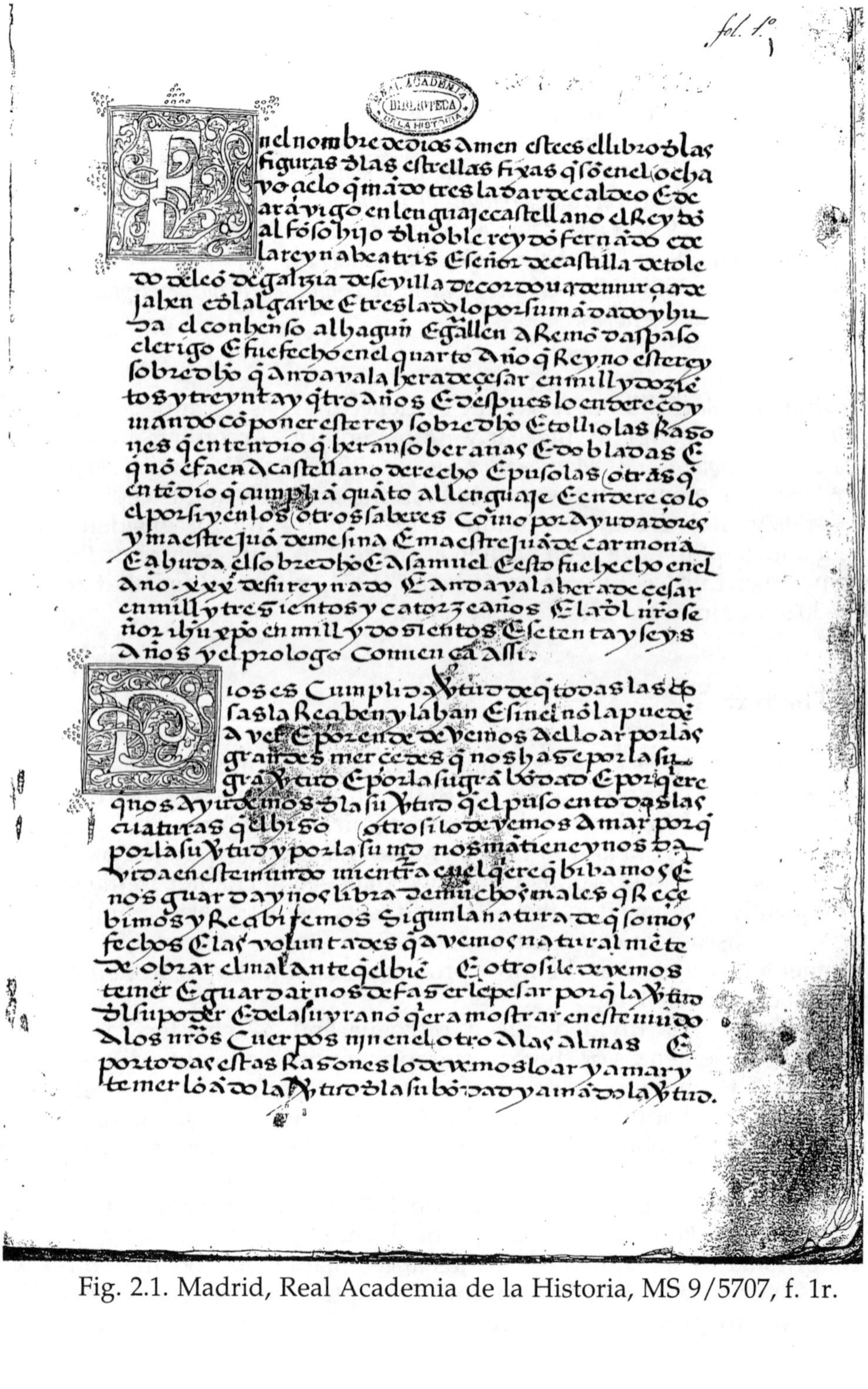

Fig. 2.1. Madrid, Real Academia de la Historia, MS 9/5707, f. 1r.

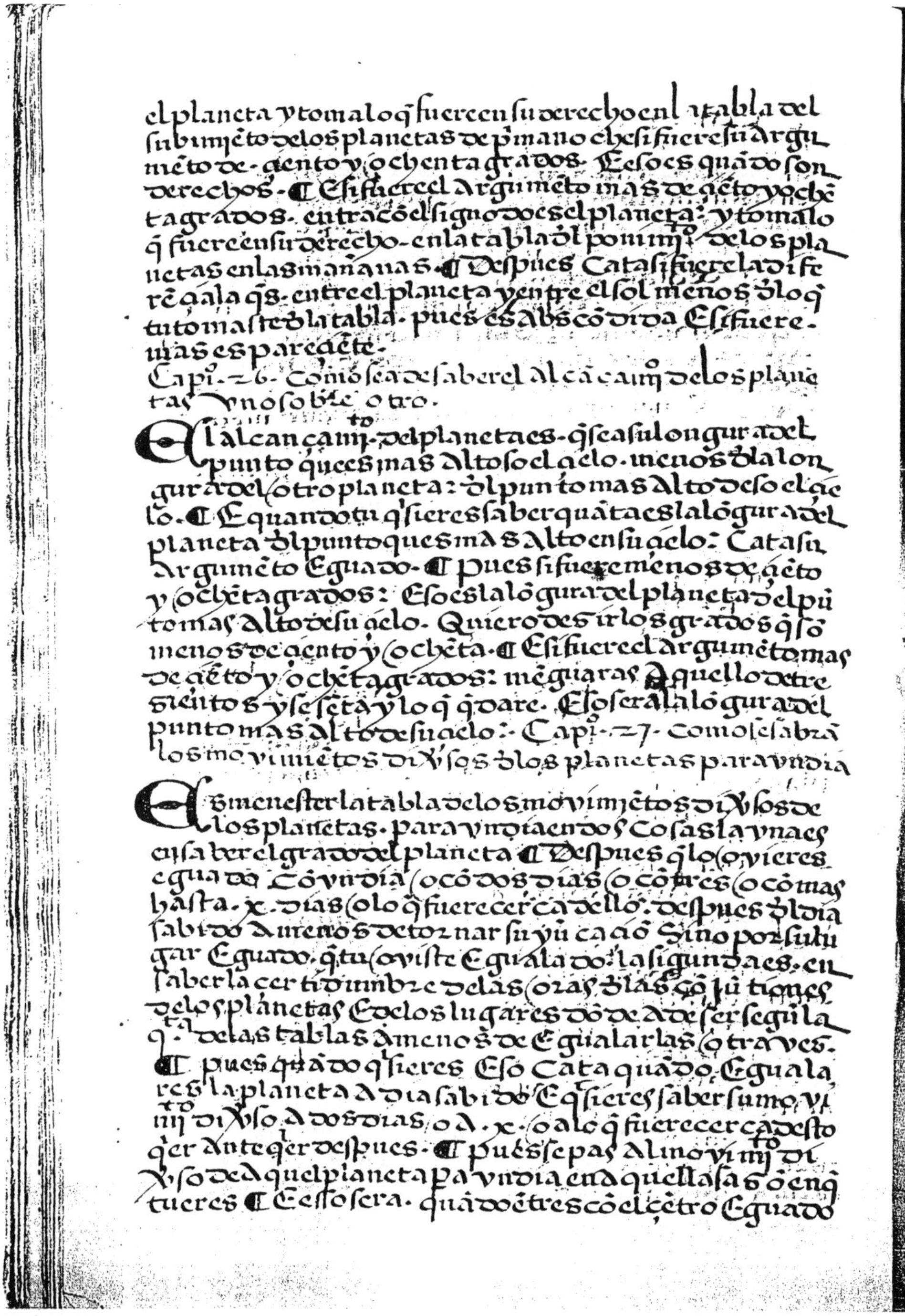

Fig. 2.2. Madrid, Biblioteca Nacional, MS 3306, f. 49v.

- Upper case is used for the initial letter of personal names.

- We have not retained the punctuation of the manuscript in the case of periods and colons which do not indicate a pause. Normally the text uses a period at the end of an abbreviation, and sometimes at the end of a line, or even to fill in a long blank space between two words. We have taken the "section markers" (¶), called *calderones* in Castilian, in the text as the graphical signs for divisions within a chapter, indicating either the beginning of a sentence or of a short paragraph, and we have numbered the sentences accordingly. This is not always the best solution from the point of view of understanding the sense of the text (as can be seen, for example, in 27:10–11), but in this we follow the decisions taken by the copyist of the manuscript.

- All editorial additions are enclosed within square brackets [ ].

2.5 A transcription of the *Libro de las tablas alfonsies* (Madrid, Biblioteca Nacional, MS 3306, ff. 34v–73v)

[f. 34v] [1] Dixo Yehuda fi de Mose fi de Mosca e Rabiçag Aben Cayut:

[2] Porque la ciençia de la astrologia es cosa que no se puede averiguar sino por rectificamientos e los rectificamientos que tienen los sabios que cumplen en esta cosa no los puede complir un hombre porque no se puede complir en vida de un hombre. mas quando se cumple cumplese por obra de muchos hombres obrando uno em pos de otro en luengos tiempos.
[3] E esto es porque en los movimientos de los çielos ay algunos movimientos que son tardios de manera que no cumplen una çircunferençia sino en millares de años.
[4] E por esto conviene de seguir los rectificamientos que en siguiendolos paresçeran y cosas pareçidas en una sazon que no eran pareçidas en otra sason.
[5] E nos agora en esta nuestra sason ques en la primera desena del quarto çentenario del sigundo millar de la hera del Çesar.
[6] E a passado del ratificar de Asarquiel aca quanto dosientos años e paresçio en algunas de las posturas que el puso diversidades

magnifiestas y paresçidas a los sentidos de manera que no puede alli caber alguna escusa.
[7] En esta sason paresçio el reynado fortunado et ayudado de Dios el reyno del muy alto y muy noble señor Rey don Alonso que Dios mantenga.
[8] E porque amava los saberes e los preçiava. mandoles haser los ynstrumentos que dixo Ptholomeo en su libro del Almagesto. sigud son las armillas y otros ynstrumentos.
[9] E mandonos retificar en la çibdad de Toledo ques [f. 35r] una de las çibdades prinçipales de España. guardela Dios.
[10] En ella fue el retificar de Asarquiel. Esto mando por endereçar y corregir las diversidades y desacordanças que paresçieron en algunos lugares de algunos de los planetas et en otros movimientos.
[11] E nos obedeçimos su mandado que de[be] ser obedesçido e rehesimos los ynstrumentos lo mejor que se pudo haser e trabajamos en rectificar una sason e siguimos en rectificar el sol quanto un año complido.
[12] E antes desto y despues rectificamoslo todavia cuanto entrava en las ygualdades y en los dos tropicos e en los otros quartos del çielo. que son el medio de tauro y descurpion. de leon y aquario.
[13] E rectificamos otrosi algunas conjunçiones de los planetas quando se allegaua una a otra e sus conjunçiones quando se llegauan con algunas de las estrellas fixas.
[14] E rectificamos munchos eclipsis de los solares y de los lunares.
[15] E retificamos otros retificamientos en que eramos dudosos e retornamoslos muchas veses por quitar la duda e no dexamos de buscar ninguna cosa ni de ynquerirla. hasta que nos paresçio enmendar lo que era rason de enmendar.
[16] E todo exsaminado dexamos por averiguado lo que es çierto o cerca de çierto. et fesimos estas tablas sobre rayzes que son sacadas de aquellos retificamientos.
[17] E ayuntamos en ellas de los capitulos que nos paresçio ser menester en sta obra.
[18] E posimos nombre a este libro el libro de las tablas alfonsies porque fue fecho y copilado por su mandado e repartimoslo en 54 capitulos los quales son estos que se siguen.

[1] Capitulo primero: Como se a de saber la hera sobre que son puestas estas tablas y su prinçipio.

[2] Todas las heras que son magnifiestas en las naçiones e usadas antiguas y nueuas son començadas de algun acaezçimiento que aconteçio e preçiose la jente de aquella naçion del tal a[f.

35v]conteçimiento. y llamole comienço de sus años y hasese quenta del tal acaesçimiento porque dure la nombradia de aquel acaezçimiento y no se olvide por luengos tiempos.

[3] E por esta forma y manera son puestas todas las heras desde que començo el mundo hasta agora.

[4] Los griegos començaron sus heras desde el Rey Alexandro. tomando el [año] quel nasçio porque fue rey muy poderoso y de quien ellos mucho se preçiaron y ansi su hera se llama la hera de Alexandro.

[5] Otrosi los romanos tomaron el año en que Çesar a reynar començo. y del nombraron su hera porque fue rey con quien se honraron mucho. y esta es la hera que se usa agora y es nombrada la hera de Çesar.

[6] Tambien los arabes tomaron por hera el año en que dixo Mahomat que hera propheta. y pusieronlo por prinçipio de su hera. porque segun ellos creen aquel fue el dador de su ley e por eso esta hera es nombrada la hera araviga.

[7] Usan otrosi los moros agora nueuamente de hera persiana ansi llamada por la victoria que ovieron del rey de Persia. y començaron a usar della en el año que los destruyeron e su reyno sujusgaron.

[8] Llamose el postrimero rey de Persia Yezdager y en memoria de tan noble hasaña et victoria los moros usan desta hera y esta es la hera persiana nueua que se usa en este nuestro tiempo desta manera. y por estas cabsas se nombran ansi estas heras.

[9] E nos vemos que en este nuestro tiempo acaezçio notable acaezçimiento e honrado e de tanta estima como todos los antepassados.

[10] Y este es el reynado del señor Rey don Alonso que sobrepujo en saber. seso y entendimiento. ley. bondad. piedad e noblesa a todos los reyes sabios.

[11] E por esto tovimos por bien de poner por comienço de hera el año en que començo a reynar este noble rey por cabsa que se use y manifieste esta hera ansi como se usaron y manifestaron las otras heras antes della porque dure et quede la nombradia deste noble rey para siempre.

[12] E posimos el co[f. 36r]mencamiento deste año sobredicho por comienço desta hera. e posimosle nombre la hera alfonsi.

[13] Pues el que quisiere obrar con estas tablas conviene que primero sepa esta hera pues que las rayses de los medios cursos y de los centros y de los argumentos e de los otros movimientos y las otras obras que son puestas en estas tablas son todas puestas sobre la rais desta hera.

[14] Esta raiz es en el medio del dia del domingo. el ques en ante del dia del lunes en que entro henero en el año primero de los años de la

hera destas tablas ques el año que reyno el rey ayudado de Dios. don Alonso a quien Dios mantenga.
[15] Pues el comienço destas tablas es deste dia sobredicho et la fin de cada dia es en su mesmo dia e assi son todos los dias e lo que se sigue de los meses y de los años.
[16] E este año en que se hisieron estas tablas fue año de mill y dozientos y noventa de la hera de Çesar e acaesçio este dia sobre que son puestas las rayses en estas tablas.
[17] E es el dia en que fue ante del primero dia de henero el sobredicho con el dia dies y seis del mes de xahuel. el que fue en año de seisçientos y quarenta y nueue de la hera de los alarabes e es el dia desiocheno del mes de asfidacme e es el mes doseno de los persianos meses el que fue en año de seisçientos y veinte de la hera de Yanzdayarr.
[18] E los años y los meses que son puestos en estas tablas son años y meses romanos como adelante diremos si Dios quisiere.
[19] E todas estas rayses sobre que se fundan estas tablas e todas sus obras son hechas y compuestas al mediodia de la çibdad de Toledo. ques la çibdad en que fue este notable acaezçimiento de la naçençia deste señor.
[20] E la longura desta çibdad sobredicha del çerco oçidental del orisonte de Arin. donde se parezçen ambos polos es veinte y ocho grados.
[21] E del cerco de orizonte deste lugar sobredicho de Arin es çiento y çinquenta y dos grados.
[22] E la longura del çerco del sol [de] mediodia desta çibdad del cerco de mediodia del lugar susodicho ques en Arin es contra occidente en sesenta y dos grados.
[23] E su latitud de la linea equinoçial contra la parte de septentrion es treinta y nueve grados y çinquenta y quatro minutos.
[f. 36v] [24] E estos son los terminos de la çibdad de Toledo sobre quien son puestas las rayses et las obras destas tablas.

[1] Capitulo 2º. De los años romanos y de sus meses sobre que son puestas estas tablas.

[2] Los romanos y las otras gentes que no hazen sus años sigun la vista de la luna. mas hasen sus años solares. tienen el comienço del dia contra ellos quando se parezçe el sol en el orizonte oriental. y por eso se adelanta el dia sobre la noche todiavia[1] contra ellos. E la

[1] *Sic,* instead of "todavia".

quenta de los dias del año es contra ellos de dos maneras. la una es segun hasen los persianos de trezientos y sesenta y çinco dias continuos sin fraçion o repartimiento ninguno.
[3] E la sigunda es segun hasen los romanos de trezientos y sesenta y çinco dias et un quarto de dia e quando se les ayunta destos quartos un dia hasen visiesto aquel año y añaden en el un dia e sera estonçes el año de trezientos y sesenta y seys dias.
[4] E son desacordados en el mes do an de añedir el visiesto porque ay algunos dellos que lo añaden en el mes de diziembre et otros ay que lo añaden en el mes de hebrero. E ansi lo hasen agora todos los romanos en este nuestro tiempo e ansi lo pusimos nos en estas nuestras tablas.
[5] E otrosi discrepan entre si en el mes primero del año. porque unos dizen ques el mes de otubre otros ques el mes de henero. como son los romanos destos nuestros tiempos.
[6] E segun esto hesimos nos estas nuestras tablas y el hordenamiento destos meses et poniendo quantos dias tiene cada uno en la manera siguiente.
[7] El primero mes es henero y tiene treynta y un dia.
[8] El segundo mes es hebrero el qual en el año simple es de veinte y ocho dias y en el año de bisiesto de XXIX.
[9] El tercer mes es março y es de treinta y un dias.
[10] El quarto mes es abril y trae XXX dias.
[11] El quinto mes es mayo y tiene XXXI dias.
[12] El sexto es junio y es de XXX dias.
[13] El septimo es jullio y es de XXXI dias.
[f. 37r] [14] El octauo mes es agosto y trae XXXI dias.[2]
[15] El X° mes es otubre y es de treynta y un dia.
[16] El onseno es noviembre y trae treynta dias.
[17] El doseno mes es diziembre y tiene treynta y uno dias.
[18] Pues estos son los años romanos sobre que son puestas estas tablas y este es el hordenamiento de sus meses.

[1] Capitulo tercero. Como se conosçeran los años de bisiestos segund la hera alfonsi o de Çesar.

[2] Quando quisieres saber quando ay bissiesto o no segun la hera alfonsi guardaras la regla siguiente.
[3] Mira los años que passaron de la hera alfonsi con el año que tu quieres saber y parte los años de la hera alfonsi por quatros tantas

---

[2] The sentence for September is missing.

quantas veses pudieres y si todos fueren parejos estonçes avra visiesto mas si quedaren en nones no avra visiesto.
[4] E si quisieres saber esto segun la hera de Çesar mengua della todavia dos y contaras lo que quedare segun que fue dicho en la hera alfonsi ni mas ni menos.

[1] Capitulo 4. Como se quentan los años aravigos meses y dias.

[2] Los moros cuentan los començamientos de sus meses desde que se paresçe la luna nueva que sale de so los rayos del sol et su fin quando se paresçe la sigunda ves. y esto no puede ser sino quando se pone el sol.
[3] Y por esto es el comienço del dia contra ellos: desde que se pone el sol. y la noche es contra ellos adelantada sobre el dia.
[4] E son sus años lunares e la contra de los dias del año contra ellos: que son tresientos y çinquenta y quatro dias y un quinto de dia y ten otro sexto de dia: esto es onse partes de treynta partes en el dia. E hasen sus meses uno XXX dias y otro de XXIX y ansi los quentan todos por horden desde el primero hasta el postrimer mes. sacando solo el mes de dulheia que tiene XXX dias en los años del bisiesto. E las revoluçiones de sus años son de treynta en treynta años e hasen visiestos de sus años en la manera siguiente. Hasen visiesto el segundo año de la revoluçion. y el quinto. y el septimo. y el deçimo. y el treseno. y el sextodeçimo. y el decimo octauo. y el vicessimo primo. y el vigessimo quarto. y el XXVI° y el vigessimo nono.
[f. 37v] [5] El primero mes de sus meses es moharran y es de treynta dias.
[6] El segundo es safar y es de XXIX dias.
[7] El 3° es rabe primo y es de treinta dias.
[8] El 4° es rabe secundo y es de XXIX dias.
[9] El 5° es jumet primo y es de treynta dias.
[10] El 6° es jumet secundo y es de XXIX dias.
[11] El 7° es raiab y es de treynta dias.
[12] El 8° es xabem y es de XXIX dias.
[13] El 9° es ramadan y es de treinta dias.
[14] El X° es xavel y es de XXIX dias.
[15] El XI° es dulqueda y es de treynta dias.
[16] El XII° es dulheyia y es de veinte y nueve dias en los años simples y de treinta dias en los años de visiesto.
[17] Pues estos son los años aravigos.

[1] Capitulo 5°. De los años persianos e como llaman e reparten sus meses.

[2] Los persianos hasen sus años solares mas no ponen en ellos ninguna fraçion.
[3] E la quenta de los dias del año açerca dellos es de tresientos y sesenta y çinco dias y non mas sin fraçio ninguna. E hasen todos sus meses de treynta dias. sacando el mes octauo non mas al qual hasen de treynta y çinco dias.
[4] E el primer mes de sus meses es afrozdiginhe y es de treynta dias.
[5] El segundo mes es adorynahseme et es de XXX dias.
[6] El tercero es hardetinhe et es de XXX dias.
[7] El 4° es tahmarmhe y es de XXX dias.
[8] El 5 es mardethmhe y es de XXX dias.
[9] El 6 es cahynrmeh y es de treynta dias.
[10] El 7 es maharmeh y es de treinta dias.
[11] El 8 es abenmeh y es de treynta y çinco dias.
[12] El 9 es adrynhmeh et es de treinta dias.
[13] El X° es dynmeh y es de treinta dias.
[14] El XI° es bethmenxibe y es de treinta dias.
[15] El XII° es azfidarinhe y es de XXX dias.
[f. 38r] [16] [E]stos pues son los años persianos y estos son los nombres de sus meses.

[1] Capitulo VI°. Señales para saber en que dia entra el año y cada uno de los meses por la quenta.

[2] Quando quisieres saber en que dia entrara el año o mes toma los años complidos que passaron de aquella hera y buelvelos todos en dias.
[3] Y despues mira que si buelues años romanos o alfonsies si te saliere con los dias alguna fraçion que qualquier que sea as de quitar un dia complido y añadillo sobre lo que as de los dias.
[4] Mas si tu bolvieres hera de Çesar y te saliere fraçion con los dias y fuere aquella fraçion medio o mas de medio dia. hasela tambien dia conplido y añadelo sobre los dias y si fuere menos de medio echalo y no lo quentes.
[5] E si bolvieres años aravigos y te saliere con los dias fraçion y fuere aquella fraçion mas de medio dia. hazlo dia complido y añadelo sobre los dias e si fuere medio dia o menos. echalo y no lo quentes. E lo que se te ayuntare de los dias despues desto añadelo sobre el a la hera alfonsi dos. e a la hera de Çesar uno. e a la hera de Persia tres. e a la hera araviga V.

latitude, deals with a subject of considerable technical difficulty in 34 sentences. On the other hand, chapter 15, on the mean position of the planets, gives a long explanation in 22 sentences on how to read mean motion tables, a topic that is rather straightforward.

New technical terms and new meanings of already existing terms seem to be concentrated not in the descriptive chapters but in those with a high degree of difficulty. We also note frequent changes in terminology: for example, in chapter 33 the expressions *movimiento de la ladesa* (33:4) and *movimiento de la latitud* (33:9) are used for "argument of latitude" instead of the expected *argumento de latitud*. It is surprising to find a variety of terms for "equation", one of the most important concepts in mathematical astronomy. In the text we find two different terms: *ynvocaçion* (and different variants of it) appears in chapters 27, 30, 31, 33, 34, and 49, whereas *equaçion* is used in the rest of the text. It is tempting to consider this vacillating terminology to be a reflection of the dual authorship, but that would require a thorough linguistic analysis of this and other texts attributed to the same authors. In the meantime, we shall focus on other aspects of this text, namely, its astronomical content, its scientific context, and its impact on the development of European astronomy.

2.4 Transcription criteria

As Sánchez-Prieto remarks (1998, p. 90), "a paleographic transcription implies an interpretation of the writing (*escritura*), at least in 'translating' the handwriting (*manuscritura*) into the stricter conventions imposed by typesetting". The transcription we present here is intended to provide a tool for research in the history of science and, for this reason, we have not cluttered it with the paleographical niceties of a diplomatic transcription. Nevertheless, we have chosen to remain faithful to the text and we have made as few modifications as possible. The editorial intervention is restricted to the following cases:

- All abbreviations have been resolved without any indication. In particular, a bar (*lineta*) over a letter has been replaced by an *n*, except before *p* and *b*, in which case we have replaced it by an *m*. In the case of an *n* with a bar over it, we have replaced it by *ñ*.

- The letters *b*, *v*, and *u* are maintained as in the manuscript, except for the articles "vn", "vna", "vno", and their plural forms, that have been transcribed by "un", "una", and "uno", etc.

[6] E lo que se te ayuntare de los dias despues desto echalos siete y siete. y lo que quedare siete o menos de siete esso sera la señal de la entrada del año. Despues mengualo del dia del domingo ansi como van los dias por horden y da a cada dia uno. e do llegare la quenta de los dias en aquel dia entrara aquel año en que tu eres. quiero desir el mes primero de sus meses.
[7] E si tu quisieres saber la señal de qualquiera de los otros meses del año. toma cada mes del año passado y mira quantos dias tiene allende de veinte y ocho.
[8] E si fuere aquel año bisiesto y fuere passado el mes del bisiesto. toma un dia de mas e lo que quedare de las diferençias de los dias añadelo sobre la señal del año. y reparte lo que quedare en siete y siete y lo que quedare menos de siete o siete aquello sera la señal del mes. Mengua despues lo del dia del domingo segun van los dias por horden dando a cada dia uno e en aquel mesmo dia entrara el mes que tu quieres saber do llegare la quenta de los dias.
[f. 38v] [9] E si quisieres haser esta quenta de otra manera toma los años complidos passados de aquella hera que tu quieres saber e echalos a dosientos y dies y dosientos y dies. si fuere la hera araviga. o veinte y ocho y veinte y ocho. si fuere la hera alfonsi o de Çesar. o siete y siete si fuere la hera persiana. E lo que quedare despues multiplicalo en uno y quarta si fuere la hera alfonsi o de Çesar o en quatro et tres deçimas e dos terçios de un die sino. si fuere la hera araviga. o en uno no mas si fuere la hera persiana e lo que se te ayuntare despues son dias.
[10] Pues si oviere fraçion has dello lo que te mostre ante et añade sobre cada hera lo que antes te mostre y cumple la obra assi como antes te dixe. y hallaras lo que tu demandas.

[1] Capitulo 7. Como se conosçeran las señales de las entradas de los años y meses por tablas.

[2] Quando quisieres saber lo sobredicho por tabla toma los años complidos passados de aquella hera y echalos veinte y ocho a 28 si fuere la hera alfonsi. o siete y siete si fuere la hera persiana o dexalos ansi como estan si fuere la hera araviga.
[3] E despues añade sobre el uno todavia lo que quedare. si fuere la hera araviga o dexalo ansi como esta. si fuere la hera alfonsi o persiana.
[4] E despues que ovieres fecho aquesto entra con el uno en las tablas de las señales de los años de aquella hera et toma lo que fuere en su derecho de las señales et guardalo.

[5] E si te sobraren algunos años lo qual no acaesçe sino en los años aravigos solamente. entra con ellos en la tabla de los años bueltos y toma lo que fuere en su derecho de las señales y guardalo tanbien.
[6] Despues toma la señal del mes primero de los meses de aquel año que tu quieres si fuere la hera arauiga.
[7] E si fuere otra hera no tomes señal del mes primero pues si fuere araviga y tomares la señal del primer mes. ansi como te dixe. añadela sobre las señales que tu guardaste. E lo que se te ayuntare de todo esto quentalo siete a siete e lo que quedare menos de siete o siete echalo del dia del domingo sigun van los dias por horden y da a cada dia [f. 39r] uno. E a do llegare la quenta de los dias. en aquel mesmo dia entrara el primer mes de los meses de aquel año cuya señal de entrada tu quieres saber.
[8] E si allende desto quisieres saber la señal de alguno de los otros meses del año. tomaras la señal que hallares en derecho del mes complido questa antes de aquel mes cuya señal tu quieres saber si fuere la hera alfonsi o persiana. e si fuere hera araviga toma lo que fuere en derecho de aquel mesmo mes cuya señal tu deseas saber.
[9] E miraras las señales para conosçer aquesto sigun la tabla del año que corriere. porque una tabla sirve para el año simple y otra para el bisestil. y vistas las tablas as de parar mientes al año en que esta el mes cuyo dia de entrada quieres saber.
[10] Y si es mes de año comun acude a la tabla del año comun e si es mes de año de bisiesto miraras la tabla del año bisestil.
[11] Pues si fuere año comun miraras las tablas de los años comunes y si bisestil las señales del año de bisiesto.
[12] E lo que fuere señal de aquel mes que tu tomaste de una destas dos tablas añadirlo as sobre la señal del año y lo que se ayuntare haslo siete y siete.
[13] Despues lo que quedare menos de siete o siete echalo[3] del dia del domingo ansi como van los dias por horden. E donde se allegare la quenta de los dias. en aquel mesmo dia entrara el mes que quisieres saber.
[14] E si allende de esto quisieres saber de qualquiera de los meses romanos con la hera alfonsi o de los meses persianos e fuere aquel mes de los meses del año primero de qualquiera destas dos heras. entra en tabla de los meses con los meses complidos e toma la señal que fuere en su derecho. et añade sobre ella dos si fuere la hera persiana. E lo que se ayuntare mengua dello siete si los oviere e lo que quedare aquello sera señal de mes que aun no entro.

---

[3] MS: echalo echalo.

[15] Comiença pues con la quenta del domingo ansi como hesiste en lo que se adelanto.

[1] Capitulo VIIIº. Como se tornaran años de qualquier hera a años de otra qualquier era por quenta.

[2] Quando ovieres años de qualquier hera y quisieres saber quantos años ay[4] de otra qualquiera hera. toma los años de la hera sabida y buelvelos todos en dias. y esto sera multiplicallos [f. 39v] en dias de un año de los años desta hera sabida. E lo que se ayuntare de dias e defeçiones si ay las oviere. y tornalos todos a años de aquella otra hera que tu quieres saber e eso sera en partir aquellos dias e aquellas defeçiones o faltas sobre dias de un año de años de aquella otra hera que tu quieres saber. E lo que saliere seran años complidos de años de aquella otra hera que tu quisiste saber e lo que quedare de mas seran dias.
[3] Pues lo que fuere de años y de dias eso es lo que tu demandaste. y essa es la quantia que ay en esos años sabidos de los años de la otra hera que tu demandauas et de dias de demas. verbigraçia.
[4] Pongamos que nos queremos saber quanto a en tresientos años aravigos de años romanos. Para esto aver de saber es menester que bolvamos estos tresientos años aravigos en dias multiplicandolos en tresientos y çinquenta y quatro y tres deçimos y dos terçios de un deçimo.
[5] De manera que hecha la quenta de dias son seys mill millares y tresientos millares y sesenta y ocho mill y seys çientos.
[6] Partiendo pues los dias de un año romano ques trezientos y sesenta y çinco dias y mas un quarto. salionos la quenta dosientos y noventa y uno años complidos y veinte y tres dias de mas.
[7] Esta pues es la quantia o quenta que ha en tresientos años aravigos de años romanos.

[1] Capitulo 9. Como se tornaran los años de las heras unas a otras por las tablas compuestas de los dias no mas.

[2] Quando quisieres saber esto entra en las tablas compuestas de los dias en la tabla de aquella hera que tu quisieres tornar con los años complidos passados en la tabla de los colectos de las tablas que son puestos a aquella hera.

---

[4] MS: ay ay.

[3] E demanda en ellas tanto como aquellos años passados o que es acerca dello con menos e toma lo que fuere en su derecho de lo que es escripto en somo de la tabla del quarto offiçio y del 3° y del 2° y del primero y guardalo.
[4] Despues mengua lo que tu entraste o acreçen[f. 40r]taste con lo que oviste em primero. y entra con la tabla de los años bueltos et busca ay lo que fuere tanto como el o lo mas cerca del con menos e toma lo que fuere en su derecho de aquellos offiçios. et ponlos so aquellos que posiste ante cada offiçio so su semejante. despues mengua lo que tu entraste con el de lo que te avia fincado. E lo que te fincare entra con el en la tabla de los meses y demanda ay tanto como el o lo que fuere mas çerca del con menos et toma lo que fuere en su derecho de aquellos offiçios. et ponlos so lo que avias escripto segunda vez. cada offiçio so su semejante.
[5] E si quedaren algunos dias de mas. ponlos so el primer offiçio de aquellos offiçios que tu tomaste.
[6] E despues ayuntalos y comiença del primer offiçio y si se te ayuntare dello sesenta o mas de sesenta alça cada sesenta en uno y ponlo en el offiçio ques ante del y guarda lo que quedare.
[7] E despues ayuntaras lo que ay en el sigundo offiçio. et si se te juntare dello sesenta o mas de sesenta alça cada sesenta en uno y ponlo en el offiçio que es antes del y guarda lo que quedare.
[8][5] E despues ayunta otrosi lo que ha en el segundo offiçio. y si se te ayuntare dello sesenta o mas que sesenta alca de cada sesenta uno y ponlo en el terçer ofiçio y guarda lo que quedare.
[9] Despues ayunta lo ques en el terçero ofiçio. y si se te ayuntare dello sesenta o mas de sesenta alça cada sesenta uno y ponlos en el quarto ofiçio y guarda lo que quedare.
[10] Despues ayunta lo que quedare en el quarto ofiçio y guardalo y lo que se te ayuntare despues desto de los quatro offiçios ponle nombre el padron y guardalo que as menester adelante.
[11] E despues entra con este padron en la tabla de los collectos de aquella hera demandada.
[12] E es la hera a que en tu quieres tornar la hera sabida et demanda ay tal como el o que es menor quel.
[13] E toma lo ques en derecho de la quantia de los años collectos y guardalo.
[14] Despues medra del padron lo que tu entraste con el et lo que fincare entra con el en la tabla de los espandidos de aquella hera demandada.

---

[5] Sentence 8 repeats the content of the previous sentence with a slightly different wording. In Rico's transcription, sentence 8 is omitted.

[15] E demanda y tal como el o lo que fuere çerca del con menos e toma lo que fuere en su derecho de la quantia de los años y guardalo.
[16] E despues mengua lo que tu entraste con el de lo que te avia quedado del padron.
[17] E lo que fincare entra con en[6] en la tabla de los meses y demanda hy tal como el o lo que fuere çerca del con menos y toma lo que fuere en su derecho de los meses y guardalo.
[18] E lo que te fincare del primer offiçio seran dias e ayuntarlos as a lo que tu guardaste de los [f. 40v] años y de los meses.
[19] E lo que fuere despues desto de años et de meses et de dias eso sera la quantia que ha en aquellos años los que heran escontra ti sabudos de años de aquella otra hera demandada que tu quesiste tornar a ella los años sabudos que tu avias et de sus meses et de sus dias.

[1] Capitulo X. De la diversidad de los tiempos entre cada era y era. cotejando la era[7] alfonsi a las otras. y de los años romanos y persianos e aravigos y sus dias.

[2] Sepas que lo que a en la hera alfonsi et la hera persiana de años persianos son seis çientos y diez y nueve años y tresientos y çinquenta y tres dias.
[3] E de años romanos seis çientos y XIX años. y C y noventa y ocho dias y un quarto de dia.
[4] E de años antiguos seys çientos y treynta y ocho años y dosientos dias.
[5] E de dias CC vezes mill dias y 26 vezes mill dias y CC y ochenta y ocho dias.
[6] E es su señal por la compostura de los dias sobre los 4 offiçios es uno y dos y çinquenta y uno y 28.
[7] E lo que ay entre la hera alfonsi y la hera araviga de años aravigos son seys çientos y quarenta y ocho años. y CC y ochenta y dos dias.
[8] E de hera persiana seys çientos y 29 años y tresientos y veinte et siete dias.
[9] E de hera romana seis çientos y 29 años y çiento y sesenta y nueve dias y tres quartos de dia.
[10] E de dias dozientas veses mill dias y veinte y nueve veses mill dias y noveçientos y 12 dias.

---

[6] *Sic*, instead of "el".

[7] MS: era era.

[11] E su señal por la compostura de los dias sobre los 4 offiçios es uno y tres LI y LII.
[12] E lo que a entre la hera arauiga y persiana es que de hera araviga son diez años y ochenta dias. y de hera persiana IX años CCC y XXXIX dias.
[13] E de hera romana IX años y CCCXXXVI dias y tres quartos de dia.
[14] E de dias tres mill dias et seys çientos y 24 dias.
[15] E su señal por la compostura de los dias por los 4 offiçios. es çifra y uno y [f. 41r] çifra y 24.
[16] E sepas que la mas antigua destas tres heras es la hera araviga y despus della la hera persiana. Despues de la hera persiana es la hera alfonsi.

[1] Capitulo XI. Como se [sa]ca una hera por la quenta de la otra.

[2] Quando tu ovieres sabido una hera y quisieres saber por ella otra hera. torna los años de aquella hera sabida et sus meses los complidos en dias et añade sobre ellos los dias del mes menguado si los oviere.
[3] E guarda lo que se ayuntare de los dias.
[4] E despues torna los años de la hera sabida a los años de la hera demandada assi como se dixo en ell octauo capitulo.
[5] E guarda lo que fuere de los años de la hera sabida despues que los tornaste a la hera demandada.
[6] E si aconteçiere que aya con ellos algunos dias de mas. ayuntalos a los dias que avias guardado.
[7] Pues si se ayuntare dellos mas de dias de un año de los años de aquella hera demandada. mengua de los dias de un año y guarda lo demas y son dias.
[8] Despues añade en la quantia de los años un año.
[9] E lo que fuere despues de los años complidos et de los dias esos son los años de la rayz et sus dias por eso guardalos.
[10] Despues sabe quanto a entre las dos heras de los años que son de la natura de aquella hera demandada et de los dias assi como diximos en el X capitulo. et lo que fuere esa es la diferençia guardala.
[11] Despues cata si fuere la hera demandada mas antigua que la sabida añade los años de la diferençia y sus dias sobre los años de la rais y sobre sus dias.
[12] E si fuere la hera sabida mas antigua de la demandada. mengua los años de las diferençias y sus dias de los años de la rais y de sus dias. y lo que se ayuntare despues o lo que quedare de años y dias guardalo.

[13] Despues cata los dias que tu aguardaste con los años.
[14] Pues si fuere de mas dias de un año. mengua[f. 41v]los dellos et añade en la quantia de los años uno. E despues toma lo que sobro de los dias y tornalos meses de aquella hera demandada. dando a cada mes dellos la quantia de sus dias.
[15] E comiença con el primero mes de los meses de aquella hera assi como lo avemos demostrado en lo passado.
[16] Pues si oviere con los dias fraçion quando tu sacares la hera alfonsi de otra hera un quarto de dia o mas de un quarto. o quando sacares la hera araviga de otra hera mas de medio dia. alça aquella fraçion a dia complido y añadelo sobre los dias. E si fuere aquella fraçion quando tu sacares la hera araviga de otra hera medio dia o menos. echalo et no lo metas en quenta.
[17] E lo que fuere despues desto de la quantia de los años y de los meses y de los dias eso sera lo passado desdel comienço de aquella hera demandada hasta el dia en que tu estas. E si quisieres hasello en otra manera. hordena la hera sabida y tornala toda en dias et añade sobre ellos los dos dias que son sobre las dos heras. et si fuere la hera demandada mas antigua que la sabida.
[18] O mengualos dellos si fuere la hera sabida mas antigua que la demandada. et lo que se ayuntare o fincare despues desto de los dias tornalos años de la natura de los años de aquella natura de los años de aquella hera demandada assi como arriba se dixo. E lo que fueren los años guardalo.
[19] E lo que quedare de los dias tornalo meses de la natura de los meses de la natura de los meses de aquella hera dando a cada mes la quantia de sus dias.
[20] E comiença con el primer mes de los meses de aquella hera e lo que fuere despues de años y de meses y de dias esso es lo que passo desdel comienço de aquella hera demandada hasta el dia en que estas.

[1] Capitulo XII. Como se saca la diferençia de las heras por las tablas de los dias compuestos.

[2] Quando quisieres esso et ovieres tomado alguna de las tres heras sobredichas. sabidos sus años. sus meses y dias. entra con los años [f. 42r] complidos en la tabla de los colectos de aquella hera sabida en las tablas de los dias compuestos assi como te demostre en el IX capitulo. E en las tablas de los años que proçeden por horden de aquella hera tanbien as de mirar e saca el padron ansi como te demostre en el nono capitulo. E guarda lo que fuere del padron e despues saca del X° capitulo ques lo que ay entre las dos eras et ques

su señal por la compostura de los dias sobre los quatro offiçios et guarda esa señal.
[3] E despues mira si fuere la hera demandada mas antigua que la hera sabida et añade la señal sobre el padron. poniendo cada offiçio sobre su semejante.
[4] E si fuere la hera sabida mas antygua que la demandada. mengua la señal del padron cada offiçio de su semejante et lo que se ayuntare despues desto o lo que fincare eso sera el padron demandado. guardalo.
[5] E despues entra con el en la tabla de los collectos de aquella hera demandada en las tablas de los dias compuestos et demanda ay otro tal como el o lo que fuere mas acerca del con menos. et lo que fuere despues desto de años et de meses et de dias esto es lo passado de la era demandada.

[1] Capitulo 13. Como se a de distinguir la era alfonsi e la hera persiana e la hera araviga y la era de Çesar cada una de la otra por las tablas compuestas de los años meses y dias.

[2] Quando quisieres eso e ovieres hera sabida de años. meses et dias. entra con sus años los complidos en la tabla de los collectos de aquella hera sabida en las tablas que son para sacar la una de la otra. la ques compuesta de años y de meses y de dias.
[3] E demanda ay tanto como aquellos años o lo que fuere mas çerca dellos con menos.
[4] E toma lo que fuere en su derecho en la tabla ques puesta hasia aquella hera demandada de años. dias et fraçiones si las oviere y guardalo.
[5] E despues mengua lo que entraste con el de lo que as.
[6] E lo que quedare entra con el en la tabla de[8] [f. 42v] los años que van por horden de aquella hera sabida otrosi por orden. et demanda ay tal como ellos o lo que fuere çerca dellos con menos.
[7] E toma lo que fuere ay en su derecho otrosi de años et de dias y de fraçiones si las oviere ay y ponlo debaxo de lo que tomaste ante cada offiçio con su semejante.
[8] E despues mengua lo que entraste con el de lo que te avia fincado de la raiz e lo que te fincare son meses.
[9] Entra otrosi con ellos en la tabla de los dias de los meses de aquella hera sabida e toma lo que fuere en su derecho de los dias en esa mesma tabla et ponlo baxo de lo que oviste de los dias.

---

[8] Mg. adds "con previlegio real".

[10] E despues mira si ovieres algunos dias de mas sobre los meses y ponlos otrosi so los dias.
[11] E despues torna et ayunta cada offiçio a su semejante e comiença en primero de las fraçiones si fuere la hera demandada araviga. E quando se te ayuntaren dellos XXX. tornalos un dia et añade un dia en los dias. et lo que quedare ponlo aparte.
[12] E despues ayunta los dias et si se te ayuntare de los dias un año de los años de aquella hera demandada o mas mengua de los dias [los] dias de un año complido.
[13] E si fuere la hera demandada de las eras que an en sus años fraçiones. mengua de los dias [los] dias de un año complido con sus fraçiones. y añade en la quantia de los años uno.
[14] E lo que quedare de los dias menos de los dias de un año de los años de aquella era demandada. guardalos.
[15] E si fuere la hera demandada araviga et fincare della en las fraçiones de XVI arriba. alçalos un dia et añade en los dias uno.
[16] E si fuere menos de XVI. echalos. y no los metas en quenta.
[17] E lo que fuere de los dias despues desto. entra con ello en la tabla de los dias de los meses que son de la hera demandada.
[18] Pues si fuere la era demandada era Alfonsi o era de Çesar entra en una de las tablas que son puestas.
[19] Por eso quiero desir la tabla de los meses comunes o ordinarios et de los meses bisiestos.
[20] Et sabras en que tabla dellas entraras en catar el año donde son los dias en que tu entras.
[21] Pues si fuere espandido o año comun entra en la tabla [f. 43r] de los dias de los meses romanos espandidos e si fuere bisiesto entra en la tabla de los dias[9] de los meses romanos bisextos. y demanda ay tanto como as de dias o lo que fuere mas çerca dellos con menos.
[22] E toma lo que fuere en su derecho de los meses e lo que fueren esos seran meses complidos passados de aquella hera demandada.
[23] E despues mengua los dias con lo que tu entraste de los dias e lo que fincare seran dias passados del mes que es despues de aquellos meses passados que tu tomaste.
[24] E despues ayunta los años e lo que fueren seran años complidos passados de aquella hera demandada de mas sobre los meses y sobre los dias que tu avias tomado ante. y esso es lo que tu demandavas.
[25] E esta opinion es despues que passare la primera revoluçion de las reuoluçiones alfonsies.
[26] Mas si ante que la primera reuoluçion fuere passada es la opinion tal qual yo te demostrare. Cata si fuere la hera sabida hera

[9] MS: dias dias.

alfonsi. entra con ella en la tabla de los años espandidos alfonsies et toma lo que hallares en su derecho en la tabla de la era demandada et guardalo.
[27] E despues entra en la tabla de los meses y cumple la opinion assi como te amostre en lo ya dicho.
[28] E lo que saliere ayuntallo as a la raiz de aquella era demandada la ques escripta ençima de su tabla en la tabla de los años collectos alfonsies.
[29] E si fuere la hera demandada arauiga haras con sus fraçiones lo que ante te mostre e lo que saliere eso es lo que tu demandavas.
[30] E si fuere la era sabida hera de Çesar o la era persiana. mengua la rayz ques escripta ençima de la tabla de lo que tu ovieres.
[31] E entra con lo que fincare en la tabla de los espandidos de aquella era sabida.
[32] E faras con el ansi como lo hesiste con la hera alfonsi.
[33] E lo que te saliere eso es lo que demandauas.
[34] E si fuere la era sabida araviga mengua della seis çientos y quarenta y ocho años. y CCLXXX y un dia y diez y ocho fraçiones. y entra con lo que fincare en la tabla de los años aravigos espandidos e haras con el ansi como hesiste con la hera alfonsi y hallaras lo que tu demandares.
[35] E esta raiz [f. 43v] es menos que la raiz ques puesta ençima de su tabla en la tabla de los collectos con XXIIII partes.

[1] Capitulo XIIII°. Para saber sacar los medios cursos de los planetas y de los otros movimientos despues del tiempo de la hera alfonsi.

[2] Quando quisieres ygualar las luminarias o cada uno de los otros planetas a mediodia de otra villa fuera de Toledo seyendo sabida su longura de aquella villa saca los medios cursos y los argumentos y los çentros de los planetas que quisieres. a la sason que quisieres al mediodia de la çibdad de Toledo. segun son puestas en estas tablas.
[3] E desi cata la longura de aquella otra villa de la parte de oçidente et si fuere mayor que la longura de Toledo. muestra que es oriental de Toledo aquella otra villa.
[4] E si fuere su longura menos que la longura de Toledo. muestra que aquella otra villa es ocçidental de Toledo e desi mengua estas longuras la una de la otra sacando la menor de la mayor.
[5] E parte la diferençia por XV y lo que saliere de la partiçion seran oras y fraçion de ora.
[6] E lo que fuere aquello es la diferençia ques entre el mediodia de aquella villa y el mediodia de Toledo.

[7] E cata quanto se mueve cada medio curso et cada argumento et cada çentro en tantas horas y en tanta fraçion de hora y nombrala diferençia.
[8] Desi cata si fuere aquella otra villa oriental de Toledo. mengua aquella diferençia de las rayses que sacaste al mediodia de Toledo. cada una de su natura. la del medio curso del medio curso. la del argumento del argumento. la del çentro del çentro.
[9] E si fuere la otra villa occidental de Toledo. añade la diferençia de las rayses que sacaste al mediodia de Toledo. cada una a su linaje.
[10] E con esto avras los medios cursos y los argumentos y los centros al mediodia de aquella otra villa.
[11] E endereça dellos los lugares e avras los lugares ygualados al mediodia de aquella otra villa que demandaste.

[1] Capitulo XV. Como se an de saber sacar los lugares de los planetas en las villas que son orientales o ocçidentales de la çibdad de Toledo.

[f. 44r] [2] Quando quisieres eso. sepas quanto ha entre el tiempo que tu demandas eso en el. et del tiempo de la raiz destas tablas. de años romanos y de meses y de dias e de horas yguales.
[3] E entra en las tablas de los movimientos de qual movimiento quisieres y de qual planeta quissieres.
[4] Primeramente en la tabla de los años collectos. la que creçe de XX en XX con lo que fuere mas çerca a la quenta de años complidos que tu as empero menos.
[5] E toma lo que hallares en su derecho de signos et de grados y de minutos y sigundos. e escrivelo cada uno por si.
[6] Et despues cata lo que te quedo de los años complidos que oviste menos de XX y entra con ello en la tabla de los años espansos.
[7] E toma lo que hallares en su derecho de signos e de grados et de minutos et de sigundos. y escrivelo so la otra quenta cada uno so su linaje.
[8] E despues cata el año en ques aquel tiempo demandado si es bisiesto o no.
[9] E si fuere bisiesto. con los meses complidos que son passados de aquel año [entra] en la tabla de los meses del año bisextil.
[10] E si el año no fuere bisiesto entra en la tabla de los meses del año [no] bisextil.
[11] E toma lo que fuere en su derecho de aquel mes en que fuere destas dos tablas sobredichas. de signos e de grados e de minutos et de sigundos e ponlo so la otra quenta. cada uno so su linaje.

[12] E despues cata los dias que son passados del mes no complido y entra con ello en la tabla de los dias.
[13] E entra otrosi en las oras en la tabla de las horas y con los minutos de la hora en la tabla de los minutos. E toma lo que fuere en derecho de cada uno y escrivelos so los otros cada uno so su linaje.
[14] E despues ayuntalo todo començando de los sigundos.
[15] E quando llegare la quenta a LX has de cada LX un minuto y ponlo so los minutos.
[16] E lo que fincare menos que LX. escrivelo en su mesmo derecho.
[17] E desi ayunta los minutos e de cada LX. haz un grado y ponlo con los grados. e lo que fincare menos que LX. ponlo en su mesmo derecho.
[18] E despues ayunta los grados y de cada XXX. has un signo y ponlo con los signos.
[19] E lo que fincare menos de XXX. escrivelo en su mesmo lugar.
[20] E desi ayunta los signos e quanto ay oviere de dosenas echalo.
[21] E lo que fuere menos de [f. 44v] dose escrivelo ay.
[22] E lo que fuere despues del ayuntar de signos y grados y minutos y sigundos aquello es la rayz de aquel movimiento y de aquella planeta señalada que tomaste. e en aquel tiempo señalado que tu demandaste.

[1] Capitulo XVI. De la equaçion del sol.

[2] Saca el medio curso del sol al tiempo que tu quieres. e lo que fuere mengua del su aux. e lo que fincare eso es el argumento.
[3] Entra con el en las reglas de la quenta en la tabla de la equaçion del sol.
[4] E toma de su equaçion lo que fuere en su derecho.
[5] E lo que fuere mengualo del medio curso del sol si fuere el argumento menos de seys signos. et añadelo sobre el si fuere mas de seys signos.
[6] E lo que fuere el medio curso despues del menguamiento del o despues del añadimiento del o despues del añadimiento sobre el eso sera el lugar çierto del sol.

[1] Capitulo 17. De la equaçion de la luna.

[2] Saca el medio curso de la luna et su argumento. e otrosi el medio curso del sol a la ora que tu quieres.
[3] E despues mengua el aux del sol de medio curso de la luna.

[4] E lo que fincare toma su signo e multiplicalo en V sigundos y medio e guarda lo que se ayuntare.
[5] E despues cata la luna.
[6] E pues si fuere entre el aux del sol et entre su opposito mengua lo que tu guardaste del medio curso de la luna.
[7] E si fuere la luna entrel opposito del aux del sol y entre su auxe. añade lo que tu guardaste sobre el medio curso de la luna.
[8] E pues lo que fuere el medio curso de la luna despues del menguamiento del o en despues del añedimiento sobre el esso sera el medio curso ygualado de la luna guardalo.
[9] E despues mengua del el medio curso del sol e lo que fincare doblalo e ponle nombre longura doblada.
[10] E despues entra con ella en la tabla de la equaçion del çentro de la luna y toma lo que fuere en su derecho de la equaçion del çentro e de los minutos proporçionales.
[11] E despues cata a la quenta con que tu entraste.
[12] Pues si fuere menos de VI signos añade la equaçion del çentro sobre el argumento.
[13] E si fuere mas de VI signos mengua la equaçion del çentro del argumento.
[14] E lo que fuere el argumento despues del [f. 45r] añedimiento sobre el o del menguamiento del. aquello sera el argumento ygualado.
[15] E entra con el en las tablas de las equaçiones de la luna. y toma lo que fuere en su derecho de la equaçion del argumento et de la equaçion de la diversidad del diametro del piçiclo.
[16] E toma de la equaçion de la diversidad del diametro tanto quanto es la proporçion de los menudos proporçionales de sesenta.
[17] E si tu lo quisieres haser de otra manera. multiplica los mynutos proporçionales en la equaçion de la diversidad del diametro.
[18] E lo que fuere añadelo sobre la equaçion del argumento et lo que se ayuntare eso es la equaçion eguada.
[19] E lo que fuere mengualo del medio curso eguado de la luna si fuere el argumento eguado menos de VI signos. e añade sobre el si fuere mas.
[20] E lo que fuere el medio curso eguado despues del menguamiento o del añedimiento aquello es el lugar eguado de la luna.

[1] Capitulo 18. De la equaçion de los V planetas.
[2] Quando tu quisieres ygualar alguno de los planetas tres de suso. saca el medio curso del que tu quieres dellos y tambien el medio curso del sol a la ora que tu quieres.
[3] E despues mengua el medio curso del planeta del medio curso del sol y lo que quedare eso es el argumento.

[4] Despues mengua del medio curso del planeta su aux e lo que quedare eso es el çentro.
[5] Entra con el en las reglas de las quentas en la tabla de la equaçion de aquel planeta.
[6] E toma lo que fuere en su derecho de la equaçion del çentro. E ponlo aparte e menguaIo del si fuere el çentro menos de VI signos.
[7] E añadelo sobre el si fuere mas.
[8] E lo que fuere del çentro despues del menguamiento del o del añadimiento del eso es çentro eguado.
[9] E toma lo que fuere en su derecho de los minutos proporçionales.
[10] E despues cata el çentro con que tu entraste en la primera.
[11] Pues si fuere menos de VI signos añade su equaçion sobre el argumento.
[12] E si fuere menos digo mas de VI signos mengua su equaçion del argumento e lo que fuere el argumento despues del añadimiento o menguamiento eso es el argumento eguado.
[13] E[10] entra con el en las reglas de las quentas en las tablas de la equaçion de aquel planeta.
[14] E toma lo que fuere en su derecho de la equaçion del argumento e de la equaçion de la diversidad [f. 45v] del diametro del epiçiclo en una de las dos longuras quiero desir. en la longura mas luenga o en la longura mas çerca.
[15] E sabras en qual dellas tomaras por escripto ques sobre la cabeca de los minutos proporçionales que tu tomaste de cerca o de lexos.
[16] E lo que fuere multiplicalo en los menudos proporçionales.
[17] E lo que fuere aquel multiplicado menguaIo de la equaçion del argumento si tomaste en la longura mas luenga o añadelo sobre el si tomaste la longura mas cercana.
[18] E lo que fuere eso es la equaçion eguada del argumento eguado.
[19] Pues si fuere el argumento eguado menos de seis signos añade la equaçion eguada sobre el çentro eguado.
[20] E si fuere mas de VI signos mengua la equaçion eguada del çentro eguado.
[21] Pues lo que fuere el çentro eguado despues del añadimiento sobre el. o el menguamiento del. esso es el lugar de aquel planeta. ayunta a el su aux y sera su lugar çierto.
[22] E la equaçion de Venus y de Mercurio es tal como la equaçion de los tres planetas de suso no a entre ellas alguna dissimilitud. fueras ende en el argumento de cada una dellas es lo que sale de las tablas. E el medio curso de cada una dellas es el medio curso del sol.

---

[10] MS: E e.

[1] Capitulo 19. De la equaçion de la cabeça del dragon.

[2] Saca su medio curso a la ora que quisieres et lo que fuere mengualo de 12 signos et lo que queda es su lugar. E la cola es todavia en su opposito de la cabeça.

[1] Capitulo 20. Como se a de saber la declinaçion del sol ygualador del dia e si es septentrional o meredional en cada sason.

[2] Quando esto quisieres saber mira el lugar natural del sol a la sason que quisieres. E lo que fuere entra con el en la tabla de la declinaçion y toma lo que fuere en derecho de la declinaçion. E lo que fuere aquella es la declinaçion del sol en aquella sason.
[3] E es su arredamiento del çerco egualador del dia a septentrion et a mediodia. Ca si el lugar con que entraste fuere de un grado hasta seis signos aquella declinaçion es meredional[11] del equinoçio e si el lugar fuere de seis signos hasta 12 aquella declinaçion es me[f. 46r]redional del equinoçio.
[4] E si el lugar fuere de un grado hasta tres signos el sol es subiente en septentrion.
[5] E si fuere de tres signos hasta VI es decendiente en septentrion.
[6] E si fuere de 6 signos hasta 9 es deçendiente en mediodia.
[7] E si fuere de 9 signos hasta 12 es subiente en mediodia.

[1] Capitulo 21. Como se a de saber la latitud de la luna ques su arredramiento del sodiaco al septentrion o a mediodia en cada sason.

[2] Quando esto quisieres saber. sepas el lugar de la luna e el lugar del geuzaar.
[3] E mengua el lugar del geuzaar del lugar de la luna. e lo que fincare es el argumento de la latitud.
[4] E entra con ello en la tabla de la latitud de la luna y toma lo que fuere en su derecho de la latitud. y aquella es la latitud de la luna.
[5] E es su arredramiento del zodiaco al septentrion o al mediodia.
[6] E si este argumento con que entraste fuere de un grado hasta tres signos aquella latitud es septentrional subiente.
[7] E si fuere de tres signos hasta VI signos la latitud es septentrional deçendiente.

[11] Instead of "septentrional".

[8] E si fuere de VI signos hasta IX la latitud es meredional deçendiente.
[9] E si fuere de IX signos. hasta 12 sera la latitud meredional subiente.

[1] Capitulo 22. Que muestra la latitud de cada uno de los V planetas e otrosi su alongamiento del zodiaco a septentrion y mediodia en todo tiempo.

[2] Quando quisieres saber la latitud de alguno de los otros planetas tres arriba dichos que son Saturno. Jupiter y Mares saca el çentro de qual quisieres dellos y entra con el en la tabla del planeta que demandas. et toma lo que hallares en el derecho de los minutos proporçionales y guardalos.
[3] E cata lo que tienes escripto de suso en la regla donde los tomaste de septentrion o de mediodia et otrosi guarda aquello.
[4] E despues cata el argumento eguado de aquel planeta en aquella sason. e entra con el en las tablas de la latitud de aquel planeta.
[5] E si los minutos proporçionales fueren en septentrion toma lo que fuere en derecho del argumento en la regla que tiene escripta de suso: latitud septentrional.
[6] E si los menudos proporçionales fueren en mediodia [f. 46v] toma lo que fuere en derecho del argumento en la linea que tiene septentrion[12] latitud meredional.
[7] E guarda a su parte lo que fuere la latitud de qualquier dellos.
[8] E guarda otrossi lo que halles en somo de la regla donde tomaste la latitud del subiente o deçendiente.
[9] Despues toma aquella latitud e la proporçion que an los minutos proporçionales de sesenta. y lo que fuere aquella es la latitud de aquel planeta en aquella sason que demandaste.
[10] E es su arredramiento del zodiaco et la su parte a mediodia o de septentrion y de subiente y deçendiente segun lo hallaste y guardaste.
[11] Mas quando quisieres saber la latitud de Venus. entra con su çentro en las tablas de su latitud y toma la que fuere en su derecho de los minutos proporçionales de la declinaçion e de los minutos proporçionales del declinamiento. et escrive cada uno dellos a su parte y escrive sobre cada uno lo que hallares en somo de la regla donde lo tomas de alto o de baxo.

---

[12] The text should probably read "escripta de suso", as in the previous sentence.

[12] E toma otrosi lo que fuere en aquel derecho del çentro en la regla de la latitud tercera et guardala otrosi et la parte que le hallares escripta de suso es siempre septentrional.
[13] E despues entra con el argumento eguado en aquellas mesmas tablas. en tabla de la declinaçion et en tabla del desviamiento.
[14] E toma lo que fuere en su derecho de la declinaçion et del desviamiento y escrive cada uno dellos a su parte.
[15] E escrive sobre cada uno lo que hallares en somo de la regla donde lo tomas de alto o de baxo.
[16] E despues toma de la declinaçion la proporçion de sus menudos proporçionales a sesenta et lo que fuere nombralo la latitud primera.
[17] E si fuere el escripto sobre la declinaçion y sobre sus minutos acordados e semejantes que sean ambos altos o ambos baxo. escrive sobre esta latitud primera meredional. E si se desacordare la declinaçion sus menudos proporçionales que sea el uno alto y el otro baxo. escrive sobre la latitud primera septentrional.
[18] E despues toma del sujamiento la proporçion que an sus menudos proporçionales a sesenta e lo que fuere nombralo latitud segunda.
[19] E otrosi si el desviamiento et sus menudos proporçionales se acordaren en el escripto de alto o de baxo. escrive sobre la latitud meredional sigunda.
[20] E si se desacordaren de manera que [f. 47r] sea el uno alto et el otro baxo. escrive sobre la sigunda latitud septentrional.
[21] E despues cata estas dos latitudes. la primera e la sigunda.
[22] E si amas las fallares septentrionales o amas meredionales ayuntalas. e lo que ende fuere nombralo latitud eguada e escrive sobre ella la parte en que se acordaron.
[23] Et si estas dos mesmas latitudines se desacordaren de manera que sea la una meredional y la otra septentrional. mengua la menor de la mayor et lo que queda nombralo latitud ygualada por qual destas dos maneras oviste.
[24] E si hallares acordada con la terçera en septentrion. ayuntalas e lo que fuere aquello es la latitud de Venus en septentrion e si la latitud eguada se discordare con la ladeza terçera. mengua la menor de la mayor. E lo que fincare aquello es la ladeza de Venus e la su parte es la parte de la mayor.
[25] E quando ovieres esta latitud de qualquiera destas dos maneras aquella es la latitud averiguada de Venus. ques su arredramiento del zodiaco. E la su parte a septentrion o a mediodia sigun lo hallaste.
[26] Mas quando quisieres saber la latitud de Mercurio entra con su çentro en las tablas de su ladeza et toma lo que fuere en su derecho de los menudos proporçionales de la su declinaçion y de los menudos proporçionales del desviamiento.

[27] E escrivelos con sus titulos segun lo diximos en Venus.
[28] E toma otrosi con el çentro su ladesa terçera e escrivela con su parte e siempre la hallaras meredional.
[29] E despues entra con su argumento eguado e toma lo que fuere en su derecho de la declinaçion y escrivela con so titulo segun que diximos en Venus.
[30] Mas en el desviamiento para mientes al centro e si fuere de un grado hasta XCIII grados o de 267 que se hasen 8 signos y 27 grados hasta el complimiento del çerco.
[31] Entra con el argumento eguado en la tabla del desviamiento de Mercurio. la qual tiene por titulo en somo estas cuentas sobredichas. y toma lo que fuere en su derecho del desviamiento y escrivelo con su titulo.
[32] E si el çentro fuere de 93 grados hasta 267 entra con el argumento eguado en la tabla del desviamiento de Mercurio. la que tiene esta quenta en somo en su titulo. E toma della el [f. 47v] desviamiento con su titulo.
[33] E quando ovieres el desviamiento con su titulo de qualquier destas tablas. has toda la obra segun hesiste en Venus hasta que ayas averiguado su latitud et su parte segun la oviste en Venus.
[34] Mas el sobimiento y el abaxamiento en [es]tos dos planetas no lo podimos determinar en las tablas. porque se diversifica por muchas partes.
[35] Mas quando lo quisieres saber saca la latitud de qual quisieres a un dia señalado segun lo avemos dicho.
[36] E despues sacala a VIII o a X dias despues.
[37] Desi cata si la ladeza fuere septentrional e si la hallares creçiendo en la ladesa sabras ques subiente en septentrion.
[38] E si hallares la ladesa descreçiendo sabras ques deçendiente en septentrion.
[39] E si la su latitud fuere meredional e hallares la ladesa que creçe sabras ques deçendiente en mediodia.
[40] E si la hallares en la primera equaçion septentrional e en la segunda meredional sabras que en la primera fue deçendiente en septentrion.
[41] E si fuere en la primera meredional y en la segunda septentrional sabras que en la primera fue subiente en mediodia.
[42] E la fin de la ladesa de Venus a septentrion o a mediodia es VIII grados y XXXVI minutos.
[43] E la fin de la ladesa de Mercurio a septentrion o a mediodia es quatro grados. y XXXVIII minutos.

[1] Capitulo 23. Como se a de saber la retrogradaçion de cada uno de los çinco planetas y de su enderecamiento.

[2] Quando esto quisieres saber sepas el çentro eguado del planeta e su argumento eguado a la sason que quisieres.
[3] E cata el çentro eguado. si fuere de un grado hasta tres signos o a nueve signos hasta dose muestra quel planeta es en el medio alto del çerco de su aux.
[4] E si el çentro eguado fuere de tres signos hasta IX. muestra quel planeta es en el medio baxo del cerco de su aux.
[5] Desi cata la tabla donde son puestos los 4 estados de cada planeta.
[6] E si el planeta fuere en el medio alto del cerco de su aux toma de estos 4 estados el primero y el segundo. y obra con [f. 48r] ellos en la siguiente manera.
[7] Mira el argumento eguado. e si fuere conveniente e ygual con el estado primero muestra que aquel planeta es su estaçion primera començando a retrogradar. E si el argumento eguado fuere mas del estado primero et menos del estado segundo muestra que aquel planeta es retrogrado.
[8] E si este argumento eguado fuere semejante y ygual al estado sigundo muestra que aquel planeta esta en la segunda estaçion començandose dese enderecar.
[9] E si este argumento eguado fuere menos del estado primero o mas del estado sigundo muestra quel planeta es derecho.
[10] Pues quando el planeta fuere retrogrado y quisieres saber quantos dias a que començo a retrogradar. mengua el estado primero del argumento eguado e lo que fincare partelo por el movimiento del argumento de aquel planeta en un dia e lo que saliere de aquella partiçion tantos dias que començo a retrogadar.
[11] E si quisieres saber a quantos dias se endereçara despues que fuere retrogrado. mengua el argumento eguado del estado segundo e lo que fincare partelo por el movimiento del argumento a un dia e salirte a lo que quisiste.
[12] E si la planeta fuere derecha e quisieres saber quantos dias a que se endereço. mengua el estado sigundo del argumento eguado y lo que quedare partelo por el movimiento del argumento a un dia e salirte a lo que quisiste.
[13] Et si quisieres saber quando se retrogradara quando fuere derecha. mengua el argumento eguado del estado primero y lo que quedare partelo por el movimiento del argumento a un dia e salirte a lo que quisiste.
[14] Mas si el planeta fuere en el medio baxo del çerco de su aux et es el ques nombrado hadit. que quiere desir ell opposito dell aux. toma

de los 4 estados del planeta el 3° e el 4° E obra con ellos sigun diximos del primer estado e del sigundo ygualmente ny mas ni menos. poniendo el terçero en ves del primero y el 4° en ves del 2° y abras lo que demandaste.
[15] Nos labramos en este capitulo por çentro eguado e no lo mostramos en la equaçion de los planetas. ca no lo ovimos ay menester segun [f. 48v] nuestras reglas.
[16] Mas porque aquello avemos menester. convienenos mostrar aqui como se yguala.
[17] Saca el çentro de las tablas e toma con ello su equaçion de su tabla apartada. E si el centro fuere mas[13] de VI signos. mengua de su equaçion e lo que fuere el çentro despues desto aquello es el çentro eguado con que obramos en este capitulo e en otros mas adelante.

[1] Capitulo 24. Como se a de saber el subimiento de los V planetas y su opponimiento.

[2] Quando quisieres saber el subimiento de los V planetas y su opponimiento. si fuere alguno de los tres arriba dichos que son Saturno. Jupiter y Mercurio[14] cata el argumento eguado de qualquier dellos.
[3] Pues si fuere de un grado hasta complimiento de VI signos. en el es el subimiento de las mañanas. Empero si fuere mas de seis signos hasta complimiento de 12 signos es en ponimiento de prima noche.
[3] Mas Venus y Mercurio porque andan todavia alrededor del sol y porques su movimiento a las veses apresurado. a las veses tardio. acaesçeles con el sol 4 estados que les yo te dire luego.
[4] Quando fuere el argumento eguado de Venus de un grado hasta 4 signos y 17 grados es subimiento de prima noche. E eso es quando se paresçe sobre el orisonte de occidente y es estonçe mas apresurado del sol.
[5] E quando es de complimiento de 4 signos y 17 grados hasta complimiento de VI signos. es en ponimiento de prima noche.
[6] E eso es quando es mas tardio del sol et retrobes e alcançale el sol.
[7] E de complimiento de 6 signos hasta 7 signos y 13 grados es en subimiento de las mañanas et es estonçe mas tardio el sol.
[8] E hasta 12 signos es en ponimiento de las mañanas.

---

[13] Between the lines, above the word "mas", the text adds "menos".

[14] *Sic*, instead of "Mares".

[9] E quando fuere su argumento de Mercurio eguado de un grado hasta tres signos y 22 grados es en subimiento de prima noche.
[10] E de 3 signos. y 22 grados hasta complimiento de VI signos es en ponimiento de prima noche.
[11] E de complimiento de VI signos. hasta VIII signos y 8 grados es en subimiento de las mañanas.
[12] E de 8 signos y 8 grados hasta complimiento de 12 signos es en ponimiento de las mañanas.
[f. 49r] [13] E es su estado en la presura y en el tardamiento ansi como lo diximos en el estado de Venus.
[14] Pues quando quisieres saber el subimiento de Saturno. Jupiter y Mares en las mañanas en oriente. quando es en el comienço de su aparesçimiento et su salida de so los rayos quando lo passa el sol. cata quando fuere el argumento eguado de qualquier dellos cerca de XX grados. pues es en el comienço de su aparesçimiento e de su salida de so los rayos.
[15] Mas su ponimiento de prima noche es quando lo alcança el sol y lo cubre.
[16] Eso es quando es su argumento eguado cerca de CCCXL grados.
[17] E Venus e Mercurio. quando fuere el argumento eguado de qualquiera dellos cerca de XX grados. es en el comienço de su paresçimiento en las mañanas en oriente.

[1] Capitulo 25. De como se a de saber el paresçimiento de los planetas e de su ascondimiento.

[2] Quando quisieres saber eso yguala aquel planeta que quieres e sepas su lugar.
[3] E otrossi egua el sol et sepas su lugar.
[4] Despues mengua el menor del mayor y sepas la diferençia de quien es.
[5] Pues si fuere la diferençia a comparaçion del planeta. as de haser la siguiente diferençia. conviene a saber que o es abscondido o pareçiente en ocçidente.
[6] E entra pues con el signo do es el planeta y toma lo que fuere en su derecho en la tabla del ponimiento de los planetas de prima noche. de grados y de minutos.
[7] E cata si fuere la diferençia que es entre el sol y el planeta. si fuere mas desta que tu tomaste de la tabla. pues el planeta es paresçiente e si fuere menos es abscondida.
[8] E si fuere la diferençia del sol. pues la planeta o es paresçiente o en oriente o en abscondido.

[9] Pues quando quisieres saber eso entra con el signo do es el planeta y toma lo que fuere en su derecho en la tabla del subimiento de los planetas en las mañanas.
[10] E despues cata si fuere la diferençia ques entre el sol y el planeta menos desto que tu tomaste de la tabla pues es abscondida. y si fuere mas es pareçiente.
[11] Mas Venus y Mercurio an subimiento de prima noche y ponimiento en las mañanas.
[12] Sepas pues lo que es entre qualquier dellos y entre el sol y de quien es la diferençia y eso guarda.
[13] Y despues entra con el signo do es [f. 49v] el planeta y toma lo que fuere en su derecho en la tabla del subimiento de los planetas de prima noche si fuere su argumento de çiento y ochenta grados. E eso es quando son derechos.
[14] E si fuere el argumento mas de çiento y ochenta grados. entra con el signo do es el planeta y toma lo que fuere en su derecho en la tabla del ponimiento de los planetas en las mañanas.
[15] Despues cata si fuere la diferençia la ques entre el planeta y entre el sol menos de lo que tu tomaste de la tabla. pues es abscondida. e si fuere mas es pareçiente.

[1] Capitulo 26. Como se a de saber el alcançamiento de los planetas uno sobre otro.

[2] El alcançamiento del planeta es que sea su longura del punto que es mas alto so el çielo menos de la longura del otro planeta del punto mas alto de so el çielo.
[3] E quando tu quisieres saber quanta es la longura del planeta del punto que es mas alto en su çielo. cata su argumento eguado.
[4] Pues si fuere menos de çiento y ochenta grados eso es la longura del planeta del punto mas alto de su çielo. Quiero desir los grados que son menos de çiento y ochenta.
[5] E si fuere el argumento mas de çiento y ochenta grados. menguaras aquello de tresientos y sesenta y lo que quedare eso sera la longura del punto mas alto de su çielo.

[1] Capitulo 27. Como se sabran los movimientos diversos de los planetas para un dia.

[2] Es menester la tabla de los movimientos diversos de los planetas para un dia en dos cosas la una es en saber el grado del planeta.

[3] Despues que lo ovieres eguado con un dia o con dos dias o con tres o con mas hasta X dias o lo que fuere cerca dello. despues del dia sabido a menos de tornar su yuncaçion sino por su lugar eguado que tu oviste egualado. La sigunda es en saber la çertidumbre de las oras de las conjuntiones de los planetas e de los lugares donde a de ser segun la quenta de las tablas a menos de egualarlas otra ves.
[4] Pues quando quisieres eso cata quando egualares la planeta a dia sabido e quisieres saber su movimiento diverso a dos dias o a X o a lo que fuere cerca desto quier ante quier despues.
[5] Pues sepas el movimiento diverso de aquel planeta para un dia en aquella sason en que tu eres.
[6] E esso sera quando entres con el çentro eguado [f. 50r] de aquel planeta en las reglas de la quenta. E toma lo que fuere en su derecho del movimiento del çentro de aquel planeta con la ynovoçaçion e guardalo.
[7] E despues entra con el argumento eguado de aquel planeta y toma lo que fuere en su derecho del movimiento del argumento con la ynvocaçion e guardalo.
[8] E despues cata si fuere el movimiento del argumento que tu tomaste de la tabla en sason de la endereçadura. E esso sabras por el escripto que es de suso en somo de la tabla endereçado.
[9] Ayunta pues ambos los movimientos e lo que se ayuntare eso es el movimiento diverso çierto de aquel planeta en un dia.
[10] Pues multiplicalo en quenta de los dias que tu quieres saber en ellos esta cuenta e lo que se ayuntare añadelo sobre el lugar eguado del planeta aquel dia sabido.
[11] Si fuere en estos dias despues de aquel dia sabido. o menguarlo dende si fuere ante.
[12] E lo que fuere despues del añadimiento o del menguamiento eso sera el lugar del planeta a aquel dia que quisieres e si fuere el movimiento del argumento en la sason de la retrogradaçion et es el ques escripto sobre el retrogrado. mengualo del movimiento del çentro. E quedara el movimiento del planeta para un dia. y es endereçado tardio obra pues con el como se adelanto.
[13] E si no fuere lo que salier del movimiento del argumento en el estado de la retrogradaçion. mas del movimiento del çentro. mengua dello el movimiento del çentro e quedara el movimiento del planeta para un dia y es retrogrado.
[14] Pues multiplicalo en la quantia de los dias e lo que se ayuntare cata si fuere en los dias que tu quieres despues del dia sabido y del eguado. y mengua lo que se ayuntare del multiplicar del lugar eguado del planeta e quedara su lugar a aquel dia.
[15] E si quisieres haser esto en el planeta en muchos dias tal como 20 dias o 30 quando fuere el planeta en el estado que se conviene esto en

ella. sepas el movimiento mediano della en el argumento y del movimiento del çentro en aquellos dias.
[16] E añade aquello sobre el argumento y sobre el çentro si fuere tu quenta a lo que a de venir o mengualo dende si fuere tu quenta a lo que passo. E lo que fuere despues del añadimiento o del menguamiento del argumento e del çentro.
[17] E entra [f. 50v] con ellos en la tabla de los movimientos diversos de los planetas a un dia et obra con ellos assi como se adelanto.
[18] Mas en Saturno. Jupiter y Mares conviene que hagas lo que se sigue en ellos a 15 dias. o a mas de XV dias hasta treynta dias quando fueren endereçados presurados en el movimiento o quando fueren retrogrados presurados en el movimiento o quando fueren con el sol en un signo o cerca del. porque entonce son mas pressurados que pueden ser et de menos diversidad en su movimiento e en su endereçamiento.
[19] E ansi hasen los planetas de suso quando fueren en el opposito del signo del sol son pressurados en movimiento en su retrogradaçion sin desvariar en su movimiento.
[20] Mas Venus y su retrogradaçion a mucha desvariança y tardase en ello.
[21] E quando fueren los planetas tardios en movimiento o estantes o en el comienço de la retrogradaçion o en su fin. no conviene que hagas esso en ellos dende a muchos dias. ende mas en Mares. Mercurio y Venus porque son de muchas variaciones en estos estados. por esso no les hagas mas de tres dias o de quatro.
[22] E Venus es de mas desvario que Mares.
[23] E Mercurio es de mucha variança por la pressura de su movimiento que ha en su epiçiclo.
[24] Pues no le hagas quenta mas de X dias.
[25] E quando quisieres açercar las horas de las conjunçiones e saber su lugar a menos de eguar o sin ygualar los planetas muchas veses. para mientes al dia en que tu piensas que sera la conjunçion et lo que fuere çerca della y eguala los dos planetas que se an de ayuntar a medio de aquel dia e despues cata si los fallares en un minuto y esso es lo que tu quisyste.
[26] E si fueren desvariados e fuere lo mas liviano de menos grados de lo mas pessado. e sepas lo que a entre ellos de grados et de minutos et guardalo.
[27] E despues sepas el movimiento de cada uno dellos para un dia et mengua el movimiento del pesado del movimiento del liviano. e lo que fincare eso sera el adelantamiento.
[28] Pues parte sobre el lo que guardaste saldran dias. e lo que fincare de los menudos sepas lo que les cabe de 24 horas.

[29] Pues añade estos dias y estas horas [f. 51r] sobre la hera en que tu yguialaste los dos planetas. E lo que fuere de la hera esa sera la hora de la conjunçion çertera.
[30] E despues multiplica el movimiento de la pesada en la quantia de los dias e añade lo que se ayuntare de grados y de minutos sobre el lugar de la pesada.
[31] O si quisieres multiplica el movimiento de la liviana en la quantia de los dias e añade lo que se ayuntare sobre el lugar de la liviana. E lo que fuere de qualquiera destas dos opiniones esso sera el lugar de la conjunçion.
[32] Et si fueren muchos los dias que tu añadiste e son los que te salieron de la partiçion. yguala los planetas a aquella sigunda era otra vez. porque hallaras diferençia en ambos lugares. E obra con ellos assi como se adelanto hasta que tengas el lugar çierto de la conjunçion.
[33] Et ansi haras en la conjunçion del sol con los planetas. E sabras el movimiento del sol a un dia entrando en la tabla con el argumento del sol el que es su alongamiento de su aux.
[34] E tomaras lo que fuere en su derecho en la tabla del movimiento del çentro de Venus e lo que fuere eso sera el movimiento diverso del sol para un dia.
[35] E si fuere el grado del planeta liviano mas que el grado de la pessada sepas que la conjunçion es passada.
[36] Mengua pues lo menos de los dos movimientos de lo mas mayor dellos et lo que fincare partelo sobre el adelantamiento del planeta e lo que te saliere seran dias. E lo que quedare de los minutos sepas quanto les monta de 24 horas.
[37] E mengua estos dias y estas horas de los dias en que ygualaste los dos planetas y esa sera la hora çierta de la conjunçion.
[38] E multiplica la quantia de aquellos dias y de aquellas oras en lo que se mueve la una de aquellas dos planetas en un dia. E lo que saliere de grados et de minutos mengualo del lugar de aquel planeta e quedara el lugar para la hora de la conjunçion.
[39] E si fueren los planetas ambos retrogrados e quisieres saber quando se juntaran ambos en el retrogradamiento. sabe el movimiento de cada uno dellos e en el retrogradamiento en un dia. E mengua lo menor de lo mayor. e sabe quanto es el adelantamiento del uno sobre el otro y parte sobre el lo que a entre ellos de grados y de minutos y guarda lo que [f. 51v] saliere.
[40] E despues cata si fuere el liviano menos grados de lo pesado. pues sepas que ya fueron ayuntados ante con quantia de aquellos dias e aquellas oras.

[41] Pues mengualo de la hera e si fuere el liviano en mas grados de lo passado[15] a la quantia de aquellas horas y dias se ayuntaran e añedildos[16] sobre la hera.
[42] E despues multiplicar la quantia de los dias y de las horas en lo que se mueve qualquiera dellas e mengualo de su mesmo lugar.
[43] E lo que fincare esso sera el lugar de la conjunçion si no es la conjunçion passada.
[44] O añadelo sobre aquel mesmo lugar si la conjunçion fuere passada e lo que fuere despues aquel sera el lugar de la conjunçion.
[45] E si fuere el uno de los planetas retrogado y el otro derecho y quisieres saber la ora de su conjunçion et sera esto a Venus o a Mercurio con el sol.
[46] Pues sepas el movimiento de cada una dellas en un dia e ayunta los dos movimientos. E lo que se ayuntare dellos parte sobre el los grados y los minutos que son entre los dos planetas et lo que saliere de los dias et de las horas guardalo.
[47] E despues cata si fuere el planeta que es retrogado en menos grados del otro ques derecho. Sepas que la conjunçion fue ante de aquella hera en que tu ygualaste los dos planetas con aquellos dias e horas que te salieron.
[48] E si fuere el planeta ques retrogado en mas grados de el otro ques derecho. la conjunçion a de ser depues de aquella hera con la quantia de aquellos dias e aquellas horas que te salieron.
[49] Multiplica pues la quantia de los dias y de las oras en el movimiento del planeta ques derecho por un dia et añadelo sobre su mesmo lugar.
[50] E lo que se ayuntare esso sera el lugar de la conjunçion. si la conjunçion no fuere passada. O mengualo de su mesmo lugar. si la conjunçion es passada. e lo que fuere esso es el lugar de la conjunçion.
[51] E ansi obraras en saber la conjunçion de la luna con qualquier planeta que sea derecha o retrogada ansi como se adelanto en saber la conjunçion de los planetas uno con otro.
[52] Mas no pares mientes [f. 52r] en la retrogradaçion de los planetas. en saber la quantia de su movimiento en la retrogradaçion con la luna por la pressura del movimiento de la luna y porque lo alcança ayna.
[53] E otrosi no pares mientes en saber el movimiento de los planetas tardios en la quenta de la conjunçion de la luna con ellas por la pressura del movimiento otrossi.

---

[15] *Sic*, instead of "pessado".

[16] *Sic*, instead of "añadelos".

[54] Mas a de ser la partiçion en esto sobre el movimiento de la luna a una hora.
[55] E saldran horas desde la hora de la conjunçion digo de la ynvocaçion çierta que haras a la luna y al planeta. hasta la ora de su çierta conjunçion.
[56] Pues esto es quando[17] as menester en la obra de la tabla de los movimientos de los planetas.
[57] E quando fuere la luna en quadradura del sol. e esto es quando ay entre ellos un quarto de çerco o lo que fuere cerca de quarto de çerco con un medio de signo. de mas o de menos. agora sea esta quadradura antes del sol agora despues del.
[58] Pues entra con el argumento eguado de la luna para saber su movimiento a una hora en las dos reglas de la quenta.
[60] E toma lo que fuere en su derredor o derecho en la tabla del movimiento de la luna en las quadraduras. E lo que fuere eso sera el movimiento de la luna a una hora sobre que se partira lo que ay entre ella y entre el planeta cuya conjunçion con la luna tu quieres saber y alcançar.

[1] Capitulo 28. Como se a de endereçar el asçendente et el medio çielo por lo que passo del dia o de la noche e para saber endereçar las otras cosas.[18]

[2] Quando esto quisieres saber mira quanto ay desdel subimiento del sol hasta aquella hora si fuere de dia o del ponimiento del sol hasta aquella ora si fuere de noche.[19]
[3] E si esto supieres por horas yguales quier de dia quier de noche. multiplicalas en 15 y los menudos si ay fueren lo que les cabe.
[4] E si lo supieres por horas temporales. e fueren de dia multiplicalas en los tiempos de un hora de aquel dia e si fuere de noche en los tiempos de un hora de aquella noche. E lo que fuere de aquella multiplicaçion por qualquier manera destas guardalo.
[5] E despues cata el lugar natural del sol. en qual signo esta [f. 52v] e en quantos grados del signo. y entra con el en la tabla de los subimientos de la villa en que esto demandas e toma lo que fuere en su derecho de los subimientos. y esto si fuere de dia.

---

[17] *Sic*, instead of "quanto".

[18] *Sic*, instead of "casas".

[19] Between the lines, above the word "noche", the text adds "de dia".

[6] Mas si fuere de noche entraras en esta mesma tabla con el grado del ponimiento opposito al lugar sobredicho del sol. E tomaras sus subimientos e lo que fueren los subimientos por qualquier destas dos maneras añadeles lo que guardaste en lo de la multiplicaçion de las horas.
[7] E lo que fuere a esto ayuntado si no allegare a 360 guardalo et si fuere mas de 360 mengua dello 360 y guarda lo que queda.
[8] E lo que fuere este guardado por qualquier manera destas. entra con ello en los subimientos de tu villa sobre que esto hesimos.
[9] E toma lo que fuere en su derecho de los grados yguales e lo que fuere aquello es el grado del asçendente en aquella sason de aquel signo donde hallaste los grados con que entraste.
[10] E despues entra con aquella mesma quenta con que ante entraste en la tabla de los subimientos del cerco derecho y toma lo que fuere en su derecho de los grados yguales. E lo que fuere aquel es el grado del medio çielo ques la X casa.
[11] E despues entra con el grado del asçendente en la tabla de la equaçion de las casas de la villa sobre questo labrares e toma lo que fuere en su derecho en la casa .2.3.4.5.6. E lo que fuere aquellos seran los çentros destas casas sobredichas. E la quarta casa es opposita de la Xª la 7ª del asçendente. la 8ª de la 2ª la 9ª de la 3ª la 2ª[20] de la 5ª la 12ª de la 6ª.
[12] Con esto avras endereçado el ascendente et las otras casas. E este asçendente et los çentros de las casas que te salen por este hecho todos son del comienço natural. Si los quisieres tornar del otro començamiento proprio mengua de cada uno dellos quanto es el movimiento del llegamiento y del tornamiento mientra que fuere de menguar.
[13] E nos en estas nuestras tablas pusimos tablas de las equaçiones de las casas sobre la çibdad de Toledo. porqueste lybro es fecho sobre esta çibdad sobredicha.

[1] Capitulo 29. De la conjunçion de los dias con sus noches.

[2] Muchos piensan que cada dia con su noche es ygual con qualquier otro dia con su noche et tienen ansimesmo que cada dia con su noche es de 24 [f. 53r] horas. mas no es ansi porque el un dia mediano con su noche es subimiento de los tiempos del çerco equinoçial. que son todos 360 quier suba sobre el çerco del orisonte quier sobre el cerco de mediodia.

---

[20] *Sic*, instead of "11ª".

[3] E lo que sube mas de los 360 grados del cerco equinoçial co[n] 59 minutos. lo ques movimiento mediano del sol en un dia et en una noche.
[4] E el dia diverso con su noche es lo que sube de los tiempos del equinoçio que son 360. E lo que sube de mas del equinoçio con el movimiento diverso del sol en un dia e en una noche. e es de todo en todo a horas mas de 59 minutos y a horas de menos. y a las veses ni de mas ni de menos.
[5] E si nos pusieremos el comienço del dia del cerco del orisonte. cambiarsenos a en cada orizonte por la diversidad de los tiempos de los subimientos de los grados del cerco del sodiaco que se desvarian en cada orison.
[6] E si nos pusieremos el començamiento del dia del çerco de mediodia sera todavia fincable en estado ni se cambiara ni se mudara.
[7] E[21] esto es por la ygualdad de los subimientos de los grados del çerco del sodiaco quando passa por medio del çielo. passan con ellos de los tiempos del equinoçio en cada çibdad una quantia ni mas ni menos.
[8] E por esto no pusieron los sabios antiguos el comienço de los dias en quanto cabe de endereçar las estrellas e en sus quentos. ni del nasçimiento del sol ni de su ponimiento. mas pusieronlo del mediodia o del medio de la noche.
[9] E todos los movimientos que son puestos en las tablas son puestos sobre dias yguales.
[10] E quando el hombre no parare mientes en lo que se ayunta entre los dias yguales y entre los dias diversos. puede ser que en el movimiento del sol e los otros planetas fuera dende la luna non le empesca. mas en la luna señera. haserle a enojo pareçido por la pressura del movimiento de la luna.
[11] E lo que se ayuntare de esta desvariança ques entre los dias yguales y los dias diversos es cerca de media hora.
[12] E muevese la luna en esta media hora en unos tiempos cerca de 18 minutos mas lo que añaden unos dias diversos sobre los dias medianos. E lo que ende mengua dellos es doble desta media hora e hasese una hora.
[13] E esta desvariança sobredicha se ayunta de [f. 53v] dos partes. la una es la diversidad del movimiento del sol y la otra es la desvariança del passamiento de los signos quando passan por medio del çielo porque no passan ende con una quantia. E el lugar del

[21] MS: E e.

menguamiento es çerca de 20 grados passados de aquario hasta el comienço de scurpion.

[14] E del lugar del añedimiento es cerca del comienço de scurpion hasta 20 grados de aquario. E lo que se ayunta del desvariamiento del movimiento del sol es çerca de tres grados y quarto.

[15] E lo que se ayunta dello por el passamiento de los signos sobre el medio çielo es cerca de 4 grados y 32 minutos.

[16] E hasese pues lo que se ayunta de ambas partes 7 grados y 48 minutos ques media ora ygual y de una parte de 48 partes de un hora y esta es la quantia de la desvariança que ay entre los dias yguales y entre los dias diversos. E nos posimos los movimientos medianos en este nuestro libro a quenta que fue el lugar del sol por su movimiento mediano en VI grados y e[22] 57 minutos de capricornio e su lugar çierto es visible en 7 grados y 29 minutos capricornus.

[17] E a estos lugares a menester que se ayunten los otros lugares en todos los dias del año los que quisieren obrar con este libro.

[18] Pues quando quisieres eguar los dias diversos e tornarlos dias medianos toma lo que ay entre el lugar del sol mediano el sobredicho y entre el lugar segundo mediano que tu quieres e cata lo que an entrambos lugares de los grados de los signos e guardalo et ponle diferençia primera.

[19] E despues toma lo que a entre su lugar çierto el sobredicho e entre el lugar çierto segundo e cata lo que a entre ellos e ponle nombre diferençia sigunda.

[20] E despues cata quantos son los subimientos del cerco derecho desta diferençia sigunda e guardalos. O si fuere la quantia destos subimientos mas de la diferençia primera. sepas que ay entre ellos e ques su proporçion un hora ygual e lo que oviere o fuere añadelo sobre aquellos dias diversos que tu oviste.

[21] E si fuere la quantia de los subimientos menos de la diferençia primera mengualo dellos. E lo que fuere dias despues del añadimiento o del menguamiento sera tiempo tornado en dias diversos a dias medianos.

[22] E si quisieres tornar dias medianos a dias diversos cata a los subimientos de la diferençia sigunda.

[f. 54r] [23] Pues si fuere menos de la diferençia primera añade lo que te salio de ora sobre los dias medianos. e si fueren mas mengualo dellos.

[24] E lo que fueren de los dias medianos despues del añadimiento o despues del menguamiento sera tiempo tornado de dias medianos a dias diversos e en esta opinion no avemos yerro alguno.

---

[22] *Sic.*

[1] Capitulo 30. Como se a de saber la conjunçion y anssimesmo la oposiçion.

[2] Quando quisieres eguar la conjunçion o la opposiçion sabe los años romanos passados de la hera alfonsi con el año en que es aquella conjunçion o aquella opposiçion. E entra con aquellos años en la tabla de las conjunçiones si quisieres conjunçion o en la tabla de las opposiçiones si quieres opposicion et toma lo que hallares en su derecho en las quatro tablas ques la una dellas dias et oras e la 2ª medio curso de las luminarias et la 3ª argumento de la luna et la 4ª el movimiento de la latitud.
[3] Et despues entra en los espandidos con lo que te finco de los años et toma lo que hallares en su derecho en las 4 tablas sobredichas e escriviras cada cosa en su linaje.
[4] E despues entra en la tabla de los meses con el mes romano cuya conjunçion tu quieres saber o su opposiçion. E tomaras lo que fuere en su derecho de las quatro tablas sobredichas e sabras en qual tabla de los meses entraras en catar el año en que es aquel mes. Si es visiesto entra en la tabla de los meses de los años de bisiesto. si simple entraras en los meses del año simple.
[5] Et quando supieres en qual de las tablas sobredichas as de entrar tomaras lo que hallares en derecho de aquel mes cuya conjunçion o opposiçion tu quieres saber por las 4 tablas que arriba te dixe.
[6] E pornas pues lo que te saliere debaxo de lo que oviste antes escripto cada uno so su linaje e mengua de los dias que tu hallares en la tabla de los años en simples tanto como aquellos dias que tu hallares en derecho del mes con que entraste.
[7] E si fueren los dias que hallares en derecho del mes mas de los dias que son en derecho del año con que tu entraste añade sobre los dias que son en derecho del año con que entraste la quantia de los dias del mes que fue antes del mes con que entraste. e mengua de lo que se ayuntare los dias que tu hallaste en derecho del mes et escrive lo que te fincare.
[8] E aquella conjunçion o aquella opposiçion sera en aquel mes ques antes del mes con que tu entraste.
[9] E avra en el año questo aconteçiere treze conjunçiones o treze opposiçiones.
[10] E nos avemos señalado [f. 54v] sobre estos años en las tablas y escrivimosles de fuera en la tabla su numero 13 porque se conosca que acaezçen en ellos 13 conjunçiones o 13 opposiçiones.
[11] E la conjunçion o la opposiçion que fuere en este tal año en el mes de diziembre se endereça por el mes trezeno que nos posimos en las tablas de los meses por el mes despues de diziembre. que nos le nombramos hy el mes soberano.

[12] E quando los dias ovieres por qualquier maneras de las sobredichas ayunta las horas et sus minutos y sus sigundos et has dellos dias et horas et minutos de hora.
[13] E lo que se ayuntare de dias et de horas et de minutos aquello sera la hera de la conjunçion o de la opposiçion mediana.
[14] Pues si tu quisieres saber la conjunçion o la opposiçion çierta media del medio curso del sol su aux. E entra con lo que finca en la tabla de la ynvocaçion del sol y toma lo que fuere en su derecho de la ynvocaçion del sol y cata si fuere esto con que tu entraste menos de 180 grados. quiero desir VI signos mengua lo que te saliere de la ynvocaçion del medio curso del sol.
[15] E si fuere mas de 180 grados añadelo sobre el medio curso del sol. E lo que fuere del medio curso del sol despues del añadimiento o del menguamiento esso sera el lugar çierto del sol.
[16] Despues egua el medio curso de la luna con su alongamiento del aux del sol e saca el medio curso eguado de la luna ansi como te lo amostre en el 17 capitulo e guarda lo que fuere.
[17] E despues entra con el argumento de la luna en la tabla de la ynvocaçion de la luna e toma lo que fuere en su derecho de la ynvocaçion del argumento.
[18] E despues cata si fuere el argumento menos de 18[23] grados. mengua lo que te saliere de la ynvocaçion del argumento del medio curso eguado de la luna. E si fuere mas de 180 grados. añadelo sobre el medio curso eguado despues del añedimiento o del menguamiento aquello sera el lugar çierto de la luna. E si anadieres la yncaçion sobre el medio curso eguado de la luna añade a esa mesma sobre el movimiento de la latitud.
[19] E si la menguaste menguala otrossi del movimiento de la latitud.
[20] Pues si hallares el sol et la luna ayuntados en un grado et en un minuto eso es lo que tu quisiste saber.
[21] E si oviere entre ellos longura añade sobre ellos todavia su dozena parte. E lo que se ayuntare partelo sobre el movimiento de la luna a una hora e lo que te [f. 55r] saliere seran horas y fraçiones de hora.
[22] Pues si fuere la longura a la luna mengua a estas horas e a estas fraçiones de la hera que oviste.
[23] E si fuere la longura al sol añade las horas et las fraçiones sobre la era que tu oviste. avras era eguada a la hora de la conjunçion et de la opposiçion.
[24] E las horas que tu ovieres de mas sepas que son despues de mediodia de aquel dia que salio.

---

[23] *Sic*, instead of "180".

[25] E despues yguala los dias con sus noches ansi como te mostre antes en el capitulo 29 e torna a la hera diversa e lo que fuere aquella sera la hera de la conjunçion o de la opposiçion visible en la çibdad de Toledo.
[26] E si fuere la longura de la luna mengua la longura et su dosena parte del lugar de la luna e del movimiento de la ladesa. E mengua la dosena parte de la longura del sol.
[27] E si fuere la longura del sol añade la longura e su dosena parte sobre el lugar de la luna e sobre el movimiento de la latitud. E añade la 12ª parte de la longura sobre el lugar del sol e sera el lugar çierto de la conjunçion o de la opposiçion.
[28] E lo que te saliere del movimiento de la ladesa despues del añadimiento e del menguamiento aquello sera el movimiento de la latitud eguada a la ora de la conjunçion o de la opposiçion.
[29] E si fuere tu cuenta a la opposiçion sepas que la luna sera en oposito de aquel grado y de aquel minuto que te salio.
[30] E toda esta opinion es en conjunçion o en opposiçion que no a en el ecliuse.
[31] Mas si oviere en el eclipse et tu lo quisieres averiguar as de tener manera que sea el retificar tal como sale de la yncaçion mesma.
[32] Pues sepas que avras de ygualar aquella conjunçion o aquella opposiçion ansi como te mostre en este capitulo sin añedimiento ni menguamiento.
[33] E quando ovieres su ynvocaçion tal e supieres su era qual yo te mostre a haser. tornate y saca el medio curso del sol et el medio curso de la luna et su argumento aquella ora cada uno dellos de su tabla. E egua cada uno de los dos luminares qual yo te e amostrado en la yncaçion de cada uno dellos en su lugar y no trueques nada dello.
[34] Pues si los hallares ayuntados en un minuto esso es lo que tu quesiste.
[35] E si oviere entre ellos longura e fuere poca haz con ella como tengo dicho en este capitulo.
[36] E si fuere aquella longura mucha egualo la 3ª vez hasta que te salgan sus lugares en un minuto e avras estonçe sus lugares bien averiguados.

[1] Capitulo 31. Como se sabra el catamiento de la luna segun su diversidad en las horas del eclipsy.

[f. 55v] [2] Quando quisieres saber la diversidad del catamiento de la luna en longura et en anchura a la hora de la conjunçion que es la hora del eclipsi. sepas el lugar eguado de la luna a la hora de la conjunçion cierta. Sabe otrosi el grado de la asçendente aquella hora e que tanta

es la longura de la conjunçion çierta del mediodia de horas yguales et de sus partes.
[3] E quando supieres todo esto entra con aquellas horas eguales en la tabla de la diversidad del catamiento en las horas que son ante de mediodia o en las horas que son despues de mediodia si la conjunçion fuere despues de mediodia.
[4] E toma lo que fuere en su derecho en el signo do es la luna de los minutos de la longura y de los minutos de la latitud.
[5] E lo que fuere aquello sera la diversidad del catamiento de la luna si la luna fuere en el començamiento de aquel signo et si no fuere en el començamiento del signo e oviere andado algunos grados del sabe quantos grados del a andado e lo ques la su proporçion de treynta e guarda aquella proporçion.
[6] E despues entra en el 2° signo con aquellas horas que antes entraste y toma lo que fuere en su derecho de la diversidad del catamiento de la luna en longura y anchura. et escrivelo sobre lo primero que tu oviste tomado ante y mengua lo menor de lo mayor e lo que fincare toma dello tanto como de aquella proporçion que te mande guardar e guarda lo que fuere.
[7] E cata si fuere aquello que tu hallaste en derecho del signo segundo en cada uno de los dos catamientos mas de lo que tu hallaste en el primer signo en su derecho. E añade lo que tu tomaste con la proporçion sobre aquello que tu hallaste en el primer signo en su derecho en qualquier acatamiento de la longura o anchura.
[8] E si fuere lo que tu hallaste en derecho del segundo signo en cada uno de los dos acatamientos menos mengua lo que tomaste con la proporçion de lo que oviste tomado del primer signo.
[9] E lo que fuere despues del añedimiento o del menguamiento esso sera la diversidad del catamiento de la luna en longura y anchura al grado de la luna que oviste egualado aquella ora.
[10] E si oviere algunas fraçiones con las horas tomaras aquellas fraçiones lo que les pertenesçe de las yncaçiones de los planetas. y esta opinion es quando fuere la luna en su alongamiento mediano.
[11] Mas si no fuere ay entra con el argumento [f. 56r] de la luna en la tabla del endereçamiento e toma lo que hallares en su derecho de los minutos e multiplicalo en los minutos de la longura y anchura cada uno aparte. E guarda lo que se ayuntare.
[12] E despues cata el argumento con que tu entraste en la tabla del endereçamiento en el qual medio es caydo.
[13] Pues cayere en la mitad de suso mengua lo que se te ayunto del multiplicar de los minutos de la longura y anchura cada uno dellos aparte.
[14] E si cayere en el medio de yuso añade lo que se te ayunto del multiplicar sobre los minutos de la longura y anchura cada uno dellos

aparte. E lo que fuere despues del añadimiento o despues del menguamiento essos seran los minutos çiertos de la longura y de la anchura de la diversidad del catamiento de la luna.
[15] E despues sepas si oviere entre el lugar de la luna e el grado del asçendente que es en aquella hora menos de noventa grados añade la diversidad del catamiento de la longura sobre el lugar de la luna.
[16] E si fuere mas de 90 grados mengualo del lugar de la luna e lo que fuere esso sera el lugar visible de la luna aquella hora.
[17] E si quisieres saber la hora de la conjunçion visible toma la diversidad del catamiento en la longura de la [conjunçion] eguada. E partelo sobre el adelantamiento que a la luna sobre el sol en la una hora e añade sobre el su dosena parte e lo que se ayuntare partelo sobre el movimiento diverso de la luna a una hora.
[18] E lo que te saliere de ora e de sus fraçiones mengualos de las oras de la conjunçion eguada si la oviere entre el asçendente y el lugar de la luna menos de 90 grados e añadelas sy oviere mas de 90 grados. E lo que fuere despues essas seran las horas de la conjunçion visible en aquella çibdad do tu obraste.
[19] E tal como esto faras quando tu quisieres saber la diversydad del catamiento de la luna a qualquier otra hora que tu quisieres.
[20] E quando oviere entre el asçendiente e el lugar de la luna menos de 90 grados añade la diversidad del catamiento de la luna sobre su lugar eguado.
[21] E si fuere mas de 90 mengualo dende y avras su lugar visible de la luna.
[22] E si quisieres saber la diversidad del catamiento de la luna a la hora de su ponimiento. sepas su longura del medio çielo a aquella hora que tanto es de horas yguales.
[23] E entra con ellas e obra ansi como se adelanto y escusarte a de esta obra en las visiones de las lunas que entres con las horas que son en la postrimera regla del signo do es la luna.
[f. 56v] [24] E toma lo que fuere en su derecho de la diversidad del catamiento de la luna en longura y en anchura.
[25] E lo que fuere esso es lo que tu quisiste.
[26] E podras escusar su yncaçion con la tabla del endereçamiento.

[1] Capitulo 32. Como se a de saber la diversidad del catamiento de la luna en longura y anchura en todo tiempo y en cada lugar. e propiamente en los lugares que tienen en anchura mas de 23 grados y 33 minutos.

[2] Quando esto quisieres sabe la anchura de la villa en que quieres labrar esto.

[3] E lo que fuere mengualo de 90 e lo que queda sera la altura de la ygualdad en aquella villa guardalas amas.
[4] E despues sepas el tiempo del dia en que esto quisieres saber por oras yguales en aquella villa.
[5] Sabe tambien el lugar de la luna en aquella hora en que grado es de los grados de los signos del comienço natural e guarda todo esto.
[6] E desi saca el arco que aqui nombramos arco del altura en la siguiente manera.
[7] Cata la hora del dia en que esto demandas e si fuere en començamiento del dia et la luna estoviere en el orizonte occidental este arco del altura sera entonçe de 90 grados.
[8] E esto determinamos desta manera. por la diversidad del catamiento de la luna porque nunca es menester sino en ora de conjunçion de las luminarias o cerca de ella.
[9] E si la hora en que esto demandas fuere en mediodia e la luna fuere en medio çielo cata la declinaçion del grado en ques la luna et su parte.
[10] E si fuere septentrional menguala de la anchura de la villa.
[11] E si fuere meredional añadela a la anchura de la villa e lo que fuere la anchura de la villa despues desto aquello sera el arco de la altura que demandaremos.
[12] Mas si la hora en questo demandas no fuere en alguna destas tres sasones terminada sabe quantas horas ygualas son passadas de aquel dia en aquella villa.
[13] E si es aquella hora antes de mediodia o despues sepas el grado del asçendente del començamiento natural en aquella hora. E sepas los sobimientos e toma su signo en la tabla de los signos e nombralo el signo primero.
[14] Despues desto sabe quanto ay del grado de la luna hasta el grado del asçendente de los grados de los signos.
[15] E si la hora fuere antes de mediodia sepas quanto a del grado de la luna hasta el opposito del grado asçendente otrossi de los grados de los signos. E lo que fuere esto por qualquiera destas dos maneras aquello es lo que ay de la luna al orisonte en aquella hora. e lo que [f. 57r] fuere sepas su signo et nombralo signo segundo.
[16] E despues sabe el signo de la altura de la ygualdad en aquella villa et lo que fuere nombralo signo 3°.
[17] E despues mira quanto ay del grado del asçendente al uno de los dos puntos de la ygualdad al que fuere mas cerca. e lo que fuere sabe su signo e nombralo 4° signo.
[18] Desto multiplica el signo primero en el signo 3 y parte por el 4° signo lo que fuere de la multiplicaçion e lo que saliere desta multiplicaçion nombralo V signo.

[19] E multiplica este signo 5° en el 2° e lo que fuere desta multiplicaçion partelo por sesenta e lo que saliere desta partiçion sera signo. Sabe pues su arco segun que saben el arco por el signo e lo que fuere este arco mengualo de 90 et lo que quedare aquello sera el arco del altura que fue demandado.

[20] E quando el arco del altura fuere sabido por qualquier destas maneras sobredichas guardalo.

[21] E despues saca lo que aqui nombramos angol del altura en la siguiente manera. Cata a la hora en questo demandas si fuere en comienço del dia o en el fin del. E si la luna fuere en el orizonte o en el occidente saca lo que ante nombramos signo 5 e lo que fuere sepas su arco e lo que fuere aquel arco aquello sera el angol de la altura que aqui demandamos.

[22] E si la ora en questo demandas fuere en mediodia et la luna en medio del çielo sabe los subimientos del grado de la luna en el çerco derecho començando del punto de la ygualdad del verano.

[23] Esto saberse a tomando los subimientos del grado en la tabla de los movimientos o subimientos del cerco ygual menguando dellos 90 . E lo que fuere estos subimientos en esta guisa sabe su signo et nombralo nota primera.

[24] E despues sabe quanto ay del grado de la luna hasta el uno de los dos puntos de la ygualdad al que mas çerca fuere e lo que fuere. Toma su signo y nombralo nota segunda y despues multiplica la nota primera en 60 e lo que fuere de la multiplicaçion partelo por la nota 2ª. E lo que saliere desta partiçion sera signo. sabe su arco e lo que fuere este arco aquello es el angol del altura que aqui demandamos.

[25] Mas si la hora en que esto demandas no fuere en una destas sasones cata el 5° signo que ante sacamos y sepas su arco.

[26] Despues sabe el signo del complimiento deste arco e lo que fuere este signo multiplicalo en 60 y lo que fuere de la multiplicaçion nombralo guardado.

[27] Despues cata el arco del altura que ante supiste e toma su signo e lo que fuere este signo parte el guardado por el e lo que saliere desta partiçion sera signo. Sepas su arco [f. 57v] e lo que fuere aquel es el angol que aqui demandamos el qual nombramos angulo del altura. Pues quando ovieres por qualquier via destas tomado el arco del altura e el angulo del altura susodicos para mientes a la luna e si la hallares en el zodiaco sin ninguna latitud a septentrion ni a meredion este arco y este angulo es con que obran en saber la diversidad del catamiento de la luna.

[28] Mas si la luna oviere la ladesa a septentrion o a mediodia este arco y este angulo an menester de eguarlos en la siguiente manera.

[29] Sabe la ladesa de la luna que tanta es aquella hora y en su parte e despues cata en[24] angulo de la altura que ante sopiste e toma su signo y nombralo la corda primera e toma el signo de su complimiento y nombralo la corda sigunda.
[30] Despues desto multiplica la corda primera en el signo de la latitud de la luna e lo que fuere de la multiplicaçion nombralo guardado primero.
[31] E multiplica la corda 2ª en el signo de la latitud de la luna e lo que fuere nombralo guardado 2º.
[32] Despues parte lo primero guardado por 60 e lo que saliere desta partiçion sera signo. sabe su arco e lo que fuere este arco es equaçion de arco o arco de equaçion guardalo.
[33] E cata la anchura de la luna e si fuere septentrional mengua el arco de la equaçion del arco del altura que ante oviste.
[34] Mas si la ladesa de la luna fuere meredional añade el arco de la equaçion al arco del altura que ante oviste.
[35] E lo que fuere el arco del altura despues desto nombralo arco endereçado e toma su signo e multiplicalo en sy mesmo y lo que ende fuere nombrado llamalo quadrado primero.
[36] E toma lo guardado segundo e partelo por 60 e multiplica en si mesmo lo que saliere de la partiçion e lo que fuere nombralo guardado[25] 2º.
[37] Despues ayunta los dos quadrados e toma su raiz de lo que fuere del ayuntado e lo que fuere esta raiz sera signo. Sepas su arco y aquel sera el arco del altura endereçado e averiguado el ques menester en saber la diversidad del catamiento de la luna. guardalo para la obra.
[38] Despues toma el guardado sigundo e partelo por el signo del arco averiguado antedicho e lo que fuere desta partiçion es signo. Sabe su arco e lo que fuere nombralo el arco del endereçamiento.
[39] Despues cata la anchura de la luna et si fuere septentrional et fuere lo que es de la luna hasta el asçendiente menos de 90 grados añade el arco del endereçamiento al angulo del altura que ante oviste.
[40] E si fuere lo ques de la luna al asçendente mas de 90 grados mengua el arco del endereçamiento del angulo del altura.
[41] E si la ladesa de la luna fuere meredional e fuere lo ques de la luna hasta el asçendente menos de 90 grados [f. 58r] mengua el arco del enderecamiento del angulo de la altura.

---

[24] *Sic*, instead of "el".

[25] *Sic*, instead of "quadrado".

[42] E si la ladesa de la luna fuere meredional y fuere lo ques de la luna hasta el ascendente menos de 90 grados mengua el arco del endereçamiento del angulo del altura e si fuere lo ques de la luna al asçendente mas de 90 grados. añade el arco del enderecamiento al angulo del altura. E lo que fuere el angulo del altura despues deste añadimiento o deste menguamiento aquello sera el angulo del altura endereçado el ques menester para saber la diversidad del catamiento de la luna.
[43] Pues quando ovieres el arco del altura y del angulo del altura endereçados y averiguados guardalos amos.
[44] Despues desto sabe el argumento eguado e saca el alongamiento del sol de sus tablas a aquella sason.
[45] E despues toma el arco del altura averiguado y entra con el en la tabla de la diversidad del catamiento del sol y toma lo que fuere en derecho de la diversidad del sol y escrivelo en su parte.
[46] Despues entra con este mesmo arco en las tablas de la diversidat del catamiento de la luna en las generales.
[47] E toma lo que aconteşçiere en derecho deste arco en la tabla primera e nombralo la primera diversidad.
[48] E toma otrosi lo que fuere en su derecho en la tabla 2ª e nombralo 2ª diversidad.
[49] Toma ansimesmo lo que fuere en su derecho en la tabla 3ª y nombralo la 3ª diversidad.
[50] E toma otrosi lo que fuere en su derecho en la tabla 4ª y lo que fuere nombralo diversidad 4ª.
[51] E ansimesmo mengua la diversidad primera de la diversidad 2ª e lo que finca nombralo la sobra primera.
[52] E mengua la diversidad 3ª de la 4ª diversidad. et nombra lo que queda la 2º sobra.
[53] Despues entra con el argumento eguado de la luna en la tabla de la proporçion entre los dos alongamientos de la luna en su epiçiclo. E toma lo que fuere en derecho de los minutos de la proporçion e de los sus segundos e multiplica los minutos de la proporçion en la sobra primera. E lo que fuere sera la diversidad del catamiento eguado primera.
[54] E despues multiplica los minutos de la proporçion en la 2ª sobra e lo que fuere desta multiplicaçion añadelo a la 3ª diversidad e lo que fuere sera la diversidad del catamiento eguado segunda.
[55] Despues desto mengua la diversidad eguada primera de la diversidad eguada 2ª e lo que queda nombralo la sobra eguada.
[56] Despues entra con el alongamiento de la luna en la del sol que ante sacamos en la tabla de la proporçion entre los dos a[f. 58v]longamientos de la luna en su çerco del çentro saliente. E toma lo que fuere en su derecho de los minutos de la sobejança e multiplica

los minutos de la sobejança en la sobra eguada. E lo que fuere desta multiplicaçion añadelo a la diversidad del catamiento eguada primera e lo que fuere sera la diversidad del catamiento de la luna sigun el arco del altura. mengua della el catamiento del sol e lo que finca nombralo diversidad del catamiento endereçada.

[57] Despues toma el angulo del altura endereçado e sepas su signo e nombralo el signo de la ladesa. E sepas el signo de su complimiento e nombralo el signo de la longura. Despues multiplica el signo de la ladesa en la diversidad del catamiento endereçada e lo que fuere de la multiplicaçion partelo por 60 y lo que saliere desta partiçion aquello sera la diversidad del catamiento de la luna endereçada y averiguada en la ladesa a aquella hora.

[58] E multiplica el signo de la longura en la diversidad del catamiento endereçada e lo que fuere desta multiplicaçion partelo por 60 e lo que saliere de la partiçion es la diversidad del catamiento endereçada y averiguada a aquella sason.

[59] Parte pues esta diversidad del catamiento en la longura por el zaba de la luna e salirte an horas o fraçiones de ora guardalo.

[60] E cata lo que ay del grado de la luna hasta el grado del asçendente. E si fuere menos de 90 grados sabras que la diversidad del catamiento en la longura es segund la horden de los signos e por esta añade la diversidad del catamiento en la longura al lugar endereçado de la luna e sera en su lugar visible. E mengua las horas que te salieron de la partiçion de la hora de aquella conjunçion e sera la hora de la conjunçion visible.

[61] E si lo ques del grado de la luna hasta el grado del asçendente fuere mas de 90 grados sabras que la diversidad del catamiento en longura es en contra de la horden de los signos. Por esto mengua la diversidad del catamiento en la longura del lugar. endereçado de la luna e sera su lugar visible. E añade las horas que te salieron de la partiçion a la hora de la conjunçion endereçada e sera la hora de la conjunçion visible.

[62] E sepas que en todos los lugares que en la ladesa mas de toda la declinaçion que en tales lugares hablamos la diversidad del catamiento en la ladesa siempre sera meredional.

[63] Saca pues el lugar del geuzaar a la [f. 59r] hora visible et sepas deste lugar e del lugar visible de la luna la ladesa de la luna y su parte segun es dicho en su capitulo. E si la latitud de la luna fuere meredional añade la diversidad del catamiento en la anchura y sera la ladesa de la luna justa y tambien sera meredional.

[64] Mas si la ladesa de la luna fuere septentrional y fuere menos que la diversidad del catamiento en la ladesa. mengua la ladesa de la diversidad del catamiento e lo que finca sera la ladesa de la luna justa y sera meredional.

[65] E si la ladesa de la luna fuere septentrional y mas de la diversidad del catamiento en la ladesa de la luna lo que fincare sera la ladesa de la luna visible e sera septentrional.
[66] E la diversidad del catamiento de la luna nunca es menester sino en conjunçion en que aya eclipsi.
[67] E en tal sason como esta el argumento de la luna no ay equaçion. ni la luna non alongamiento del sol. nin la luna otrosi non a ladesa del zodiaco. E si alguna cosa oviere de todo esto sera muy poco de manera que nin empeçe nin puede acaesçer por ello daño alguno en ninguna destas obras deste hecho.
[68] E quando esto fuere no avras menester de eguar el arco del altura ni el angulo del altura e complir tanto a la obra el arco. el angulo que te sale de la primera ves a menos de las equaçiones que les posimos.
[69] Otrosi complirte a el argumento de la luna que te sale de las tablas por argumento eguado.
[70] E otrosi complirte an de las 4° tablas de la diversidad del catamiento las dos señeras endereçando la primera y la 2ª endereçandolas con los minutos de la proporçion del epeciclo solamente.
[71] E la que ante nombramos la diversidad del catamiento eguado primera aquella mesma sera la diversidad del catamiento endereçada con que obran.
[72] E lo que ante nombramos corda primera aquello mesmo sera el signo de la ladesa.
[73] E lo que ante nombramos corda 2ª aquello mesmo sera el signo de la longura que por esto les camiamos nombres por ser señalado.
[74] E quando tu obrares abreviando todos estos sigund diximos fallaras lo que demandas sy Dios quisiere.

[1] Capitulo 33. Como se a de saber endereçar el eclipse solar.

[2] Quando quisieres saber la opposiçion del eclipsi solar egua la conjunçion que fuere de dia y el sol cerca de la cabeça del dragon o de su cola e eso es [f. 59v] quando fuere el movimiento de la ladesa de un grado hasta 12 grados. o de 168 hasta 190[26] porque la ladesa de la luna sera entonçe septentrional.
[3] E si fuere mas o menos de lo que avemos dicho no avra eclipse en todo el clima de Toledo. E en quanto fuere su ladesa cerca del anchura de Toledo desde oriente a oçcidente [no le avra] sino muy poco.

[26] *Sic*, instead of "180".

[4] Mas en las çibdades que son sus anchuras desde un grado hasta 3° grados puede ser que aya en ellas eclipsi solar quando fuere la ladesa de la luna meredional y esto sera quando fuere el movimiento de la ladesa a menos de un çerco con menos de 6 grados.
[5] E despues sabe el lugar de la conjunçion eguada y el movimiento de la ladesa ques yguada con la yucaçion de la luna y con la longura y su dosena parte.
[6] E sepas la hora de la conjunçion eguada con la diversidad de los dias con sus noches en la çibdad de Toledo a qual çibdad tu quisieres.
[7] E sepas que tanto ay entra la hora de la conjunçion y entre el mediodia de oras yguales et de sus fraçiones e si son aquellas horas antes del mediodia o despues.
[8] E despues sepas por aquellas horas las que son entre la hora de la conjunçion y entre mediodia la diversidad del catamiento de la luna en la longura e saca con ellas las horas de la conjunçion visible anssi como se dixo en el 31 capitulo.
[9] E lo que fuere en aquellas horas sepas su longura del mediodia y entra con ellas la sigunda ves et saca la diversidad del catamiento en la longura. E has como hesiste ante e saca con la diversidad que te sale la 2ª ves las horas de la conjunçion visible sigunda ves y lo que fueren aquellas horas y sus fraçiones en la 2ª ves sepas otrossi su longura del mediodia y entra con ellas terçera vez en la tabla de la diversidad del catamiento y toma lo que fuere en su derecho en la diversidad del catamiento en la longura y anchura y escrive cada uno aparte y saca las horas de la conjunçion visible por la diversidad del catamiento en la longura ansi como hesiste ante. E todavia las horas que ovieres de añedir o de menguar en estas tres veses sobredichas añadelas si fueren de añedir sobre las horas de la conjunçion eguada las que tu eguaste a primas quando eguaste la conjunçion o mengualas dellas si fueren de menguar. [f. 60r] Pues lo que fueren las horas de la conjunçion visible en la 3ª ves essas seran las eguadas horas con la diversidad del catamiento a medio del eclipsi. E essas seran las horas de la conjunçion visible y çierta e añade la diversidad del catamiento con su dosena parte sobre el movimiento de la latitud sy añediste las horas e mengualas si las menguaste.
[10] E despues toma los menudos de la diversydad del catamiento en la ladesa y es todavia meredional en Toledo y en cada çibdad cuya anchura fuere mas de 24 grados multiplicados en 11 grados y medio. y guarda lo que se juntare de grados y de menudos.
[11] E despues cata si fuere el lugar de la conjunçion con la cabeça del dragon mengua los grados y los minutos que te salieron del multiplicar del movimiento de la ladesa que tu oviste eguado con la

yucaçion de la luna y con la longura y su dosena parte y con la diversidad del catamiento en la longura.
[12] E si fuere el lugar de la conjunçion con la cola del dragon añade los grados y los minutos sobre el movimiento de la ladesa eguado. E lo que fuere el movimiento de la ladesa despues del añadimiento o del menguamiento esso sera el argumento de la ladesa a medio del eclipsi.
[13] Pues entra con el en la tabla del ecliuse solar en la longura mayor si fuere la luna en su longura a la luenga.
[14] E en la tabla del eclipsi solar en la longura menor si fuere la luna en su longura la çercana e toma lo que fuere en su derecho de los dedos y de los minutos del caymiento.
[15] Pues lo que fueren los dedos eso sera lo que se ecliusara del diametro del sol.
[16] E lo que fueren los menudos del caymiento añade sobrellos su dosena parte. E lo que fuere partelo sobre el movimiento diverso de la luna a una hora e mengua lo que te saliere de partes de hora de las horas de la conjunçion visible las que son en medio del eclipsi e lo que quedare essas seran las horas del comienço del ecliuse e añadelas sobre la mitad del eclypsi. E lo que fuere essas seran las horas de la fin de la claridad.
[17] E si no fuere la luna en su longura la mas luenga ni en su longura la mas çerca. toma lo que fuere en derecho del argumento de la ladesa en cada una de ambas las tablas. E toma la diferençia que ay entre ellas de los dedos y de los minutos del caymiento e guardalo cada uno aparte.
[18] E despues entra con el argumento de la luna en la tabla de la proporçion y toma lo que fuere en su derecho de los menudos sobejanos e sepas su proporçion de 60. E [f. 60v] toma la quantidad de aquella proporçion de la diferençia que tu guardaste e lo que fuere añadelo sobre lo que tu fallaste en la tabla de la longura mayor e lo que se allegare essos seran los dedos eguados del eclipsi e los menudos eguados.
[19] Obra pues con ellos assi como se adelantro.
[20] E quando quisieres saber que tanto se eclipsara de la sobrehas del cerco del sol entra con los dedos del diametro en la tabla ques puesta para esto e toma lo que fuere en su derecho de los dedos del çerco del sol e lo que fuere esso sera el eclipsado.
[21] E cata si fuere el argumento de la ladesa con lo que tu entraste en la tabla del eclipsi mas de un grado y menos de 180 grados pues el eclipsi sera de parte de septentrion.
[22] E si fuere de mas de 180 y menos de 360 pues el eclipsi sera de parte de mediodia.

[1] Capitulo 34. Como se a de saber el eclipsy o la opinion del eclipsi lunar.

[2] Quando quisieres saber el eclipsi lunar endereça la opinion que fue de noche y la luna cerca de la cabeça o de la cola e sepas el argumento eguado de la ladesa a la ora de la oppinion çierta.
[3] Cata si fuere la luna en su longura la mas luenga y entra con el argumento de la ladesa en la tabla de los eclipsis lunares en la longura mas luenga et toma lo que fuere en su derecho de los dedos y de los minutos del caymiento e de medio de la mora si oviere mora.
[4] E si fuere la luna en su longura la mas çerca entra con el argumento de la ladeza en la tabla de los eclipses lunares en su longura la mas cerca e toma lo que fuere en su derecho de los dedos y de los minutos del caymiento e de medio de la mora.
[5] E si no fuere la luna en uno destos dos lugares ya nombrados y toma lo que fuere en su derecho del argumento de la ladesa en cada uno de los dos lugares e mengua lo menor de lo mayor cada uno de su linaje. y sepas la diferençia que ay entre ellos.
[6] E despues entra con el argumento eguado de la luna en una de las dos reglas de la quenta en la tabla de la proporçion.
[7] E toma lo que fuere en su derecho de los minutos sobejanos y multiplica aquellos menudos en la diferençia que tu guardaste de los dedos y de los minutos del caymiento y del medio de la mora si ay la oviere. E añadiras lo que te saliere sobre lo que hallaste en derecho del argumento de la ladesa en la longura mas luenga cada uno sobre su linaje y seran dedos eguados en medio de la mora eguada a medio del eclipsi.
[8] E despues añade sobre los minutos del caymiento su dosena parte e partela sobre el movimiento diverso de la luna a una hora. E lo que saliere seran las horas que son del començamiento del eclipsi hasta començamiento de la mora sy ay la oviere o hasta el medio del eclipsi sy no oviere ay mora.
[f. 61r] [9] E despues añade sobre los menudos del medio de la mora su dosena parte e partela sobre el movimiento diverso de la luna a una hora e lo que saliere sera el medio de la mora de la luna.
[10] Dobla pues las horas del medio de la mora y las horas del caymiento e ayuntalas. E lo que se ayuntare sera lo que ay entre el comienço del eclipsi hasta el fin de su claridad.
[11] E mengua las horas del caymiento y las horas del medio de la mora de las del medio del eclipsi las que son las horas de la mesma opinion o de la opinion mesmas. e lo que fincare sera su començamiento del eclipsi.
[12] E mengua las horas de en medio de la mora de las horas de emedio del eclipsi e sera el comienço del fin de la tiniebla.

[13] E añade las horas de en medio de la mora sobre el medio del eclipsi e sera el començamiento de la claridad.
[14] E añade sobre el medio del eclipsi las horas del medio de la mora y las horas del caymiento e sera la fin de su claridad.
[15] E tomaras los minutos del caymiento y del medio de la mora e añade sobre el su dosena parte e mengua lo que se ayuntare del lugar de la luna a medio el eclipsi. sera su lugar al començamiento del eclipsy e añadelo sobre el lugar de la luna al medio el eclipsi. sera su lugar al fin de la claridad.
[16] E si quisieres saber que tanto se eclipsara de la sobrehaz del cerco de la luna. entra con los dedos del eclipsi en la tabla de la yucaçion de los dedos y toma lo que fuere en su derecho en la tabla de la luna de dedos y de minutos. y eso sera lo que se eclipsara de la sobrehas del çerco de la luna.

[1] Capytulo 35. De que color sera el eclipsi.

[2] Quando quisieres saber la color del eclipsi cata la ladesa de la luna.
[3] E si fuere de un minuto hasta X minutos sera la color del eclipsi bien prieto.
[4] E si fuere de X hasta XX sera prieto y avra en ello verdura.
[5] E si fuere de 20 hasta 30 sera prieto y avra en ello bermejura.
[6] E si fuere de 30 hasta 40 y tyrara a amarillo.
[7] E si fuere de 40 hasta 50 sera prieto color abierto. E quanto mas fuere de L sera su color mas abierto.
[8] E a en esto otra carera. buelvenla con la primera y de ambas carreras puede aver en ello mas çertidumbre y es que pares mientes al lugar de la luna. E si fuere mas su aux con 6 signos sera prieto.
[9] E si fuere mas de su aux con 5 signos o con 4 sera prieto y avra en el verdura.
[10] E si fuere mas de su aux con 3 signos o con 9 signos sera prieto e [f. 61v] tirara a amarillo.
[11] E si fuere mas de su aux con 2 signos o con X sera color abierto.
[12] E si fuere mas de su aux con un signo o con XI sera color abierto que tirara a claridad.

[1] Capitulo 36. Como se sabra el aparesçimiento de la luna en cada villa.

[2] Quando esto quisieres saber toma el lugar del sol y el lugar natural de la luna e estos lugares sean a terçia de una hora despues del ponimiento en la noche que se sigue a la conjunçion.
[3] E sepas el lugar visible de la luna y la su ladesa visible otrosi endereçados por la diversidad del catamiento en longura y en ladesa a la ora del poniente o ponimiento. como es dicho en su capitulo.
[4] Despues mengua el lugar natural del sol del lugar natural de la luna e lo que fincare es el alongamiento.
[5] Despues cata si fuere el lugar de la luna del començamiento de capricornio hasta la fin de geminis. E si fuere el alongamiento 9 grados o menos no te cumple de obrar con ello que la luna no paresçera en aquella noche.
[6] E si el alongamiento fuere 15 grados o mas no te cale tampoco obrar. que sin duda no paresçera.
[7] E si el lugar de la luna fuere del començamiento de cançer hasta la fin de sagitario y fuere el alongamiento X grados o menos ansimesmo no te aprovechara obrar con ello que en ninguna manera paresçera.
[8] E si el alongamiento fuere 24 grados o mas. otrossi no labres en ello que sin duda [no] paresçera.
[9] Mas quando la luna fuere del començamiento de capricornio hasta la fin de geminis e fuere el alongamiento mas de 9 grados y menos de 15 puede ser que paresçera y puede ser que no.
[10] E este termino es el que a menester obra.
[11] E otrossy quando la luna fuere de començamiento de cançer hasta el fin de sagitario e el alongamiento fuere mas de X grados o menos de 24 otrosi puede ser que paresca e puede ser que non.
[12] E otrosi este termino a menester obra.
[13] Pues quando hallares la luna en uno destos terminos y quisieres averiguar si paresçe o no paresçe. toma el su lugar visible y entra con ello en la tabla del desviamiento de la luna la ques apartada para este fecho y toma lo que fuere en su derecho de los minutos del desviamiento. E lo que fueren estos minutos multiplicallo as en la ladesa visible de la luna e lo que fuere desta multiplicaçion guardalo e nombralo guardado.
[14] Despues cata el lugar vi[f. 62r]sible de la luna con que entraste en la tabla e si fuere de començamiento de capricornio hasta la fin de geminis y fuere la ladesa visible de la luna septentrional. mengua lo guardado del lugar visible de la luna.
[15] E si la ladesa visible fuere meredional añade el guardado al lugar visible de la luna.
[16] E si el lugar visible de la luna con que entraste fuere de començamiento de cançer hasta el fin de sagitario e fuere la ladesa visible de la luna septentrional añade el guardado al lugar visible de la luna.

[17] E si la ladesa visible fuere meredional mengua el guardado del lugar visible de la luna e lo que fuere el lugar despues deste anadimiento o deste menguamiento aquello es el lugar endereçado de la luna para este hecho.
[18] E sepas los subimientos del opposito del grado del sol en aquella villa otrosi. E mengua los subimientos del opposito del grado del sol y de los subimientos del opposito del lugar de la luna e lo que finca es el arco del ponimiento.
[19] Despues toma el signo de la ladesa de la villa y partelo por el signo del complimiento del anchura de la villa e lo que fuere desta partiçion multiplicalo en la ladesa visible de la luna. E lo que fuere del multiplicar añadelo al arco del ponimiento si la ladesa de la luna fuere septentrional o mengualo dello si la ladesa de la luna fuere meredional.
[20] E lo que fuere el alco[27] del ponymiento despues desto aquello es el arco de la vista.
[21] Pues si este arco fuere menos de X grados no paresçera por ninguna guisa.
[22] E si este arco fuere de 14 grados complidos o mas sin duda paresçera.
[23] Mas si este arco fuere de los X hasta los 13 cata el alongamiento que ante diximos.
[24] E si fuere este arco X grados o mas e fuere el alongamiento 13 grados o mas paresçera. e si fuere menos no paresçera.
[25] E otrossi sy fuere el arco de la vista 12 grados o mas e fuere el alongamiento XI grados o mas paresçera. si no no paresçera.
[26] E otrosi si fuere el arco de la vista 13 grados o mas e si fuere el alongamiento X grados o mas paresçera. si no no paresçera.

[1] Capitulo 37. Como se a de saber el signo y el complimiento y la querda y la saeta cada uno dellos por el arco y como se sabra el arco por qualquiera dellos por quenta y por tabla.

[2] Los sabios antiguos de geometria discordan mucho en mostrar la proporçion que al diametro a su çerco.
[3] E la cabsa por que desacordaron en esto es [f. 62v] porque saber esta cosa en si çiertamente es cosa que no se puede saber mas fueron açercandola a la verdad lo mas que pudieron.
[4] E este açercamiento es que en la linea del çerco a tres tanto del diametro y ocho partes et media de LX^{a}.

[27] *Sic*, instead of "arco".

[5] E esto es sobre lo que hizo Ptholomeo.
[6] Empero en todas las obras de astrologia dixo Ptholomeo esta proporçion e desembargose della. E partio el diametro por si y el çerco por si e partio el diametro por 120 partes porques quenta que se parte en muchas partes sin fraçiones.
[7] E partio todo el çerco por si por 160[28] de la siguiente manera.
[8] Otrosi asigno todos los arcos por las partes de todo el arco e nombro la quantia de cada un arco de las partes de todo el cerco e asigno las cordas a todo el diametro. E nombra la quantia de cada una corda e de las partes de todo el diametro.
[9] Pues todas las quantias e las partes que son puestas en los arcos son de las partes que hasen el çerco 360. E todas las quantias de las partes que son puestas en las cordas y en los signos y en las saetas todas son de las partes por que se departe todo el diametro por 160.[29]
[10] E es provado que la corda de la 6ª parte del çerco es quanto el medio del diametro.
[11] E la 6ª parte de todo el çerco es sesenta partes de las partes del çerco e el medio diametro es 60 partes de las partes del diametro.
[12] Pues claro esta que todo arco que es de sesenta partes que su corda sera 60 partes e si multiplicares la corda de los 60 en sipse e lo menguares de la multyplicaçion de todo el diametro en sipse e tomares la rayz de lo que finca. sera corda del terçio de todo el çerco.
[13] E la 3ª parte de todo el çerco es 120 e la su corda que sale por este hecho que diximos sera 103 grados e 55 minutos 23 segundos açerca.
[14] Si multiplicares el medio del diametro en si e ayuntares estos dos quadrados e tomares la raiz deste ayuntado que es 67 grados y 4 minutos y 55 sigundos açerca e menguares desto la quarta parte del diametro fincaran 37 grados 4 minutos 55 sigundos.
[15] Esto sera corda del deçimo del çerco ques el arco de 36 grados.
[16] E si multiplicares esta corda del arco de 36 en si e menguares lo que ende fuere de la multiplicaçion de todo el diametro en si e tomares la rayz de lo que finca que es 114 grados y 7 menudos e 36 segundos açerca. esta sera corda del arco con que se cumple el arco de 36 a medio çerco. E es arco de 144 e con este hecho quando alguna corda de algun arco ovieres sabido podras saber la corda de arco que se cumple [f. 63r] a medio çerco aquel arco sabido.

---

[28] *Sic*, instead of "60".

[29] *Sic*, instead of "120".

[17] E si menguaren la corda del arco de 120 la que ante es dicha de todo el diametro y tomares el medio de lo que finca e lo multiplicares en todo el diametro e tomares la rayz de lo que ende se ayunto. sera 31 grado y 3 minutos 30 sigundos. Esto sera corda del arco de 30 e el arco de 30 es el medio del arco de 60.
[18] E con este ho quando corda de algun arco ovieremos sabido podremos saber la corda de su medio. podemos [saber] tambien con esto la corda del arco de 15 y tambien podremos saber la corda del arco de 7 grados y medio.
[19] E porque ante avemos sabido la corda del arco de 36 podremos saber la corda del arco de 18 y la corda del arco de 9 y la corda del arco de quatro y medio.
[20] E quando la corda de siete y medio ovieremos sabido e otrosi la corda de quatro y medio ovieremos sabido. por lo que agora dyremos podremos saber la corda del arco de 12 ques ayuntada del arco de los 7 y medio y de los 4 y medio.
[21] Que quando dos cordas de dos arcos oviere sabidas e quisieres saber la corda del arco que se ayunta de ambos estos arcos. sepas la corda del arco con que se cumple cada uno de ellos a medio çerco de lo que ante diximos.
[22] E despues multiplica la corda del un arco de lo que ante supiste en la corda del otro e lo que fuere guardalo e nombralo el guardado primero. E despues multiplica la corda del arco con que se cumple el uno dellos a medio çerco en la corda del arco con que se cumple el otro a medio çerco e guarda lo que ende fuere y nombralo el guardado segundo.
[23] E cata estos dos guardados e mengua el menor del mayor e lo que fincare partelo por todo el diametro.
[24] E lo que ende saliere sera la corda del arco con que se cumple el arco compuesto a medio çerco.
[25] E este arco es en esta nuestra postura 168 ques lo que cumple sobre 12 a medio çerco e la su corda sale por este fecho 119 grados 20 minutos 33 segundos açerca.
[26] E si multiplicares esta corda en si e lo que saliere menguares de la multiplicaçion del diametro en si y tomares la rayz de lo que finca. sera la corda de 12 y sera por esta quenta 12 grados y 32 minutos y 36 segundos. y por lo que ante es dicho podremos saber la corda del arco de seis ques medio de 12. Fallarla emos por este fecho 6 grados e 16 minutos 49 segundos.
[27] E podremos otrosi saber [f. 63v] la corda del arco de tres ques el medio de seis e fallarla emos por este fecho 3 grados 8 menudos e 28 segundos.
[28] E podremos otrosi saber la corda de un grado. y medio y fallarla emos por esta quenta un grado 34 menudos 15 segundos.

[29] Podremos ansimesmo saber la corda del arco de 45 menudos e fallarla emos por este hecho 47 menudos e 8 segundos açerca.
[30] E todo arco que no pudiere entrar en ninguno destos fechos sobredichos ni por partir ny por ayuntar sigund diximos ni por ninguna de las otras cosas que diximos no se puede saber su corda por quenta de geometria.
[31] E destos tales arcos es el arco de un grado. que no puede entrar en ninguna destas carreras que diximos. Mas Ptholomeo quando quiso poner la corda de un grado supo la corda de los 45 minutos. por la carrera que diximos e fallola sigun diximos 47 minutos e 8 sigundos e añadio a esta su 3ª parte e hizo la corda del arco de un grado un grado 2 menudos 5[30] segundos.
[32] Empero que este hecho no sea verdad ni venga por carrera de demostraçion paresçe porque la proporçion que es entre los arcos no es ygual a la proporçion ques entre las cordas.
[33] Mas la duda y el desviamiento de la verdad que puede acaesçer en tamaño arco como este es tan poca cosa que no hasla daño en ningun hecho ni es cosa que se siente.
[34] E por esta carera compuso Ptholomeo la tabla de las cordas que a todos los arcos que pueden entrar en las carreras de la demostraçion que diximos. puso sus cordas verdaderas segun la quenta de geometria.
[35] E a los otros arcos que acaesçen entre los arcos verdaderos pusoles sus cordas por la partiçion de la proporçion e cumplio su tabla en esta guisa.
[36] E nos pusimos otrosi en esta tabla de las cordas en esta manera mesma en estas nuestras tablas. E estas son las que son nombradas cordas complidas.
[37] E las cordas complidas no son menester ni entran en ningun fecho de las obras de astrologia. mas ques menester y de lo que se serue en sus hechos es el signo. y el signo del complimiento y la saeta.
[38] Mas estos no se pueden saber si no supieren ante las cordas complidas pues que las cordas complidas son como çimientos y rayzes a estotros.
[39] E por esto hizo Tholomeo hiso la tabla de las cordas complidas maguer que no son menes[f. 64r]ter y dexo las otras que son menester que pues la raiz ovo mostrado. escuso lo al.
[40] E nos en estas nuestras tablas pusimos tablas de signo. y del signo del complimiento y de la saeta por ser ligeras de hallar al que las oviere menester.

---

[30] *Sic*, instead of "50".

[41] El signo es quando sacaren del un cabo una linea derecha que passe por el çentro del çerco cuyo es aquel arco e despues sacaren del otro cabo del arco otra linea que llegue hasta la linea primera e que haga con ella angulo derecho.
[42] Esta linea segunda es nombrada el signo de aquel arco.
[43] E la pieça de la linea primera que passo por el çentro que se hase del cabo del arco hasta la linea que nombramos signo aquella es la ques dicha saeta del arco.
[44] E el signo del arco que cunple con este otro arco 90 aquello es lo ques dicho signo del complimiento y todo esto lo mostramos aqui en figura.[31]
[45] Quando quisieres saber el signo por quenta cata el arco cuyo signo quieres saber si fuere menos de 90 doblalo y lo que fuere del doblamiento sepas su corda complida de lo que antes es dicho e lo que fuere esta corda toma su medio y aquel es el signo de aquel arco que quisiste saber. Mas si en arco cuyo signo quieres saber fuere mas de 90 e menos de 180 mengualo de 180 y lo que fincare nombralo argumento y has con ello lo que ante es dicho.
[46] E si el arco que ovieres fuere mas de 180 y menos de 27[32] mengua dello 180 y lo que fincare sera el argumento. has dello lo que antes es dicho y con esto avras el signo de todo el arco por carrera de quenta.
[47] E nos posimos todos estos signos en tabla.
[48] E si quisieres saber el signo del complimiento por quenta cata el arco cuyo signo del complimiento quisieres saber. E si fuere menos de 90 mengualo de 90 y lo que fincare sepas su signo de lo ante dicho e lo que fuere este signo aquello es el signo del complimiento de aquel arco que oviste.
[49] E si el arco cuyo signo de complimiento quisieres saber fuere mas de 90 sepas su argumento sigun que antes es dicho e lo que fuere aquel argumento mengualo de 90. E sepas el signo de lo que finca e aquello es el signo del complimiento del arco que oviste.
[50] E estos signos de los complimientos otrossi posimos en tablas.
[51] E si quisieres saber la saeta por quenta cata el arco cuya saeta quieres saber e sepas su corda cumplida e su signo e multiplica la corda complida en si e lo que [f. 64v] fuere nombralo quadrado de la corda. E multiplica el signo en si e lo que fuere nombralo el quadrado del signo. E mengua el quadrado del signo del quadrado de la corda y toma su raiz de lo que finca y aquello sera la saeta de aquel arco que oviste.

---

[31] This figure is missing in the manuscript.

[32] *Sic*, instead of "270".

[52] E nos pusimos otrosi todas estas saetas en tabla.
[53] Mas quando todo esto quissieres saber por las nuestras tablas quando ovieres arco sabido si quisieres saber su corda complida si el arco fuere menos [de] 180 entra con ello e si fuere mas de 180 mengualo de 360 y entra con lo que finca y toma lo que fuere en su derecho de las cordas en la tabla de las cordas complidas e lo que fuere aquello sera la corda cumplida de aquel arco.
[54] E si ovieres sabido la corda entra con ella en esta tabla en las cordas y hallaras el arco en su derecho.
[55] E quando quisieres saber el signo de algun arco sabido o el signo de su complimiento en el arco si fuere menos de 90 pues aquello mesmo es el argumento. E si fuere mas de 90 saca su argumento de lo que antes es dicho.
[56] E quando el argumento ovieres entra con ello en la tabla de los signos e de los signos del complimiento e toma lo que fuere en su derecho de aquel argumento en la regla de los signos si signos demandares o en la de los signos del complimiento si signo de complimiento demandares e lo que fuere aquello sera lo que demandaste.
[57] E si el signo ovieres sabido o el signo del complimiento y quisieres arco entra con qual quisieres dellos en su linea e hallaras el arco en su derecho.
[58] E si quisieres saber la saeta de algun arco sabido cata el arco si fuere menos de 180 entra con ello e si fuere mas de 180 mengualo de 360 y entra con lo que finca en la tabla de las saetas e toma lo que fuere en derecho de las saetas e lo que fuere aquello es la saeta del arco que oviste.
[59] E si la saeta fuere sabida y quisieres saber el arco por ella entra con ella en esta mesma tabla de las saetas y hallaras el arco en su derecho.

[1] Capitulo 38. Como se a de saber la anchura de las villas y quanto se alça el sol a mediodia en cada villa.

[2] Quando fueres en alguna villa cuya anchura no supieres y la quisieres saber cata el sol en qualquier dia y toma su altura con ynstrumento çierto y cata quanto es la mayor altura que puede aver en aquel dia y guardala.
[3] Despues endereca el sol y sepas su lugar natural a mediodia de aquel dia.
[f. 65r][4] E sepas la declinaçion de aquel grado en que hallaste el sol e si la parte de la declinaçion fuere septentrional menguala del altura que ante guardaste e si fuere meredional añade la altura que ante

guardaste e lo que fuere despues desto aquella altura aquella sera la altura de la ygualdad en aquella villa. menguala de 90 y lo que fincare sera la ladesa de la villa.
[5] E si esto quisieres saber por otra carrera de noche cata alguna de las estrellas que paresçen siempre en aquella villa y toma su altura quando es mas alta en medio del çielo e toma otrosi su altura la mas baxa que puede ser. y ayunta estas dos alturas y toma el medio de las alturas ayuntadas y lo que fuere aquello es la ladesa de aquella villa.
[6] E quando ovieres sabido la ladesa de la villa y quisieres saber quanto se alça el sol en mediodia en aquella villa mengua la ladesa de la villa de 90 y lo que finca aquello es lo que se alça el sol en aquella villa quando fuere el sol en uno de los puntos de la egualdad. E añade toda la declinaçion a esta altura de la ygualdad e aquello sera lo que se alça quando fuere en el punto del tropico estiual.
[7] E mengua toda la declinaçion de la altura de la egualdad e lo que finca [es] lo que se alça quando fuere en el punto del tropico yvernal.
[8] Mas quando quisieres saber quanto se alça el sol en medio de qualquier dia no siendo el sol en alguno destos puntos señalados endereça el sol a aquel dia y sepas su lugar natural. E sepas la declinacion de aquel lugar del sol e la parte de la declinaçion. E si la declinaçion fuere septentrional añadela a la altura de la ygualdad.
[9] E si la declinaçion fuere meredional menguala de la altura de la ygualdad. E lo que fuere el altura de la ygualdad despues desto aquello es lo que se alça el sol en medio de aquel dia que demandaste.

[1] Capitulo 39. Como se an de saber los alongamientos de los planetas y de las estrellas fixas del eguador del dia a septentrion o a mediodia.

[2] Sepas que en una partida de las estrellas fixas avemos nos puesto esto en tabla.
[3] Pues quando lo quisieres saber en alguna de aquellas estrellas que ay son puestas cata aquella estrella en aquella tabla y hallaras en su derecho quanto es su arredramiento del equinoçio con los otros sus estados.
[4] Mas si esto quisieres saber en la luna o en alguno de los otros planetas o en algunas de las otras estrellas fixas que son puestas en la tabla sepas el lugar natural de aquella estrella [f. 65v] o de aquel planeta e sepas otrosi su ladesa.
[5] E despues desto cata este lugar del estrella y si fuere en la cabeça de capricornio e no oviere ladesa en su arredramiento es tanto del equinoçio como toda la declinaçion y es meredional.

[6] E si oviere ladesa e fuere esta ladesa meredional añade esta ladesa a toda la declinaçion e lo que fuere el ayuntado aquello sera el arredramiento de la estrella del equinoçio y sera meredional.
[7] E si fuere la ladesa del estrella septentrional y fuere menos quanto da la declinaçion mengua la ladesa de toda la declinaçion y[33] finca aquello sera el arredramiento de aquella estrella del equinoçio y es meredional. E si la ladesa fuere septentrional y mas de toda la declinaçion mengua toda la declinaçion de la ladesa e lo que finca aquello es el arredramiento de la estrella del equinoçio y es septentrional.
[8] E si la ladesa fuere septentrional e ygual a toda la declinaçion aquella estrella no a arredramiento ninguno del equinoçio mas es en el.
[9] E si el lugar natural de la estrella fuere en la cabeça de cancro e no oviere ladesa el su arredramiento del equinoçio es tanto como toda la declinaçion e es septentrional.
[10] E si oviere ladesa e fuere septentrional añade la ladesa a toda la declinaçion. E lo que fuere el ayuntado aquello sera el su arredramiento del equinoçio y sera septentrional.
[11] E si la su ladesa fuere meredional y fuere menos de toda la declinaçion menguala de toda la declinaçion e lo que finca sera el arredramiento de la estrella del equinoçio y sera septentrional.
[12] Si la ladesa fuere meredional y mas de toda la declinaçion mengua toda la declinaçion de la ladesa. y lo que finca sera el arredramiento del equinoçio y sera meredional.
[13] E si la ladesa fuere meredional e egual a toda la declinaçion aquella estrella no a ningun arredramiento del equinoçio mas es en el.
[14] E si el lugar natural de la estrella fuere en la cabeça de aries o en cabeça de libra et no oviere alguna ladesa otrosi non a algun arredramiento del equinoçio mas es en el.
[15] Si oviere ladesa en su arredramiento del equinoçio es tanto como la su ladesa ni mas ni menos e la su parte es la mesma su parte de la ladesa.
[16] E si el lugar natural del estrella no fuere en ninguno destos 4 puntos terminados y no oviere ladesa cata la determinaçion de aquel grado de su lugar natural. E aquello mesmo sera el arredramiento del equinoçio e la su parte. la mesma parte de la declinaçion.
[17] Mas quando la estrella oviere ladesa e el [f. 66r] su lugar natural fuere sabido de los 4 puntos terminados y quisieres saber su arredramiento del equinoçio cata el su lugar natural y cata quanto a de la cabeça de capricornio a aquel su lugar natural. E lo que fuere esta quenta busca sus semblantes en los subimientos del çerco

---

[33] Between the lines the text adds "lo que".

derecho e toma lo que fuere en su derecho de los grados yguales e lo que fuere aquel grado e del signo cuyo fuere sepas su declinaçion e lo que fuere essa declinaçion guardalo y nombralo guardado.
[18] Despues cata la ladesa de la estrella e si fuere su parte en la mesma de su declinaçion que ante tomaste ayunta la ladesa y la declinaçion e lo que fuere nombralo ladesa eguada. E la parte desta ladesa sera la parte donde fueren ambas.
[19] E si la ladesa e la declinaçion fueren en diversas partes cata si las hallaras ambas yguales. en quenta sabras que aquella estrella no a endereçamiento del equinoçio mas es en el.
[20] E si no fueren yguales mengua el mayor del menor e lo que finca es la ladesa eguada. E la parte desta ladesa eguada es la parte del que fuere mayor.
[21] Pues quando ovieres la ladesa eguada por qualquier carrera destas toma su signo y nombralo signo primero.
[22] Despues toma el signo del complimiento de la declinaçion que ante guardaste el que nombraste guardado y nombralo signo sigundo. E toma el signo del complimiento de toda la declinaçion y nombralo signo terçero.
[23] E multiplica el signo primero por el terçero e lo que ende se ayuntare partelo por el segundo. E lo que saliere de la partyçion sera signo. sepas su arco e lo que fuere este arco aquello es el arredramiento del estrella e la parte deste arredramiento sera la parte donde fue la ladesa eguada.

[1] Capitulo 40. Como se a de saber el arco de qualquier dia y el arco de qualquier noche y el arco del dia de qualquier planeta y de su noche en cada tiempo y en cada villa y el arco de qualquier estrella y de su noche y en cada tiempo y villa.

[2] Quando quisieres saber el arco de algun dia o de alguna noche en la çibdad de Toledo o en qualquiera otra villa de quien tengas tabla de los subimientos de los signos en el orison. cata el grado del lugar natural del sol en aquel dia y entra con ello en las tablas de los subimientos de aquella villa y toma lo que fuere en su derecho de los subimientos y guardalo.
[3] Despues entra con el grado opposito del grado natural del sol en estas mesmas tablas y toma los subimientos que fueren en su derecho y lo que fuere mengua dellos los subimientos que ante guardaste.
[f. 66v][4] E si los subimientos primeros del lugar fueren mas que los subimientos sigundos del opposito añade a los subimientos del opposito 360 y mengua destos grados los subimientos del lugar. E lo

que fincare despues por qualquier manera de estas aquello sera el arco de aquel dia. e mengualo de 360 y sera el arco de aquella noche.
[5] E si esto quisieres saber en lugar do no ayas tabla de subimientos en su orizonte. sepas los subimientos del grado del lugar natural del sol y los subimientos del grado de su opposito sigun es dicho. E haz con estos subimientos segun ante es dicho y avras el arco de aquel dia o de aquella noche en aquella villa.
[6] Mas quando quisieres saber el arco del dia de la luna o del dia o de otro qualquier planeta o de estrella fixa. e esto es el tiempo que atura sobre la tierra desde que sube hasta que se pone y esto es lo ques nombrado su dia.
[7] Sepas el lugar natural de aquella estrella y la su ladesa si la oviere y el su arredramiento del equinoçio sy tambien lo oviere.
[8] E si hallares aquella estrella que non a arredramiento del equinoçio y es en el quier aya ladesa quier non el arco de su dia sera 180 y el arco de la noche otro tal.
[9] E si oviere arredramiento del equinoçio e no oviere ladesa savras el arco de su dia y de su noche y el su lugar natural sigun supiste lo del sol ny mas ny menos.
[10] E si la estrella oviere arredramiento del equinoçio y la latitud del zodiaco cata su arredramiento del equinoçio y la parte do es.
[11] E si fuere este arredramiento menos de toda la declinaçion busca su semblante en la tabla de la declinaçion. E toma el grado que fuere en su derecho de los signos septentrionales si el arredramiento fuere septentrional o de los meredionales si fuere meredional.
[12] E lo que fuere aquel grado de qualquier signo fuere. sepas el arco del dia de aquel grado y de su noche segund supiste lo del sol y lo que fuere aquello sera el arco del dia de aquella estrella y de su noche.
[13] Mas si el arredramiento de la estrella del equinoçio fuere mas de toda la declinaçion sabras el arco de su dia y de su noche en la siguiente manera.
[14] Toma el signo de la ladeza de la villa y nombralo primero. e toma el signo del complimiento de la latitud de la villa y nombralo sigundo. e toma el signo del complimiento del arredramiento del equinoçio y nombralo terçero. e toma el signo del arredramiento del estrella del equinoçio y nombralo quarto.
[15] Despues desto multiplica el segundo en el terçero [f. 67r] y lo que ende fuere nombralo padron.
[16] E despues multipl[i]ca el primero en el quarto y lo que ende fuere multiplicalo en 60 e lo que ende fuere partelo por el padron. E lo que saliere desta partiçion sera signo. sabe su arco e lo que fuere este arco doblalo y lo que ende fuere mengualo de 180.
[17] E si el arredramiento de la estrella del equinoçio fuere meredional añadelo a 180 si fuere septentrional e lo que fuere despues desto

menguar o deste añadir aquello sera el arco del dia de aquella estrella o de la luna o de otra planeta a que obraste e mengua el arco de su dia de 360 y sera el arco de su noche.

[1] Capitulo 41. Como se a de saber el grado con que sube la luna o qualquier otro planeta o estrella e el grado con que se ponen y el grado con que se paran en medio del çielo en cada tiempo e en cada villa.

[2] Quando esto quisieres saber sabe el lugar natural del estrella y su ladesa y su parte e su arredramiento del equinoçio y su parte.
[3] Despues para mientes a la estrella e si vieres que no a ladesa ninguna sabras que se para en la linea del medio çielo con aquel mesmo grado donde es su lugar natural.
[4] E si oviere ladesa sepas quanta es y su parte e cata el su lugar natural e si fuere de començamiento de cançer hasta la fin de sagitario. si la su ladesa fuere septentrional sabras que se para en medio del çielo despues que se para ay el grado de su lugar.
[5] E si la su ladesa fuere meredional sabras que aquella estrella se para en medio çielo ante que se pare ay su grado.
[6] E si el lugar del estrella fuere de començamiento de capricornio hasta la fin de gemenis y la su ladesa fuere septentrional sabras que se para en medio çielo antes que su grado.
[7] Mas si la ladesa fuere meredional sabras que se para en medio çielo despues de su grado.
[8] Pues quando quisieres saber aquel mesmo grado con que se para en medio çielo. si aquella estrella fuere en el equinoçio y no oviere arredramiento nynguno del. cata quanto ay del començamiento de capricornio hasta el grado de su lugar natural. E lo que fuere esta quenta busca su semblante en los subimientos del çerco derecho y toma lo que fuere en su derecho de grados yguales e lo que fuere aquello es el grado con que se para en medio çielo aquella estrella.
[9] E si la estrella oviere arredramiento del equinoçio cata quanto ay del començamiento de capricornio hasta el grado de su lugar natural. E lo que fuere busca su [f. 67v] semblante en los subimientos del çerco derecho y toma lo que fuere derecho en sus grados yguales. E lo que fuere este grado cata quanto a del començamiento de capricornio hasta el e lo que fuere nombralo el alongamiento primero.
[10] Despues cata lo que ovo de començamiento de capricornio hasta el grado del lugar y toma su signo y nombralo el signo del lugar.
[11] Despues cata la declinaçion del grado del lugar de la estrella e lo que fuere esta declinaçion toma el signo de su complimiento y

multiplica este signo en el signo del lugar e lo que fuere partelo por 60 e lo que saliere desta partiçion nombralo primero.
[12] Despues sepas el signo del arredramiento del equinoçio e nombralo sigundo. E sepas el signo del complimiento deste arredramiento y nombralo terçero. E sepas el signo de toda la declinaçion e nombralo quarto. E sepas el signo del complimiento de toda la declinaçion y nombralo quinto.
[13] Despues multiplica el primero en el segundo y lo que ende fuere partelo por el terçero y lo que saliere desta partiçion multiplicalo en el quarto. E lo que ende fuere partelo por el quinto e lo que saliere desta partiçion sera signo. Sepas su arco e lo que fuere este arco es el arco de la diversidad. guardalo.
[14] E cata el lugar natural del estrella si fuere del comienço de cançer hasta la fin de sagitario. E si fuere su alongamiento del equinoçio septentrional añade al arco de la diversidad el alongamiento primero.
[15] E si el lugar de la estrella fuere de comienço de capricornio hasta la fin del geminis y el arredramiento del equinoçio fuere septentrional. mengua el arco de la diversidad del alongamiento primero.
[16] E si el arredramiento del equinoçio fuere meredional añade el arco de la diversidad al alongamiento primero e lo que fuere el alongamiento primero despues deste añadir o menguar es el alongamiento eguado. busca su semblante en los subimientos del çerco derecho y toma lo que fuere en su derecho de los grados yguales e lo que fuere aquello es el grado con que se para la estrella o la luna o el planeta en medio çielo.
[17] E si quisieres saber el grado con que sube sepas el grado con que se para en el medio çielo sigun es dicho e sepas el arco de su dia otrosi segun ante es dicho y toma el arco medio de su dia y guardalo.
[18] Despues toma los sobimientos del grado con que se para en el medio çielo en el çerco derecho y este mesmo es el que ante nombramos alongamiento [f. 68r] eguado. E lo que fuere en estos subimientos mengua de ellos el medio del arco de su dia que ante guardaste. [19] E si no oviere en los subimientos tanto añadelos 360 e lo que fincare de los subimientos despues desto busca en su semblante en los subimientos de la villa y toma lo que fuere en su derecho de los grados yguales e lo que fuere aquello es el grado con que sube el estrella o la luna o el planeta en aquella villa.
[20] E si quisieres saber el grado con que se pone toma el medio del arco dese dia e añadelo a sus subimientos del grado con que se para en medio çielo en el arco derecho e si llegare la quenta a mas de un çerco complido echa ende un çerco complido y toma lo al. E lo que fuere despues desto busca su semblante en los subimientos de aquella villa y toma lo que fuere en su derecho de los grados yguales. E lo que

fuere aquel grado toma el grado de su opposito e aquello sera el grado con que se pone aquella estrella o la luna o el planeta en aquella villa. [21] E sepas que la tabla que nos hysimos de las estrellas fixas posimos ay aquellas estrellas que ay nombramos el grado con que se paran en el medio çielo y el grado con que suben y el grado con que se ponen en la çibdad de Toledo.
[22] Pues quando lo quisieres saber catalo ay y fallarlo as. mas si lo quisieres saber en alguna otra villa haslo sigun la obra que ante diximos.

[1] Capitulo 42. Como se a de saber quanto se alça cada planeta o cada estrella en la linea del medio çielo.

[2] Quando quisieres saber la altura de algud planeta o de alguna estrella fixa en el çerco del medio çielo sepas su arredramiento del equinoçio y su parte e añadelo o mengualo a la altura del equinoçio en aquella villa. segund hases en la declinaçion del sol para saber la altura del sol en medio dia. E aquello sera la altura de aquel planeta o de aquella estrella en la linea del medio çielo en aquella villa. E en la nuestra tabla do posimos las estrellas fixas posimos ay la altura de cada estrella que ay nombramos en la linea del medio çielo en la çibdat de Toledo. Pues quando lo quisieres saber en alguna estrella de las que ay son nombradas. buscalo en aquella tabla y hallarlo as.

[f. 68v][1] Capitulo 43. Como se a de saber quanto passo del dia por la altura del sol tomada.

[2] Quando catares la altura del sol en qualquiera sason e quisieres saber quantas horas del dia son passadas de las eguales o de las temporales. sepas la altura del sol en la linea del medio çielo en aquel dia y en aquella villa y lo que fuere toma su signo y nombralo primero.
[3] Despues toma el signo del altura que cataste e nombralo sigundo.
[4] E luego sabe el arco de aquel dia y toma su medio e lo que fuere en su medio toma su saeta y nombrala el terçero.
[5] Despues multiplica el sigundo en el tercero e lo que fuere partelo por el primero e lo que saliere desta partiçion mengualo del terçero. E lo que fincare sera saeta sabe su arco e lo que fuere este arco mengualo del medio del arco de aquel dia si la altura del sol que tomaste fuere oriental mas si la altura fuere occidental añadelo a el. E lo que fuere el medio del arco de aquel dia despues deste menguamiento o deste añadimiento aquello es lo que se movio el çielo desque subio el sol hasta la sason del altura que tomaste lo que fuere

partelo por XV y seran horas y fraçiones de hora e aquellas seran las horas yguales que passaron de aquel dia.
[6] E si aquello partieres por los tiempos de una hora temporal de aquel dia seran las horas temporales y sus fraçiones que son passadas de aquel dia.

[1] Capitulo 44. Como se a de saber quanto passo de la noche por la altura de alguna estrella fixa tomada.

[2] Quando catares altura de alguna estrella fixa y quisieres saber por ella quanto passo de aquella noche. sepas el medio del arco dese dia de aquella estrella et el grado con que sube en aquella villa y tambien la su altura en la linea del medio çielo lo qual sabras por los capitulos que son passados.
[3] Despues toma el signo de su altura en la linea del medio çielo e nombralo primero.
[4] E toma luego el signo de la su altura en la hora que la cataste y nombralo segundo. Ansimesmo toma la saeta del medio del arco de su dia y nombralo 3° Despues multiplica el sigundo y el terçero e lo que fuere partelo por el primero desta partiçion e lo que saliere mengualo del terçero e lo que fincare sera [f. 69 r] saeta. toma su arco e lo que fuere este arco nombralo guardado primero.
[5] Despues cata si fuere la altura de la estrella que cataste oriental. mengua el guardado primero del medio del arco del dia de aquella estrella.
[6] E si la altura que tomaste fuere occidental añade el guardado primero al medio del arco del dia de aquella estrella e lo que fuere el medio del arco de la estrella despues del añedir o del menguar aquello es lo que se movio del çielo desque subio aquella estrella hasta la ora de la altura que tomaste. guardalo y nombralo el guardado segundo.
[7] Despues cata el grado con que sube aquella estrella en aquella villa e si fuere entre el grado del sol y entre el grado de su oposito sigund la horden de los signos sabras que aquella estrella subio de dia.
[8] E si fuere el grado del subimiento de la estrella entre el grado del opposito del sol y entre el su grado del sol segud la horden de los signos sabras que aquella estrella subio de noche.
[9] Pues si hallares que la estrella subio de dia toma los subimientos del grado con que sube la estrella en la villa de aquel grado y mengualo de los subimientos del grado del opposito del sol en aquella villa otrossi. E lo que fincare destos subimientos mengualo del guardado segundo.

[10] E si hallares que aquella estrella subio de noche mengua los subimientos del grado del opposito del sol de los subimientos del grado con que sube aquella estrella e lo que fuere destos subimientos añadelo al grado segundo e lo que fuere el guardado segundo. Despues deste añedir o menguar aquello es lo que se movio el çielo desque se puso el sol hasta la hora del altura que tomaste. lo que fuere partelo por XV y salirte a las horas yguales y sus fraçiones las que son passadas de aquella noche. Si lo partieres por los tiempos de una hora temporal de aquella noche salirte a las horas temporales y sus fraçiones las que son passadas de aquella noche.

[1] Capitulo 45. Como se a de saber quantas horas yguales ay en cada dia y en cada noche.

[2] Quando esto quisieres saber en la çibdad de Toledo entra con el lugar natural del sol en la tabla de las horas del dia y hallaras en su derecho las horas yguales de aquel dia. mengualas de 24 y seran las horas yguales de aquella noche. E si lo quisie[f. 69v]res saber en otra villa despues de sabida la ladesa. saca el arco de aquel dia segun antes es dicho y partelo por 15 e lo que fuere esso seran las horas yguales de aquel dia. mengualas de 24 y seran las horas yguales de aquella noche.

[1] Capitulo 46. Como se an de saber los tiempos de una hora temporal en qualquier dia y en qualquier noche y como se sabran los subimientos de qualquier grado de los grados de los signos en cada villa.

[2] Quando quisieres saber los tiempos de un ora temporal de qualquier dia quisieres en la çibd[a]d de Toledo. entra en el lugar natural del sol en la tabla de los tiempos de las [horas] e lo que fuere en su derecho aquellos son los tiempos de una hora temporal de aquel dia.
[3] E si entrares en esta mesma tabla con el opposito del lugar natural del sol hallaras en su derecho los tiempos de una hora temporal de aquella noche.
[4] E otrossi quando quisieres saber los subimientos de qualquier grado que sea en la çibdad de Toledo entra con el lugar natural del sol en la tabla de los subimientos de los signos[34] en la çibdad de Toledo e

[34] The phrase "en la tabla de los subimientos de los signos" is repeated.

lo que fuere en su derecho aquellos son los subimientos de aquel grado en Toledo.
[5] E desta guisa lo hallaras en qualquier villa cuya tabla ovieres.
[6] Mas quando quisieres saber los tiempos de una [hora] temporal en qualquier otro lugar de sabida ladesa. saca el arco de aquel dia sigun ante es dicho. E lo que fuere partelo por 12 e lo que saliere de la partiçion aquellos son los tiempos de una hora temporal de aquel dia. mengualos de 30 aquellos seran los tiempos de una hora temporal de aquella noche en aquella villa.
[7] E quando quisieres saber los subimientos de qualquier grado en qualquier villa por quenta. toma los subimientos de aquel grado en el çerco derecho e guardalos.
[8] Otrosi despues sabe la declinaçion de aquel grado y su parte. y guardala tambien.
[9] Despues toma el signo de la ladesa de la villa y nombralo primero.
[10] E toma el signo de la declinaçion del grado y nombralo 2°.
[11] E toma el signo del complimiento de la ladesa de la villa y nombralo 3°.
[12] Toma aun allende desto el signo del complimiento de la declinaçion del grado y nombralo 4.
[f. 70r] [13] Despues multiplica lo primero en el 2° e lo que fuere partelo por el 3° y lo que saliere desta partiçion multiplicalo en 60 y lo que fuere partelo por el 4°. E lo que saliere desta partiçion sera signo. Sepas su arco e lo que fuere este arco mengualo de los subimientos del çerco derecho que antes guardaste si la declinaçion fuere septentrional o añadelo a ellos si la declinaçion fuere meredional.
[14] Lo que fueren los subimientos del çerco derecho despues deste añedir o deste menguar aquellos son los subimientos de aquel grado en el orisonte de aquella villa.

[1] Capitulo 47. Como se sabran tornar las horas yguales en temporales y las temporales en yguales.

[2] Quando ovieres horas temporales de qualquier dia o de qualquier noche e quisieres saber quantas horas yguales ay en ellas. multiplica aquellas horas temporales en los tiempos de una hora temporal de aquel dia sy fuere de dia o de aquella noche si fuere de noche.
[3] E parte lo que fuere esta multiplicaçion por 15 y lo que saliere desta partiçion aquello es lo que cabe de horas yguales a las oras temporales que ante oviste.
[4] E si ovieres horas yguales de qualquier dia o de qualquier noche y quisieres saber quantas horas temporales ay en ellas de aquel dia o de aquella noche multiplica las horas yguales por 15 y parte lo que fuere

de la multiplicaçion por los tiempos de una hora temporal de aquel dia o de aquella noche si fuere de noche. E lo que fuere de la partiçion aquellas son las horas temporales que caben con aquellas horas yguales que ante oviste.

[1] Capitulo 48. Como se an de saber haser las reuoluçiones de los años del mundo y de los naçimientos.

[2] Quando quisieres saber quando allegara el sol a algun punto señalado por haser reuoluçion de año de nasçimiento o de año del mundo o de algun acaesçimiento cata el lugar del sol en la raiz de aquella cosa.
[3] Despues cata el año cuya revoluçion quieres haser y el dia que es mas çerca de ser el sol en aquel punto de la raiz y endereça el sol al medio de aquel dia. E cata este lugar con el lugar de la rais y mengua el menor del mayor y lo que fuere la diferençia guardalo e cuya es de ellos.
[4] E despues toma esta diferençia e partela por [f. 70v] el buc del sol ques su movimiento a una hora e lo que saliere de la fraçion seran horas y fraçion de hora.
[5] E lo que fuere añadelo al medio del dia a que endereçaste el sol sy la obra fuere al lugar de la rayz. o mengualo de aquel mediodia sy la obra fuere al lugar de la revoluçion que endereçaste.
[6] Lo que fuere la hera despues de aquel mediodia a que el sol endereçaste o antes del aquella es la hera de aquella revoluçion en aquel año. Tornala diversa endereçandola con la equaçion de los dias con sus noches segund es dicho en su lugar.
[7] E saca el asçendente a aquella hera e los çentros de las otras cosas[35] e endereça los otros planetas a la hera ygual y con esto avras lo que demandaste.
[8] E si esto quisieres hazer con la tabla que nos hesimos para esto en nasçimiento o en otro acaezçimiento que fuere despues de la hera alfonsi. sepas aquel año en que fue esta cosa quanto es de la hera alfonsi y entra en la tabla de la hera de las revoluçiones en los años colectos con lo que fuere mas çerca e menos del. E toma lo que fuere en su derecho de las horas y de sus partes.
[9] Despues cata el año de que es la revoluçion que demandas quanto es de la hera alfonsy. E has con ello lo que hiziste con el año de la raiz e lo que ende ovieres nombralo la diversidad de la reuoluçion.

---

[35] *Sic*, instead of "casas".

[10] Mira tambien estas dos diversidades y mengua la menor de la mayor y sabe la diferençia quanta es y cuya es. Sy la diferençia fuere a la diversidad de la rayz mengua la aquella diferençia de la hera de la raiz si fuere la diversidad de la revoluçion añade la aquella diferençia de la hera de la raiz. E lo que fuere la era de la rayz despues desto aquello sera la hera de la rayz en aquel año.
[11] Despues entra con el año de la rayz en la tabla de los subimientos de la revoluçion en los collectos e en los espansos. E toma lo que ay hallares de los subimientos e ayuntalos e lo que fuere nombralo la diversidad de los subimientos de la rayz.
[12] Despues entra con el año donde es la revoluçion que demandas e toma otrossi lo que fuere ay en los collectos y en los espansos de los subimientos e ayuntalos otrosi y nombra lo que ende fuere la diversidad de los subimientos de la reuoluçion.
[13] Despues cata estas dos [f. 71r] diversidades menguando la menor de la mayor y mengua esta su diferençia de los subimientos del asçendente de la rayz sy menguaste de la hera de la raiz o añadelo a ellos sy añediste a la hera de la raiz. E lo que fueren los subimientos despues desto tornalos a los grados yguales en los subimientos de tu villa. y aquello sera el asçendiente de la revoluçion de aquel año.
[14] Mas quando quisieres hazer las revoluciones de los años del mundo ques la entrada del sol en el punto de la ygualdad del verano o la su entrada en el punto del comienço del signo de aries. qualquier dellas quisieres. conviene que labres para esto con la otra carrera que ante diximos nos con esta tabla. porquesta tabla esta compuesta a los acaezçimientos que son despues de la hera alfonsi.

[1] Capitulo 49. Como se a de saber endereçar el movimiento de la cabeça de aries en alongado y en tornado.

[2] Quando esto quisieres endereçar por saber los lugares naturales de los planetas y es el su alongamiento del començamiento natural ques punto de la ygualdad de verano en el çielo alto raso no estrellado. E este es el lugar con que se obran todos los retificamientos e todas las obras eguadas e rectificadas.
[3] Quando quisieres eso saca el movimiento mediano de la cabeça de aries a aquel tiempo que tu quieres con la hera alfonsi ansi como hases en sacar las rayzes de los movimientos de los planetas. E lo que fuere esso sera el argumento.
[4] Entra pues con el en la tabla de la yucaçion del allegamiento y del tornamiento e toma lo que le cupiere de la yucaçion. E lo que fuere aquello es la quantia del alongamiento o del tornamiento.

[5] Pues si el argumento fuere menos de seys signos es aquella equaçion en allegamiento e si fuere el argumento mas de seys signos es la equaçion en torneamiento.
[6] Pues si fuere el allegamiento añadelo a los lugares de las luminarias y de los planetas que te salieron de las equaçiones de sus tablas.
[7] E si fuere el tornamiento mengualo destos mesmos lugares sobredichos. E lo que fueren estos lugares despues de este añadimiento o deste menguamiento aquellos son sus lugares naturales en el çielo alto no estrellado. Que los [f. 71v] lugares que salen de las tablas de sus equaçiones son los suyos nombrados propios y es su alongamiento de la cabeça de la forma de aries en el çielo estrellado. son las formas.
[8] Pues quando fuere este argumento sobredicho 6 signos ygualmente o 12 signos ygualmente no avra alli alongamiento ni tornamiento. Digo estonces sera el punto de la cabeça de la forma de aries del çielo estrellado en derecho del punto de la ygualdad del verano del çielo raso no estrellado sin ningun desviamiento. E seran los lugares proprios de los planetas ellos mes los lugares naturales sin ninguna diferençia.

[1] Capitulo L. Como se sabra la declinaçion universal del cerco de los signos.

[2] Quando quisieres eso saca el movimiento del polo al tiempo que tu quisieres con la era alfonsi e despues entra con el en la tabla de las declinaçiones universales del sodiaco. Entraras con los signos en la ladesa de la tabla e con los grados en su longura y tomaras lo que hallares en su derecho de los grados de la declividad y de sus minutos y de sus segundos e guardalo.
[3] E si oviere algunos minutos con los grados del movimiento del polo egualos ansi como hases quando sacas los medios cursos. E lo que saliere despues esso sera la declividad universal del zodiaco a aquel tiempo que tu quisiste.

[1] Capitulo LI. Como se a de saber la sombra espandida y la sombra menguada por la altura y la altura por qualquier de las dichas sombras.

[2] Quando quisieres esso e ovieres altura toma su signo y ponle nombre el padron.

[3] E toma el signo del complimiento del altura y partelo sobre el padron y lo que saliere multiplicalo en 12 y lo que fuere esso sera la quantia de los dedos de la sombra espandida.
[4] Pues si ovieres dedos de la sombra espandida y quisieres saber su altura multiplica aquellos dedos en si e la que se ayuntare añade sobre el 144. E lo que se ayuntare toma su rayz y ponle nombre padron.
[5] E despues parte sobre el 720 y lo que saliere arquealo arqueamiento de los signos y lo que fuere el arco esso sera la altura de aquella sombra espandida.
[5] E si tu ovieres altura e quisieres saber [f. 72r][36] su sonbra la menguada toma el signo del complimiento del altura y ponle nombre el padron. y despues parte sobre el signo del altura y lo que saliere multiplicalo en 12 y lo que se ayuntare esso sera la quantia de los dedos de la sombra menguada de aquella altura.
[6] E si tu ovieres dedos de sombra menguada y quisieres saber por ellos la altura multiplica aquellos dedos en si y lo que se ayuntare añade sobre el 144. E lo que se ayuntare toma su rayz y ponle nombre padron.
[7] E despues multiplica los dedos de la sombra menguada en 60. E lo que se ayuntare partelo sobre el padron y lo que saliere arquealo arqueamiento de los signos y lo que fuere el arco aquello sera la altura de aquellos dedos de la sombra menguada.
[8] E si tu ovieres dedos de qualquiera sombra destos y quisieres saber por ella los dedos de la otra sombra. parte sobre los dedos de qualquiera sombra [que] tu ovieres 144 y lo que te saliere esso seran los dedos de la otra sombra.

[1] Capitulo LII. En fazer echamiento de los rayos de los planetas en el sodiaco sigund opinion del Bactegui.

[2] Quando quisieres saber el rayo sextil de la planeta o de su quadradura en el zodiaco. cata si fuere el planeta en el zodiaco mesmo sin ladesa ninguna pues sepas que su rayo sextil caera sobre 60 grados de los grados yguales e su triangulo sobre 12[37] grados de los grados yguales. Mas la quadradura sera todavia sobre 90 grados de los grados yguales quier aya el planeta ladeza quier no la aya e

---

[36] This folio has been numbered erroneously: "73" instead of "72".

[37] *Sic*, instead of "120".

ansi haze el opposito ques todavia en 180 agora el planeta tenga latitud agora no la tenga.
[3] Pues si tuviere el planeta latitud del zodiaco parte todavia 1800 sobre el signo del complimiento de la ladeza del planeta e lo que te saliere arquealo arqueamiento de los signos e lo que fuere el arco aquel sera argumento del sextil. Pues añadelo sobre el grado del planeta y mengualo dende e lo que se ayuntare esso sera el rayz sextil siniestro en el zodiaco e lo que fuere esso sera rayz sextil diestra en el zodiaco.
[4] Despues mengua el argumento del sextil de 180 todavia e lo que fincare eso sera el argumento del trino. Pues añadelo sobre el grado del planeta e mengualo dende e lo que se ayuntare esso sera el rayo siniestro del trino. E lo que fincare esso sera el rayo diestro del trino. Mas el rayo [f. 72v] de la quadradura cae todavia en el zodiaco quier a diestro quier a siniestro lueñe del grado del planeta con 90 grados como arriba se dixo.
[5] E el rayo del opposyto cae otrossi todavia lexos del grado del planeta con 180 grados.

[1] Capitulo LIII. Como se an de saber las horas del subimiento del alvor y del ponimiento de los rayos de prima noche por la baxura del sol so el orizonte en el çerco del altura e otrossi su converso.

[2] Quando quisieres esso toma el opposito del grado del sol el que es contado del punto de la ygualdad tropica ansi como lo diximos en el 49 capitulo. E lo que fuere sepas la saeta del arco de su dia y el signo de su altura en medio el dia e despues multiplica la saeta del arco del dia en signo de 19 todavia. E lo que se ayuntare partelo sobre el signo del altura de mediodia e lo que saliere mengualo de la saeta del arco del dia e lo que fincare arquealo arqueamiento de las saetas. E lo que fuere el arco mengualo del medio arco del dia e lo que fincare partelo sobre tiempos de una hora de las horas del grado sobredicho e lo que saliere seran horas y minutos de hora.
[3] E lo que fuere aquellas horas y sus minutos essas seran las horas del subimiento del alvor y del ponimiento de los [ra]yos de prima noche.
[4] Pues si quisieres saber que tanta es la baxura del sol so el orizonte en el çerco del altura a la hora del subimiento del alvor o del ponimiento de los rayos de prima noche por sus horas que son sabidas. multiplica la quantia de las horas del subimiento del alvor o del ponimiento del rayo de prima noche en sus tiempos. E lo que se ayuntare mengualo del medio arco del dia del opposito del grado del sol el sobredicho. E lo que fincare sepas su saeta e lo que fuere

mengualo de la saeta de mediodia e lo que fincare multiplicalo en el signo del altura de mediodia.
[5] E lo que se ayuntare partelo sobre la saeta del arco del dia e lo que saliere arquealo arqueamiento de los signos.
[6] E lo que fuere el arco esso sera la baxura del sol so el orizonte [f. 73r][38] en el cerco del altura quanto sube al alvor y quando se pone el rayo de prima noche.

[1] Capitulo LIIII°. Como se a de saber sacar en que dia cae el miercoles de la ceniza y de pasqua por quenta y tabla.

[2] Quando esto quisieres saber por la tabla cata la quantia de la era alfonsi con el mesmo año de que es el ayuno y la pasqua que quieres saber y echalos de 19 en 19 y lo que fincare menos de 19 entra con ello en la tabla de saber el ayuno y la pasqua por la hera alfonsi en la linea primera de aquella tabla que comiença de uno y acaba en 19 de arriba abaxo.
[3] E despues sabe la noche de henero en aquel año y entra con ella en las notas que son puestas en somo de la tabla en la ladesa y toma lo que hallares en la linea de la nota en derecho del año con que ante entraste e aquel sera el dia del miercoles de la çeniza en la una linea y el dia de pasqua en la otra a su lado.
[4] E si aquel año fuere bisiesto e hallares el dia del ayuno en el mes de março sepas ques un dia ante de lo que hallas en la tabla. E la pasqua siempre es en el año del bisiesto un dia antes de lo que la hallas en la tabla.
[5] E si esto quisieres saber por la era de Çesar haz en la era de Çesar lo que hiziste con la era alfonsy. entrando en la tabla de la hera de Çesar que es fecha para esto sigun entraste en la otra y hallaras lo que demandas.
[6] Mas si esto quisieres saber por quenta cata los años de la era alfonsi con el año mesmo en que fuere lo que demandas e ayunta siempre a ellos 4°. E lo que se hiziere echalo de 19 multiplicalo en 19 e lo que fuere la multiplicaçion añadele siempre 19 e lo que fuere echalo de 30. E lo que fuere menos de 30 nombralo guardado e ayuntale siempre 7 e lo que fuere aquel guardado despues desto nombralo raiz.
[7] Cata la nota de henero en aquel año e ayuntala a la rayz e lo que fuere mengualo de 55 siempre e lo que fincare echalo de 7 en VII. E lo que fincare menos de 7 ayuntalo a la rayz primera [f. 73v] e sera la rayz eguada e mengua della siempre quatro y lo que fincare comiença

[38] This folio has been numbered erroneously: "72" instead of "73".

de contar del primero dia de hebrero aquella quenta e donde la quenta quedare aquel dia sera miercoles de la ceniza en aquel año.
[8] E si aquel año fuere visiesto e fuere la raiz eguada depues que quitaste della los quatro sobredichos mas de 28 da a hebrero desta quenta 29 y sacarte a el dia del ayuno sobredicho.
[9] E este dia siempre te saldra por esta obra miercoles.
[10] E quando quisieres saber el dia de la pasqua despues desto sepas que el dia 47 deste dia adelante es la pasqua.
[11] Quenta pues 47 dias despues deste e adonde llegare la quenta aquel dia es la pasqua.
[12] E este dia siempre te saldra en domingo por esta quenta e aquello es lo que demandaste.

CHAPTER 3

A GLOSSARY OF TECHNICAL TERMS

This glossary contains some 300 technical terms in Castilian selected from those appearing in the text as preserved in Madrid, Biblioteca Nacional, MS 3306.

In general, only the first occurrence of each term is indicated and we have used the following notation: numbers separated by a colon refer to chapter and sentence, respectively. For each term a translation in modern Castilian (ES) and in English (EN) is given. For a few technical expressions a definition or an explanation is also given.

The terms in this glossary have been compared to similar terms found in four texts written in Castilian, or translated into it, by the astronomers in the service of Alfonso X. Three texts are uniquely preserved in Paris, Bibliothèque de l'Arsenal, MS 8322, a late 13th-century manuscript according to Millás (1943–1950, p. 149): (i) the Castilian translation of al-Battānī's canons (ff. 1v–27v) [henceforth B]; (ii) the set of tables called Almanac of Azarquiel (ff. 94r–135r) [henceforth A]; and (iii) the *Tratado del quadrante sennero* (ff. 136r–142r) [henceforth Q]. The author of the third work, Isaac ben Sid, is mentioned explicitly in the introduction (f. 136ra), but the names of the translators of the first two items are not given. The fourth text is called the *Lapidario* and it is extant in another manuscript, also written at the royal scriptorium: Escorial, Biblioteca del Monasterio, MS h-I-15 [henceforth L]. It was translated from Arabic by Judah ben Moses ha-Cohen between 1243 and 1250.

From a comparison of the terminology in these texts with that of the Castilian canons to the Alfonsine Tables, it is clear that the text of these canons, as they appear in MS 3306, is written in 13th-century Castilian, even though the manuscript itself is of a later date.

It should be stressed that in the process of constructing a new vernacular scientific language for astronomy/astrology, the translators of these Arabic texts and the authors of original works had to face the difficult task of creating new terminology and new phraseology in these domains. Most of the technical terms appearing in the *Libro de las tablas alfonsies* were obtained by adaptations from Latin, but we also find several examples of neologisms derived from an Arabic term.

As for the terms themselves, we note that they are often spelled in different ways as is the case in other texts of the same period. This vacillation is especially noticeable in terms containing "c" or "ç", "d" or "t", "s" or "z", and "v" or "u". Examples of polysemy are also frequent: *alongamiento*, with the values "distance" and "elongation" (the difference between the solar and lunar longitudes), or *signo*, meaning both "sign" (30°) and "sine" (the trigonometric function). On the other hand, it is easy to find examples of the use of different terms to describe the same concept: *ladesa*, *latitud*, and *anchura* for "latitude".

It is worth noting the peculiar usages of some terms and expressions in our text, many of which are also found in other texts in Castilian of the same period: the expression *arquear arqueamiento del signo* for "to find the arcsine"; the use of *verano* for "spring" (rather than "summer", as in modern Castilian); a multiplicity of forms for the term "equation": *equaçion, ynvocaçion, yncaçion, ynovoçaçion, yucaçion*, and *yuncaçion*; the frequent use of *catar* and its derivatives; the use of *tropico* for both "solstice" and "vernal equinox", and many others.

The only foreign language in evidence in the text is Arabic. Among the terms listed here we find some calques on Arabic, for example, *forma* for "constellation" (derived from *ṣūra*), and *çentro* for "cusp" (derived from *markaz*). The only terms, other than month names, transliterated from Arabic are: *aux, buc, çifra, hadit, geuzaar, zaba*. Moreover, the text provides several examples of literal translations from Arabic, as in the case of *lugar propio*, meaning "sidereal", as opposed to *lugar natural*, meaning "tropical". We have not noted the many cases where the Castilian is based on the corresponding term in Latin.

Since Millás did not edit the complete text of the Almanac of Azarquiel, we sometimes refer to text A by folio only.

ABAXAMIENTO (22:34)
See also *sobimiento*.
ES: descenso; EN: descent.
B (Bossong 1978, p. 114).
Comment: Together with *sobimiento*, these terms are used in the context of ascent and descent of planets, that is, their motion in latitude with respect to their nodes. In L (Rodríguez Montalvo 1981, p. 231) we find the form *abaxar* indicating that a star is close to the horizon. According to the Real Academia Española (1960–1981, 1:30), this term appears for the first time in Castilian in the *Libro de las tablas alfonsíes*.

ABENMEH (5:11)
A (f. 96v): *Ebenme*.
Comment: This is the month *Ābān Māh* in the Persian calendar. For the Persian months, see al-Bīrūnī (trans. Sachau 1879, p. 52).

ABIERTO (COLOR –) (35:7)
ES: color claro; EN: bright color.
Comment: This term appears in discussions of colors of eclipses. According to the Real Academia Española (1960–1981, 1:84), this term appears for the first time in Castilian with this meaning in the *Libro de las tablas alfonsíes*.

ABRIL (2:10)
ES: abril; EN: April.
A (f. 94v).

ABSCONDIDO (PLANETA –) (25:5)
See also *ascondimiento*.
ES: (planeta) invisible; EN: invisible (planet).
Comment: As opposed to *pareçiente/paresçiente*.

ACATAMIENTO (31:7)
See also *catamiento*.
Comment: According to the Real Academia Española (1960–1981, 1:314–315), this term, with the meaning "contemplation", "look", or "sight", was already used in Castilian in the Alfonsine text called *Siete Partidas*, prior to the *Libro de las tablas alfonsíes*.

ADELANTAMIENTO (27:27)
ES: adelantamiento; EN: advance.
Comment: In this context, *adelantamiento* means the difference between the daily velocities of two planets. According to the Real Academia Española (1960–1981, 1:675–676), it was already used in Castilian with other meanings in Alfonsine texts (*Setenario*, *Siete Partidas*) prior to the *Libro de las tablas alfonsíes*.

ADORYNAHSEME (5:5)
A (f. 96v): *Ardaxarme.*
Comment: This is the month *Ardībahisht Māh* in the Persian calendar.

ADRYNHMEH (5:12)
A (f. 96v): *Aderanme.*
Comment: This is the month *Ādhār Māh* in the Persian calendar.

AFROZDIGINHE (5:4)
A (f. 96v): *Afrozdinme.*
Comment: This is the month *Farwardīn Māh* in the Persian calendar.

AGOSTO (2:14)
ES: agosto; EN: August.
A (f. 94v).

ALCANÇAMIENTO (26:1)
ES: alcanzamiento, tránsito; EN: transit.
Comment: This term describes the situation when the difference in longitude between two planets vanishes. According to the Real Academia Española (1960–1981, 2:182), it appears for the first time in Castilian in the *Libro de las tablas alfonsies.*

ALÇAR (13:15) / ALCAR (9:8)
ES: convertir en una unidad sexagesimal superior; EN: convert into a higher sexagesimal unit.
Comment: This term is used in sentences explaining computations and transformations, such as transforming 60 minutes into 1 degree. See *offiçio.* It is also found in Q in the form *alçalo un officio*, as opposed to *abaxalo un officio* (Millás 1960, p. 252), and means "multiply by 60".

ALONGADO (49:1)
See also *tornado.*
ES: acceso; EN: access.
Comment: Together with *tornado*, these terms refer to the motion of "access and recess", that is, trepidation.

ALONGAMIENTO (22:1)
ES: alejamiento, distancia; EN: distance.
A (f. 131v; Millás 1943–1950, p. 227); B (Bossong 1978, p. 125).

— (36:4)
ES: elongación; EN: elongation.
Comment: This term is also used specifically for the distance between the Sun and the Moon.

ALTURA (38:2)
ES: altura; EN: altitude.

A (f. 130v; Millás 1943–1950, p. 230); B (Bossong 1978, p. 127).
Comment: This term refers to a coordinate in the horizontal system.

ALLEGAMIENTO (49:4)/LLEGAMIENTO (28:12)
See also *movimiento del llegamiento y del tornamiento*.
ES: acceso; EN: access.
B (Bossong 1978, pp. 123–124).
Comment: Together with *tornamiento*, these terms refer to the motion of "access and recess", that is, trepidation.

ALVOR (53:1)
See also *subimiento del alvor*.
ES: alba, amanecer; EN: dawn, morning twilight.

AMARILLO (35:6)
ES: amarillo; EN: yellow.
L (Rodríguez Montalvo 1981, p. 238): *amariello*.
Comment: This term appears in discussions of colors of eclipses.

ANCHURA (32:2)
See also *latitud*.
ES: latitud; EN: latitude.
B (Bossong 1978, pp. 128–129).
Comment: This term refers to both the geographical and the ecliptic coordinate. In Q it is used in the sense of *distancia* (length) (Millás 1960, p. 248).

ANGULO (32:27) / ANGOL (32:21)
ES: ángulo; EN: angle.
B (Bossong 1978, p. 132) and Q (Millás 1960, p. 242): *angulo*.

AÑADIMIENTO (16:6) / AÑEDIMIENTO (17:8)
ES: adición; EN: addition.
Comment: As opposed to *menguamiento*. This term is found in the form *annadimiento* both in B (Bossong 1978, pp. 134–135) and Q (Millás 1960, p. 241).

AÑADIR (2:3) / AÑEDIR (2:4)
ES: añadir; EN: to add.
B (Bossong 1978, pp. 133–134) and Q (Millás 1960, p. 243): *annader*.
Comment: As opposed to *menguar*.

AÑO (0:3)
ES: año; EN: year.
B (Bossong 1978, p. 135), Q (Millás 1960, p. 239), and L (Rodríguez Montalvo 1981, p. 239): *anno*. A (f. 94v): *año* and *anno*.

AÑOS DEL MUNDO (48:1)
ES: años del mundo; EN: world-years.

Comment: A world-year refers to the time between the entrance of the Sun in Aries and its next entrance in Aries a year later. Its astrological significance goes back to Abū Maʿshar's *Book of revolutions of world-years*, translated into Latin in Toledo in the 12th century under the title *De revolutionibus annorum mundi*.

APAREÇIMIENTO (24:14) / APARESÇIMIENTO (36:1)
See also *paresçimiento*.
ES: aparición; EN: appearance.

APRESURADO (MOVIMIENTO –) (24:3)
See also *presurado*.
ES: (movimiento) rápido; EN: swift, fast (motion).
Comment: As opposed to *tardio*.

AQUARIO (0:12)
ES: Acuario; EN: Aquarius.
B (Bossong 1978, p. 29); A (f. 97r; Millás 1943–1950, p. 158).
Comment: A also has the form *Aquarius* (f. 98r).

ARCO (32:6)
ES: arco; EN: arc.
B (Bossong 1978, pp. 136–137); Q (Millás 1960, p. 240); A (f. 127r; Millás 1943–1950, p. 229).

ARCO DE LA DIVERSIDAD (41:13)
ES: paralaje; EN: parallax.
Q (Millás 1960, p. 246): *arco de la diuerssidat*.
Comment: In the text, this expression, meaning parallax, is used incorrectly for ascensional difference.

ARCO DE LA NOCHE (40:1)
See also *arco del dia*.
ES: arco nocturno; EN: nocturnal arc.
Comment: This term refers to the arc corresponding to the length of night; see 40:4 in chapter 4.

ARCO DE LA VISTA (36:20)
ES: arco de la vista; EN: arc of vision.
Comment: For a definition of this term, see 36:19–20 in chapter 4.

ARCO DEL DIA (40:1)
See also *arco de la noche*.
ES: arco diurno; EN: diurnal arc.
Comment: This term refers to the arc corresponding to the length of daylight; see 40:4 in chapter 4.

ARCO DEL PONIMIENTO (36:18)
ES: arco del ocaso; EN: arc of setting.
Comment: For a definition of this term, see 36:18 in chapter 4.

ARGUMENTO (– DE UN PLANETA) (1:13)

ES: argumento (de un planeta); EN: argument (of a planet).

A (f. 104r; Millás 1943–1950, p. 172); B (Bossong 1978, pp. 137–138); Q (Millás 1960, p. 256).

Comment: In the case of the Sun, this term is used for the solar anomaly (see 16:3 in chapter 4); for the Moon it is defined as the difference between its longitude and that of the lunar node (see 21:3 in chapter 4); and for the planets it refers to the argument of anomaly.

ARGUMENTO DE LA LATITUD (21:3)

ES: argumento de latitud; EN: argument of latitude.

Comment: This term refers to the angular distance on the ecliptic between the lunar ascending node and the projection of the Moon onto the ecliptic.

ARIES (39:14)

ES: Aries; EN: Aries.

A (f. 97v; Millás 1943–1950, p. 159); B (Bossong 1978, p. 29); Q (Millás 1960, p. 241); L (Rodríguez Montalvo 1981, p. 241).

ARMILLA (0:8)

ES: (esfera) armillar; EN: armillary (sphere).

B (Bossong 1978, p. 139) and Q (Millás 1960, p. 239): *armella*.

ARQUEAR ARQUEAMIENTO DEL SIGNO (51:5)

B (Bossong 1978, pp. 139–140).

Q (Millás 1960, p. 241): *arquear lo arqueamiento*, and Q (p. 247): *arquar lo arqueamiento*.

Comment: In context, this expression means "to find the arcsine".

ARREDAMIENTO (20:3) / ARREDRAMIENTO (21:1)

ES: distancia; EN: distance.

B (Bossong 1978, pp. 140–141); Q (Millás 1960, p. 240). A (f. 133v; Millás 1943–1950, p. 231): *arredramiento*.

Comment: This term is used, for instance, in the expression *arredamiento del çerco egualador* (20:3), meaning declination, and in *arredramiento del sodiaco*, meaning latitude (21:1).

ASÇENDENTE (28:1) / ASÇENDIENTE (31:20)

ES: ascendente; EN: ascendant.

B (Bossong 1978, pp. 142–143). L (Rodríguez Montalvo 1981, p. 242) and Q (Millás 1960, pp. 240 and 252): *ascendent* and *ascendente*.

Comment: This term refers to the rising point of the ecliptic at a given time which is also the cusp of the first house of an astrological chart.

ASCONDIMIENTO (25:1)

ES: desaparición; EN: disappearance.

Comment: As opposed to *paresçimiento*. This term appears in chapter 25 with regard to the visibility of the planets.

ASFIDACME (1:17) / AZFIDARINHE (5:15)

A (f. 96v): *Azfidarme*.

Comment: This is the month *Isfandārmadh Māh* in the Persian calendar.

ASTROLOGIA (0:2)

ES: astronomía; EN: astronomy.

B (Bossong 1978, p. 145); Q (Millás 1960, p. 239).

Comment: Although the word used is *astrologia*, in all three occurrences—0:2, 37:6, and 37:37—the context is astronomical. This is not surprising because the boundary between astronomy and astrology was not sharply defined. The term *astronomia* does not appear in the text.

AUX (16:2) / AUXE (17:7)

ES: apogeo; EN: apogee.

Comment: As opposed to *hadit*. This term was borrowed from the Arabic *awj*. Cf. Nallino 1903–1907, 2:322.

AYUNO (DIA DEL –) (54:4)

ES: (día del) ayuno; EN: fast (day).

AYUNTAR (2:3)

ES: juntar; EN: to gather.

B (Bossong 1978, p. 152); L (Rodríguez Montalvo 1981, p. 235); Q (Millás 1960, p. 241).

BAXURA DEL SOL (53:1)

ES: (arco de) depresión del Sol; EN: solar depression (arc).

Comment: This term refers to the arc between the Sun and the horizon when the Sun is below the horizon, measured on a circle from the zenith to the Sun.

BERMEJURA (35:5)

ES: coloración roja; EN: red coloration.

Comment: This term appears in discussions of colors of eclipses.

BETHMENXIBE (5:14)

A (f. 96v): *Baharanme*.

Comment: This is the month *Bahman Māh* in the Persian calendar.

BISIESTO (AÑO DE –) (2:8) / BISESTIL (6:9) / BISEXTIL (15:9) / BISEXTO (13:21) / BISSIESTO (3:2) / VISIESTO (2:3)

ES: (año) bisiesto; EN: leap (year).

A (f. 94v; Millás 1943–1950, p. 154) and B (Bossong 1978, p. 153): *bissiesto*.

BUC (48:4)

ES: velocidad del Sol en una hora; EN: hourly velocity of the Sun.

A (f. 105r; Millás 1943–1950, p. 174): *buth*.

Comment: This term only appears once in the text. In A it is applied to the daily velocity of planets as well. It was borrowed from the Arabic *buht* (ultimately of Sanskrit origin: *bhukti*): see Nallino 1903–1907, 2:324; Suter 1914, p. 22; Neugebauer 1962a, p. 57.

CABEÇA DE ARIES (39:14)

ES: comienzo de Aries; EN: beginning of Aries.

A (f. 126v; Millás 1943–1950, p. 219); B (Bossong 1978, p. 154); Q (Millás 1960, p. 241).

CABEÇA DEL DRAGON (19:1)

ES: nodo ascendente; EN: ascending node.

A (f. 105v; Millás 1943–1950, p. 175).

Comment: As opposed to *cola del dragon*. This term refers to the ascending node of the lunar orb, i.e., the point on the lunar orb where the Moon passes from south of the ecliptic to the north.

CAHYNRMEH (5:9)

A (f. 96v): *Xahararme*.

Comment: This is the month *Shahrēwar Māh* in the Persian calendar.

CANÇER (36:7) / CANCRO (39:9)

ES: Cáncer; EN: Cancer.

A (f. 97v; Millás 1943–1950, p. 159), B (Bossong 1978, p. 29), and Q (Millás 1960, p. 256): *Cancer*. L (Rodríguez Montalvo 1981, p. 249): *cancro*.

CAPRICORNIO (29:16) / CAPRICORNUS (29:16)

ES: Capricornio; EN: Capricorn.

B (Bossong 1978, p. 29); Q (Millás 1960, p. 256). A (f. 97r; Millás 1943–1950, p. 158): *Capricorno*; and A (f. 98r): *Capricornus*.

CASA (28:11)

ES: casa (astrológica); EN: (astrological) house.

B (Bossong 1978, pp. 157–158); L (Rodríguez Montalvo 1981, p. 250); Q (Millás 1960, p. 240).

Comment: This term is applied to each of the 12 divisions of the ecliptic that serve astrological purposes. There were several methods of domification but the text does not describe any of them. Generally, the intersections of the ecliptic with the horizon and the meridian were taken to be the cusps of the 1st, 4th, 7th, and 10th houses (see North 1986).

CATAMIENTO (31:1) / ACATAMIENTO (31:7)

See also *arco de la diversidad* and *diversidad del catamiento.*

ES: paralaje; EN: parallax.

B (Bossong 1978, pp. 158–159).

Comment: This term refers to the angle subtended by the radius of the Earth as it would be seen at the distance of the Moon. In L it is used to mean "astrological aspect" (Rodríguez Montalvo 1981, p. 251).

CATAR (11:13)

ES: ver, considerar; EN: to see, to note, to consider.

B (Bossong 1978, pp. 159–160); L (Rodríguez Montalvo 1981, p. 251); Q (Millás 1960, p. 241).

CAYMIENTO (33:14)

ES: inmersión; EN: immersion.

A (f. 133v; Millás 1943–1950, p. 231); B (Bossong 1978, p. 162).

Comment: This term appears in the context of eclipses.

ÇENTENARIO (0:5)

ES: centuria, siglo; EN: century.

ÇENTRO (– DE UN PLANETA) (14:2) / CENTRO (1:13)

ES: centro; EN: center.

A (f. 133r; Millás 1943–1950, p. 233). B (Bossong 1978, p. 162): *centro* and *çenptro*; Q (Millás 1960, p. 257): *çenptro*.

Comment: This term refers to the center of a planet, i.e., the difference between the mean position of the planet and its apogee.

ÇENTRO DE LA CASA (28:11)

ES: cúspide; EN: cusp.

Comment: In the text, this term refers to the beginning, not the center, of an astrological house. Thus, *çentro* literally translates the Arabic word *markaz*, whose primary meaning is "center", but also has the technical sense of "cusp", that is, the beginning of a house.

ÇERCO (1:20) / CERCO (1:21)

ES: círculo; EN: circle.

A (ff. 130r, 132v; Millás 1943–1950, pp. 230, 232); B (Bossong 1978, pp. 163–164); Q (Millás 1960, pp. 239, 241) (both forms in all of them).

CERCO DE MEDIODIA (1:22)

ES: meridiano; EN: meridian.

Q (Millás 1960, p. 241): *cerco del mediel dia*.

Comment: This term is used for the great circle on the celestial sphere which passes through the observer's zenith and nadir, and the celestial poles.

ÇERCO DE LOS SIGNOS (50:1)
ES: eclíptica; EN: ecliptic.
B (Bossong 1978, p. 164); Q (Millás 1960, p. 241).
Comment: This term is used for the great circle resulting from the intersection of the celestial sphere and the plane of the solar orb.

ÇERCO DEL SOL (1:22)
ES: disco solar; EN: solar disk.

CERCO DERECHO (28:10)
ES: esfera recta; EN: Sphaera recta.
Comment: This term refers to the celestial sphere as it would be seen by an observer on the geographical equator: the pole lies on the horizon and the celestial equator is perpendicular to the horizon.

ÇERCO EGUALADOR (20:3)
See also *eguador and equinoçial.*
ES: ecuador; EN: equator.
Comment: This term refers to the celestial equator, a great circle resulting from the intersection of the celestial sphere and the plane of the terrestrial equator.

ÇERCO EQUINOÇIAL (29:2)
ES: ecuador; EN: equator.

CERCO YGUAL (32:23)
See *subimiento del cerco ygual.*
ES: ecuador; EN: equator.

ÇIBDAD (0:9)
ES: ciudad; EN: city.
B (Bossong 1978, p. 167): *çipdat*.

ÇIELOS (0:3)
ES: cielos; EN: heavens.
B (Bossong 1978, pp. 164–165).
Comment: In A it appears in the expression *cielo derecho* (f. 128r; Millás 1943–1950, p. 221).

ÇIELO ALTO (RASO) NO ESTRELLADO (49:2, 49:7)
ES: cielo no estrellado, novena esfera; EN: starless sphere, 9th sphere.
Comment: Here this term designates the 9th sphere.

ÇIELO ESTRELLADO (49:7)
ES: cielo estrellado, octava esfera; EN: sphere of the fixed stars, 8th sphere.
Comment: In L we find the expression *ochauo cielo* (Rodríguez Montalvo 1981, p. 231).

CIENÇIA (0:2)
ES: ciencia; EN: science.

B (Bossong 1978, pp. 307–308) and Q (Millás 1960, p. 239): *sciencia*.

ÇIERTO (LUGAR –, MOVIMIENTO –, HORA, etc.) (16:6, 27:9, 27:37)
ES: (posición, etc.) verdadera; EN: true (position, etc.).
B (Bossong 1978, p. 166): *cierto*; Q (Millás 1960, p. 253): *çierto* and *cierto*.
Comment: As opposed to *medio* or *mediano*. This refers to true positions, motions, etc., as opposed to mean positions, motions, etc.

ÇIFRA (10:15)
ES: cero; EN: zero.
Comment: This term is borrowed from the Arabic: *ṣifr*.

ÇIRCUNFERENÇIA (0:3)
ES: circunferencia; EN: circumference.

CLARIDAD (33:16)
ES: claridad; EN: brightness, clarity.
A (f. 133r; Millás 1943–1950, p. 233), B (Bossong 1978, p. 168), and L (Rodríguez Montalvo 1981, p. 252): *claridat*.

CLIMA (33:3)
ES: clima; EN: clima.
B (Bossong 1978, pp. 168–169); L (Rodríguez Montalvo 1981, p. 252).
Comment: This term refers to a zone of geographical latitude. In A *clima* is feminine (f. 128v; Millás 1943–1950, p. 222).

COLA DEL DRAGON (19:2)
ES: nodo descendente; EN: descending node.
Comment: As opposed to *cabeça del dragon*. This term refers to the descending node of the lunar orb, i.e., the point on the lunar orb where the Moon passes from north of the ecliptic to the south.

COLLECTO (AÑO –) (9:11) / COLECTO (9:2)
ES: (año) colecto; EN: collected (year).
A (f. 95v).
Comment: As opposed to *espandido*. Tables of mean motions usually have lists for collected or accumulated years, e.g., at intervals of 20 years, as well as lists for single or expanded years.

COLOR (35:1)
ES: color; EN: color.
B (Bossong 1978, p. 169); L (Rodríguez Montalvo 1981, p. 253).

COMENÇAMIENTO (4:1) /COMIENCAMIENTO (1:12) / COMIENÇO (1:2)
ES: comienzo; EN: beginning.

A (f. 101r; Millás 1943–1950, p. 166), B (Bossong 1978, p. 169), and Q (Millás 1960, p. 256): *començamiento*.

COMENÇAMIENTO NATURAL (49:2)

Comment: This term is used to designate the origin for tropical longitude, i.e., the vernal equinox.

COMPLIDO (AÑO –, DIA –, CORDA –) (0:11, 6:5, 37:51)

ES: (año, día, cuerda, etc.) completo; EN: full (year, day, chord, etc.).

B (Bossong 1978, pp. 171–172).

COMPLIMIENTO (SIGNO DEL –) (37:1)

ES: coseno; EN: cosine.

B (Bossong 1978, pp. 172–173); Q (Millás 1960, p. 241). A (f. 127r; Millás 1943–1950, p. 229): *sinu del complimiento*.

Comment: This term refers to the trigonometrical function (defined in 37:1 in chapter 4).

COMPOSTURA (10:11)

ES: composición; EN: composition.

Comment: In context, the expression *compostura de los dias* means "number of days expressed in sexagesimal form".

COMPUESTOS (DIAS –) (12:1)

See also *compostura*.

Comment: This term refers to days expressed in sexagesimal form.

COMUN (AÑO –) (7:10)

See also *simple*.

ES: (año) común; EN: common (year).

Comment: As opposed to *bisiesto*.

CONJUNÇION (0:13) / CONJUNTION (27:3)

ES: conjunción; EN: conjunction.

B (Bossong 1978, p. 174); L (Rodríguez Montalvo 1981, p. 255).

Comment: As opposed to *oposiçion*. This term refers to the moment when the Sun and the Moon have the same celestial longitude.

CORDA (32:29)

See also *querda*.

ES: cuerda; EN: chord.

Comment: Here chord is a line segment joining two points of a circumference, and it is used as the basic trigonometrical function (see 37:1 in chapter 4).

CORREGIR (0:10)

ES: corregir; EN: to correct.

CREÇER (15:4)

ES: crecer, aumentar; EN: to grow, to increase.

CUBRIR (24:15)
  ES: cubrir; EN: to cover.

CURSO (MEDIO –) (1:13)
  ES: movimiento (medio); EN: (mean) motion.
  B (Bossong 1978, p. 181). A (f. 101r; Millás 1943–1950, p. 166): *cursso*.

DEÇENDIENTE (20:6) / DECENDIENTE (20:5)
  ES: descendente; EN: descending.
  B (Bossong 1978, p. 190): *descendiente*.
  Comment: As opposed to *subiente*.

DEÇIMO (4:4)
  ES: décimo; EN: tenth.

DECLINAÇION (20:1) / DECLINACION (38:8)
  ES: declinación; EN: declination.
  Comment: This term refers to the equatorial coodinate.

— (22:11)
  See also *declinamiento* and *desviamiento*.
  ES: desviación; EN: slant.
  A (f. 105r; Millás 1943–1950, p. 174), B (Bossong 1978, pp. 182–185), and Q (Millás 1960, p. 241): *declinacion*.
  Comment: There are two components in Ptolemy's tables for planetary latitude, called "inclination" and "slant" in the English translation of the *Almagest* by Toomer (1984, p. 601). The Latin terms for these two components are *declinatio* and *reflexus* in the Toledan Tables (Toomer 1968, p. 71). The Castilian terms used in A are *declinacion* and *desviamiento* (Millás 1943–1950, pp. 217–218).

DECLINAMIENTO (22:11)
  See also *declinaçion* and *desviamiento*.
  ES: inclinación [?]; EN: inclination [?].
  Comment: This term only appears once in the text, and it is distinguished from *declinaçion*, whereas in sentence 22:13 another distinction is made between *declinaçion* and *desviamiento*.

DECLIVIDAD (50:2)
  ES: oblicuidad variable ; EN: variable obliquity.
  Comment: This term only appears twice, in chapter 50, and refers to the motion of the pole of the ecliptic.

DEDO (33:14)
  ES: dedo; EN: digit.
  A (f. 133r; Millás 1943–1950, p. 233); B (Bossong 1978, p. 185).
  Comment: This term refers to the unit used to measure the magnitude of an eclipse.

DEFEÇION (8:2)
ES: parte; EN: part.
Comment: In context, this term means "what is missing to complete a full day". Sentence 8:2 gives *falta* as a synonym.

DEMOSTRAÇION (37:32)
ES: demostración; EN: demonstration, proof.
L (Rodríguez Montalvo 1981, p. 259): *demostration*.

DEPARTIR (37:9)
ES: subdividir; EN: to subdivide.
B (Bossong 1978, pp. 187–188); L (Rodríguez Montalvo 1981, p. 260).

DERECHO (ANGULO –) (37:41)
ES: ángulo recto; EN: right angle.
Q (Millás 1960, p. 245).

DERECHO (PLANETA –) (23:9)
ES: (movimiento) directo (de un planeta); EN: direct (motion of a planet).
B (Bossong 1978, pp. 188–189).
Comment: As opposed to *retrogrado*.

DESACORDANÇA (0:10)
ES: discordante; EN: discordant.

DESENA (0:5)
ES: decena; EN: ten.
A (f. 94r): *dezena*.
Comment: This term appears in Q in the form *dezena*, but there it means "tenth" (Millás 1960, p. 245).

DESVARIANÇA (27:20)
Comment: This term, together with *variança*, is used here to mean "variation", or "difference".

DESVARIO (27:22)
B (Bossong 1978, pp. 191–192).
Comment: This term seems to be used as an equivalent of *desvariança*. In Q it appears in the form *desvariamiento* (Millás 1960, p. 245), and it is used as a synonym of *diverssidat*, that is, "difference", or "deviation".

DESVIAMIENTO (22:13)
See also *declinación* and *declinamiento*.
ES: desviación; EN: slant.
A (f. 126; Millás 1943–1950, p. 218) and B (Bossong 1978, p. 193): *desuiamiento*.

DIA (1:14)
ES: día; EN: day.
B (Bossong 1978, pp. 193–195); Q (Millás 1960, p. 239); A (f. 94v); L (Rodríguez Montalvo 1981, p. 262).

DIAMETRO (17:15)

ES: diámetro; EN: diameter.

A (f. 133r; Millás 1943–1950, p. 233); B (Bossong 1978, p. 195); Q (Millás 1960, p. 248).

DIESTRO (52:4)

ES: derecho; EN: right.

Comment: As opposed to *siniestro*.

DIFERENÇIA (11:10)

ES: diferencia; EN: difference.

A (f. 130r; Millás 1943–1950, p. 226) and Q (Millás 1960, p. 247): *differencia*. B (Bossong 1978, pp. 195–197): *differencia* and *differentia*.

DIVERSIDAD (0:6) / DIVERSIDAT (32:46) / DIVERSYDAD (31:19)

ES: diversidad; EN: diversity.

B (Bossong 1978, pp. 198–199): *diuerssidad* and *diuerssida*. L (Rodríguez Montalvo 1981, p. 262): *diuersidat*. Q (Millás 1960, p. 244): *diverssidat*.

DIVERSIDAD DEL CATAMIENTO (31:2)

See also *arco de la diversidad* and *catamiento*.

ES: paralaje; EN: parallax.

B (Bossong 1978, p. 199). A (f. 130r; Millás 1943–1950, p. 226): *diversidad*; and (f. 132v): *diverssidad*.

DIVERSIDAD DEL DIAMETRO (17:15)

Comment: This expression refers to the angular difference between the mean and true apogees on the lunar epicycle, usually called in Latin *diversitas diametri*. There is no standard term for it in Spanish or English, but in Toomer (1984, p. 238) this angle is called "increment in epicyclic equation", and in F. S. Pedersen (2002, p. 1250) it is called "variation of epicyclic diameter".

DIVERSIDAD DE LOS DIAS CON SUS NOCHES (33:6)

See *equaçion de los dias con sus noches*.

ES: ecuación del tiempo; EN: equation of time.

B (Bossong 1978, p. 198).

DIVERSO (MOVIMIENTO –, DIA –) (27:1, 29:10)

ES: diverso, desigual; EN: diverse, unequal.

DIZIEMBRE (2:4)

ES: diciembre; EN: December.

Comment: This term is found in the form *deziembre* in A (f. 94v).

DOBLAMIENTO (37:45)

ES: doble; EN: double.

DOBLAR (17:9)

ES: doblar; EN: to double.

B (Bossong 1978, pp. 199–200); Q (Millás 1960, p. 246).

DOMINGO (1:14)
ES: domingo; EN: Sunday.

DOSENA (15:20)
ES: docena; EN: dozen.

DOSENO (2:17) / DOZENA (30:21)
ES: duodécimo; EN: twelfth.

DULQUEDA (4:15)
A (f. 94r).
Comment: This is the month *Dhu l-qa*ᶜ*da* in the Arabic calendar.

DULHEIA (4:4) / DULHEYIA (4:16)
A (f. 94r): *Dulheia*.
Comment: This is the month *Dhu l-ḥijja* in the Arabic calendar.

DYNMEH (5:13)
A (f. 96v): *Dijme*.
Comment: This is the month *Dai Māh* in the Persian calendar.

ECLIPSE (30:31) / ECLIPSI (0:14) / ECLIPSY (31:1) / ECLIUSE (30:30)
ES: eclipse; EN: eclipse.
B (Bossong 1978, p. 202). A (f. 133r; Millás 1943–1950, p. 233): *eclipsi*.

ECHAMIENTO (DE LOS RAYOS) (52:1)
ES: proyección (de los rayos); EN: projection (of the rays).
B (Bossong 1978, pp. 201–202).
Comment: This term refers to a doctrine in medieval astrology.

EGUADOR (39:1)
See also *çerco egualador, çerco equinoccial, çerco ygual,* and *equinoçial*.
ES: ecuador; EN: equator.
B (Bossong 1978, pp. 203–205); Q (Millás 1960, p. 240).

EGUAL (31:3)
See *yguales*.
ES: igual; EN: equal.
A (f. 127r; Millás 1943–1950, p. 229); B (Bossong 1978, pp. 205–206); Q (Millás 1960, p. 239); L (Rodríguez Montalvo 1981, p. 263).

EGUALAR (27:3) / EGUAR (20:18)
See *ygualar*.
ES: igualar; EN: to equate.
B (Bossong 1978, p. 206).

EGUALDAD (38:6)
See *ygualdad*.
ES: equinoccio; EN: equinox.

B (Bossong 1978, p. 206) and Q (Millás 1960, p. 243): *egualdat*.
Comment: The expression *puntos de la egualdad* refers to the equinoxes, when the astronomical day and night are equal.

ENDEREÇADURA (27:8)
ES: corrección; EN: correction.

ENDEREÇAMIENTO (32:38) / ENDERECAMIENTO (23:1)
ES: corrección; EN: correction.
B (Bossong 1978, p. 209). A (f. 104v; Millás 1943–1950, p. 173): *endreçamiento*; and (f. 134v): *endereçamiento*.

ENDEREÇAR (0:10) / ENDERECAR (23:8)
ES: corregir; EN: to correct.
B (Bossong 1978, pp. 209–210); L (Rodríguez Montalvo 1981, p. 266). Q (Millás 1960, p. 240): *endereçar* and *endreçar*.

EPIÇICLO (18:14) / EPECICLO (32:70) / PIÇICLO (17:15)
ES: epiciclo; EN: epicycle.
A (f. 133r; Millás 1943–1950, p. 233): *episciclo*.

EQUAÇION (16:1)
See *ynvocaçion*.
ES: corrección, ecuación; EN: correction, equation.
B (Bossong 1978, pp. 202–203). Q (Millás 1960, p. 240): *equacion*.
Comment: This term normally refers to the correction to be added to, or subtracted from, a mean quantity in order to obtain the true quantity.

EQUAÇION DE LOS DIAS CON SUS NOCHES (48:6)
See *diversidad de los dias con sus noches*.
ES: ecuación del tiempo; EN: equation of time.
A (f. 127v; Millás 1943–1950, p. 220): *eguacion de los dias con sus noches*.

EQUINOÇIAL (LINEA –) (1:23)
See also *çerco egualador, çerco equinoccial, çerco ygual,* and *eguador*.
ES: ecuador; EN: equator.
B (Bossong 1978, pp. 214–215).

EQUINOÇIO (20:3)
ES: equinoccio; EN: equinox.

ERA (10:1)
See *hera*.

ESCURPION (0:12) / SCURPION (29:13)
ES: Escorpión; EN: Scorpio.
A (f. 97r; Millás 1943–1950, p. 158) and B (Bossong 1978, p. 29): *Scorpio*. L (Rodríguez Montalvo 1981, p. 268): *escurpion* and *escorpion*.

ESPANDIDA (SOMBRA –) (51:1)
ES: cotangente; EN: cotangent.
B(Bossong 1978, p. 217).
Comment: As opposed to *sombra menguada*. This term corresponds to the Latin *umbra extensa seu expansa* (see Nallino 1903–1907, 1:180–182, 2:343).

ESPANDIDO (AÑO –) (9:14) / ESPANSO (15:6)
ES: (año) expandido; EN: expanded (year).
A (f. 94v).
Comment: As opposed to *collecto*.

ESTAÇION (23:7) / ESTADO (23:5)
ES: estación; EN: station.
B (Bossong 1978, pp. 219–220).
Comment: In the text, both terms refer to the 4 values for the stations of each planet, that is, the positions of a planet in celestial longitude where it shifts from direct to retrograde motion, or *vice versa*.

ESTADO (39:3)
ES: estado, condición; EN: state, condition.
B (Bossong 1978, p. 220).
Comment: In context, this term refers to the various coordinates given in a star list. In L this term means "position of the stars" (Rodríguez Montalvo 1981, p. 270).

ESTANTE (PLANETA –) (27:21)
ES: (planeta) estacionario ; EN: stationary (planet).
Comment: This term refers to a planet when it shifts from direct to retrograde motion, or *vice versa*.

ESTIUAL (TROPICO –) (38:6)
See *tropico* and *yvernal (tropico –)*.
ES: (solsticio) de verano; EN: summer (solstice).
Comments: This term is found in B in the form *estiual* (Bossong 1978, p. 221) and appears together with *mudamiento* (solstice). In Q it appears as *estival* (Millás 1960, p. 241), where it is also used together with *mudamiento*.

ESTRELLA (39:3)
ES: estrella; EN: star.
B (Bossong 1978, p. 221): *estrella* and *strella*.
Comment: Note that the word *estrella* is feminine, but in this text it also appears together with a masculine article (see 39:5). Both genders for *estrella* are also found in Q (Millás 1960, pp. 239–240).

FINCAR (9:14)
ES: quedar, resultar; EN: to remain, to result.
B (Bossong 1978, p. 228).

FIXAS (ESTRELLAS –) (0:13)
ES: estrellas fijas; EN: fixed stars.
B (Bossong 1978, p. 229); L (Rodríguez Montalvo 1981, p. 274); Q (Millás 1960, p. 239).

FORMA (49:7)
ES: constelación; EN: constellation.
B (Bossong 1978, pp. 229–230).
Comment: The Arabic word for "constellation" is *ṣūra* whose primary meaning is "form" or "shape"; the authors of our text have clearly depended on a translation of this Arabic term. Note, however, that in L the term used is *constellation* (Rodríguez Montalvo 1981, p. 255).

FRAÇION (2:2) / FRAÇIO (5:3)
ES: fracción; EN: fraction.
B (Bossong 1978, pp. 230–231): *fraction*.

GEMINIS (36:5) / GEMENIS (41:6)
ES: Géminis; EN: Gemini.
B (Bossong 1978, p. 29).
A (f. 97v; Millás 1943–1950, p. 159), L (Rodríguez Montalvo 1981, p. 278), and Q (Millás 1960, p. 256): *Gemini*.

GEOMETRIA (37:2)
ES: geometría; EN: geometry.
B (Bossong 1978, p. 232); L (Rodríguez Montalvo 1981, p. 278); Q (Millás 1960, p. 239).

GEUZAAR (21:3)
ES: nodo lunar; EN: lunar node.
Comment: This term is a transliteration of Arabic *jawzahar* and can apply to the node of any planet, but in our text it refers exclusively to the lunar node.

GRADO (1:20)
ES: grado; EN: degree.
A (f. 97r; Millás 1943–1950, p. 158); B (Bossong 1978, pp. 233–234); L (Rodríguez Montalvo 1981, p. 278); Q (Millás 1960, p. 240).

GUARDAR (9:8)
ES: guardar; EN: to keep.
B (Bossong 1978, pp. 233–234); L (Rodríguez Montalvo 1981, p. 252); Q (Millás 1960, p. 246).

HADIT (23:14)
ES: perigeo; EN: perigee.
Comment: As opposed to *aux*. This term is a transliteration of Arabic *ḥaḍīḍ*.

HARDETINHE (5:6)
A (f. 96v): *Hardarme*.

Comment: This is the month *Khurdādh Māh* in the Persian calendar.
HEBRERO (2:4)
ES: febrero; EN: February.
A (f. 94v): *febrero*.
HENERO (1:14)
ES: enero; EN: January.
A (f. 94v): *enero*.
HERA (0:5) / ERA (10:1)
ES: era; EN: era.
A (f. 95v): *era*.
Comment: This term is used for dates in a calendar counted in years from an origin at a certain time, called an "epoch".
HORA (14:7) / ORA (14:5)
ES: hora; EN: hour.
A (f. 127v; Millás 1943–1950, p. 220); B (Bossong 1978, p. 236); L (Rodríguez Montalvo 1981, p. 296): Q (Millás 1960, p. 240).
HORDEN (6:12) / HORDENAMIENTO (2:6)
ES: orden; EN: order.
B (Bossong 1978, pp. 267–268) and Q (Millás 1960, pp. 267–268): *ordenamiento*.
JUNIO (2:12)
ES: junio; EN: June.
A (f. 94v).
JULLIO (2:13)
ES: julio; EN: July.
A (f. 94v): *julio*.
JUMET PRIMO (4:9)
A (f. 94r): *Jumeth I*; and (f. 96r): *Jumet I*.
Comment: This is the month *Jumādā I* in the Arabic calendar.
JUMET SECUNDO (4:10)
A (f. 94r): *Jumet II*; and (f. 96r): *Jumeth II*.
Comment: This is the month *Jumādā II* in the Arabic calendar.
JUPITER (22:2)
ES: Júpiter; EN: Jupiter.
LADEZA (22:24) / LADESA (22:37)
ES: latitud; EN: latitude.
A (f. 104v; Millás 1943–1950, p. 173), B (Bossong 1978, pp. 237–238), and Q (Millás 1960, p. 239): *ladeza*.
LATITUD (1:23)
ES: latitud; EN: latitude.
Comment: Note that the terms *ladeza*, *ladesa*, *latitud*, and *anchura* are all used to designate "latitude". The term used

most often in the text is *ladesa* (111 times), followed by *latitud* (55 times), then *anchura* (25 times), and finally *ladeza* (8 times). All four terms are used to mean both celestial and geographical latitude, with the following frequencies:

	geo.	cel.
ladesa	103	8
latitud	53	2
anchura	13	12
ladeza	7	1

The terms *latitud* and *ladeza* may be used to designate geographical latitude (40:14) or celestial latitude (22:26), and even in a single chapter *latitud* may designate both kinds of latitude (as in 40:10 and 40:14).

LEON (0:12)

ES: Leo; EN: Leo.

A (f. 97v; Millás 1943–1950, p. 159) and B (Bossong 1978, p. 29): *Leo*.

LIBRA (39:14)

ES: Libra; EN: Libra.

A (f. 97r; Millás 1943–1950, p. 158); B (Bossong 1978, p. 29); Q (Millás 1960, p. 253).

LINAJE (14:9)

ES: linaje; EN: rank.

Comment: In context, this term is used for the addition of numbers whose digits have to be put one under another "of the same lineage", meaning of the same rank or order of magnitude.

LIVIANO (PLANETA –) (27:26)

ES: (planeta) ligero, rápido; EN: light, swift (planet).

B (Bossong 1978, p. 241); L (Rodríguez Montalvo 1981, p. 285).

Comment: As opposed to *pesado*. This term is used in the discussions of the relative velocity of the planets.

LONGURA (1:20)

ES: longitud; EN: longitude.

B (Bossong 1978, pp. 243–244); Q (Millás 1960, p. 239).
Comment: This term designates both celestial and geographical longitude; as indicated below, it is also used to mean "elongation" and "distance", as was the case with *alongamiento*.

LONGURA DOBLADA (17:9)
ES: doble elongación; EN: double elongation.

LONGURA (LA) MAS ÇERCA (18:14, 33:17) / LONGURA LA ÇERCANA (33:14) / LONGURA MENOR (33:14)
ES: diferencia sustractiva en el apogeo; EN: subtractive difference at apogee.
B (Bossong 1978, p. 244). A (f. 133v–134r; Millás 1943–1950, pp. 231, 234): *longura (la) mas cerca*.

LONGURA (LA) MAS LUENGA (18:14, 33:17)
ES: diferencia aditiva en el perigeo; EN: additive difference at perigee.
B (Bossong 1978, p. 244). A (f. 133v–134r; Millás 1943–1950, pp. 231, 234): *longura (la) mas luenne*.

LUGAR (0:10)
See also *natural (lugar –)* and *propio (lugar –)*.
ES: posición; EN: position.
A (f. 101r; Millás 1943–1950, p. 166) and B (Bossong 1978, p. 244): *logar*. Q (Millás 1960, p. 240): *lugar*; and (p. 239): *logar*.

LUMINARIAS (14:2) / LUMINARES (30:33)
ES: luminarias; EN: luminaries.
B (Bossong 1978, p. 246).

LUNA (2:2)
ES: Luna; EN: Moon.

LUNA NUEVA (4:2)
ES: luna nueva; EN: New Moon.

LUNES (1:14)
ES: lunes; EN: Monday.

LLEGAMIENTO (28:12)
See *allegamiento*.

MAHARMEH (5:10)
A (f. 96v): *Maharanme*.
Comment: This is the month *Mihr Māh* in the Persian calendar.

MARÇO (2:9)
ES: marzo; EN: March.
A (f. 94v).

MARDETHMHE (5:8)
A (f. 96v): *Mardarme*.
Comment: This is the month *Murdādh Māh* in the Persian calendar.

MARES (22:2)
ES: Marte; EN: Mars.
A (f. 108r; Millás 1943–1950, p. 196) and L (Rodríguez Montalvo 1981, p. 289): *Mars*.

MAYO (2:11)
ES: mayo; EN: May.
A (f. 94v).

MEDIANO / MEDIO (MOVIMIENTO –, DIA –, SOL –, etc.) (27:15, 28:2, 29:16)
ES: (movimiento, etc.) medio; EN: mean (motion, etc.).
B (Bossong 1978, p. 249). A (f. 101r; Millás 1943–1950, p. 166): *medio*.
Comment: As opposed to *çierto*.

MEDIO (1:13)
See also *mediano*.

MEDIO ALTO (23:3)
ES: mitad superior; EN: upper half.
Comment: As opposed to *medio baxo*.

MEDIO BAXO (23:4)
ES: mitad inferior; EN: lower half.
Comment: As opposed to *medio alto*.

MEDIO (DEL) ÇIELO (28:1, 29:7)
ES: punto culminante, medio cielo; EN: midheaven.
B (Bossong 1978, p. 250): *mediel cielo*. Q (Millás 1960, p. 254): *medial çielo*.
Comment: This expression refers to the point on the ecliptic where it crosses the meridian above the horizon at a given time.

MEDIO DEL DIAMETRO (37:11)
ES: radio; EN: radius.
B (Bossong 1978, p. 250).

MEDIO DEL ECLIPSE (33:12)
ES: mitad del eclipse; EN: mid-eclipse.
Comment: This expression refers to the moment in time halfway between first and last contacts.

MEDIODIA (1:19) / MEDIO DEL DIA (1:14)
ES: mediodía; EN: noon.
B (Bossong 1978, p. 251) and Q (Millás 1960, p. 239). Q (p. 241): *mediel dia*.

MEDRAR (9:14)
ES: restar; EN: to subtract.

MENGUAMIENTO (16:6)
ES: sustracción; EN: subtraction.
B (Bossong 1978, pp. 252–253); Q (Millás 1960, p. 241).

Comment: As opposed to *añadimiento*.

MENGUADA (SOMBRA –) (51:1)

ES: tangente; EN: tangent.

B (Bossong 1978, p. 253).

Comment: As opposed to *sombra espandida*. The corresponding Latin expression is *umbra versa* that derives from the Arabic *ẓill maᶜkūs* (*lit*.: reversed or inverted shadow): see Nallino 1903–1907, 1:22–23, 180–182, and 2:343. The term in our text seems to be a mistranslation. Nallino (1903–1907, 1:22–23) has *umbra extensa* in Latin for cotangent and *umbra erecta* for tangent. The translation of al-Battānī's canons has *sombra espandida* and *sombra leuantada* (Bossong 1978, pp. 56–58).

MENGUAR (3:4)

ES: restar; EN: to subtract.

B (Bossong 1978, pp. 253–254); Q (Millás 1960, p. 241).

Comment: As opposed to *añadir*.

MENUDO (17:16)

See *minuto*.

MERCURIO (18:22)

ES: Mercurio; EN: Mercury.

MEREDIONAL (20:1)

ES: meridional; EN: southern.

A (f. 125v; Millás 1943–1950, p. 217) and Q (Millás 1960, p. 241): *meridional*. L (Rodríguez Montalvo 1981, p. 291): *miridional*.

Comment: As opposed to *septentrional*.

MES (0:15)

ES: mes; EN: month.

A (f. 94r); B (Bossong 1978, p. 255); L (Rodríguez Montalvo 1981, p. 290).

MIERCOLES (54:1)

ES: miércoles; EN: Wednesday.

MILLAR (0:3)

ES: millar; EN: thousand.

MINUTO (1:23) / MENUDO (17:16) / MYNUTO (17:16)

ES: minuto; EN: minute.

A (f. 95v) and B (Bossong 1978, p. 254): *menudo*.

Comment: This term refers both to the unit of arc and the unit of time.

MOHARRAN (4:5)

A (f. 94r): *Moharram*.

Comment: This is the month *Muḥarram* in the Arabic calendar.

MORA (34:3)

ES: duración; EN: duration.

A (f. 133v; Millás 1943–1950, p. 231); B (Bossong 1978, pp. 255–256).
Comment: This term appears in the context of eclipses, usually in the expression *medio de la mora* (half-duration [of totality]).

MOVIMIENTO (0:10)
ES: movimiento; EN: motion.
A (f. 105r; Millás 1943–1950, p. 174); B (Bossong 1978, p. 257); L (Rodríguez Montalvo 1981, p. 292); Q (Millás 1960, p. 247).

MOVIMIENTO DEL LLEGAMIENTO Y DEL TORNAMIENTO (28:12)
ES: trepidación; EN: trepidation.
Comment: This expression refers to the motion of "access and recess".

MOVIMIENTO EN ALONGADO Y TORNADO (49:1)
ES: trepidación; EN: trepidation.
Comment: This expression refers to the motion of "access and recess".

MULTIPLICAÇION (28:6) / MULTYPLICACION (37:12)
ES: multiplicación, EN: multiplication.
B (Bossong 1978, p. 257) and Q (Millás 1960, p. 248): *multiplicacion*.

MULTIPLICAR (6:9)
ES: multiplicar; EN: to multiply.
B (Bossong 1978, pp. 258–259).
Comment: Note that it is usually followed by the preposition "*en*" (rather than "*por*", as in modern Castilian).

NAÇENÇIA (1:19)
ES: nacimiento; EN: birth.
B (Bossong 1978, pp. 259–260): *nacencia*.

NAÇIMIENTO (48:1) / NASÇIMIENTO (48:2)
ES: nacimiento; EN: nativity.
Comment: These terms are used in chapter 48 in an astrological context, meaning nativity, but *nasçimiento* is used once meaning "sunrise" (29:8), as opposed to *ponimiento*.

NATURA (11:10)
ES: naturaleza, tipo; EN: nature, kind.
B (Bossong 1978, p. 260); L (Rodríguez Montalvo 1981, p. 293).

NATURAL (LUGAR –) (20:2)
See also *propio*.
ES: longitud trópica; EN: tropical longitude.
Comment: This refers to celestial longitude reckoned from the vernal equinox, a fixed point on the ninth sphere.

NOCHE (2:2)
See also *prima noche.*
ES: noche; EN: night.
B (Bossong 1978, p. 261); L (Rodríguez Montalvo 1981, p. 294); Q (Millás 1960, p. 239).

NONES (3:3)
ES: (números) impares, nones; EN: odd (numbers).
Comment: As opposed to *parejos.*

NOVIEMBRE (2:16)
ES: noviembre; EN: November.
A (f. 94v).

OCÇIDENTAL (14:4) / OÇIDENTAL (1:20)
ES: occidental; EN: occidental, western.
B (Bossong 1978, p. 262). L (Rodríguez Montalvo 1981, p. 295) and Q (Millás 1960, p. 253): *occidental.*
Comment: As opposed to *oriental.*

OCÇIDENTE (1:22) / OCCIDENTE (24:4) / OÇIDENTE (14:3)
ES: occidente; EN: occident, west.
B (Bossong 1978, pp. 264–265) and Q (Millás 1960, p. 255): *occidente.*
Comment: As opposed to *oriente.*

OCTAUO (2:14)
ES: octavo; EN: eighth.
L (Rodríguez Montalvo 1981, p. 252): *ochauo.*

OFFIÇIO (9:3)
See also *alçar.*
ES: columna, orden de magnitud; EN: rank, order of magnitude.
B (Bossong 1978, p. 265) and Q (Millás 1960, p. 252): *officio.*

ONSENO (2:16)
ES: undécimo; EN: eleventh.

OPOSIÇION (30:1) / OPPOSIÇION (30:2) / OPPOSICION (30:2)
ES: oposición; EN: opposition.
B (Bossong 1978, p. 266).
Comment: As opposed to *conjunçion.* This term refers to the moment when the Sun and the Moon are 180° apart in longitude.

OPPONIMIENTO (24:1)
See *ponimiento.*

OPPOSITO (17:6) / OPOSITO (30:29)
ES: opuesto; EN: opposite.
B (Bossong 1978, pp. 266–267) and Q (Millás 1960, p. 250): *opposito.*

ORIENTAL (2:2)
ES: oriental; EN: oriental, eastern.
B (Bossong 1978, p. 268); Q (Millás 1960, p. 253).
Comment: As opposed to *occidental*.

ORIENTE (33:3)
ES: oriente; EN: orient, east.
B (Bossong 1978, pp. 268–269); Q (Millás 1960, p. 253). L (Rodríguez Montalvo 1981, p. 252): *oryente*.
Comment: As opposed to *occidente*.

ORIZONTE (1:21) / ORISON (40:2) / ORISONTE (1:20)
ES: horizonte; EN: horizon.
B (Bossong 1978, p. 269) and L (Rodríguez Montalvo 1981, p. 297): *orizon*. Q (Millás 1960, pp. 240, 259): *orizon* and *horizon*.
Comment: This term refers to the great circle resulting from the intersection of the observer's horizontal plane and the celestial sphere.

OTUBRE (2:5)
ES: octubre; EN: October.
A (f. 94v): *ochubre*.

PAREJOS (3:3)
ES: (números) pares; EN: even (numbers).
Comment: As opposed to *nones*.

PAREÇIENTE (PLANETA –) (25:5) / PARESÇIENTE (25:7)
ES: (planeta) visible; EN: visible (planet).
B (Bossong 1978, p. 271): *paresciente*.
Comment: As opposed to *abscondido*.

PARESÇIMIENTO (24:17)
ES: aparición; EN: appearance.
A (f. 130r; Millás 1943–1950, p. 226); B (Bossong 1978, pp. 271–272).
Comment: As opposed to *ascondimiento*.

PARESÇER (36:13) / PAREZÇER (1:20)
ES: aparecer; EN: to appear.
B (Bossong 1978, p. 271): *paresçer* and *parecer*. Q (Millás 1960, p. 239) and L (Rodríguez Montalvo 1981, p. 298): *parescer*.

PARTE (4:4)
ES: parte; EN: part.

—— (22:33)
ES: dirección (N o S); EN: direction (N or S).

PARTIÇION (14:5) / PARTYÇION (39:23)
ES: división; EN: division.
B (Bossong 1978, p. 273); Q (Millás 1960, p. 243).

PARTIR (– POR) (23:11)
ES: dividir; EN: to divide.

B (Bossong 1978, pp. 273–274); Q (Millás 1960, p. 243).
Comment: Note that it is followed by the prepositions "*por*" (23:11) and "*sobre*" (27:36), whereas in modern Castilian "*por*" is required.

PASQUA (54:1)
ES: pascua; EN: Easter.

PESADO (PLANETA –) (27:27) / PESSADO (27:26)
ES: (planeta) pesado, lento; EN: heavy, slow (planet).
Comment: As opposed to *liviano*. This term is used in discussions of the relative velocity of the planets.

PLANETA (1:10) (MASCULINE) (15:22) (FEMININE)
ES: planeta; EN: planet.
Comment: Note that the word *planeta* is masculine (1:10), but in this text it is sometimes feminine (15:22). Both genders are found in A (f. 130r; Millás 1943–1950, p. 226), B (Bossong 1978, pp. 279–280), L (Rodríguez Montalvo 1981, p. 301), and Q (Millás 1960, p. 239).

POLO (1:20)
ES: polo; EN: pole.
B (Bossong 1978, p. 281); Q (Millás 1960, p. 246).

PONIMIENTO (28:2) / PONYMIENTO (36:20) / OPPONIMIENTO (24:1)
ES: ocaso; EN: setting.
B (Bossong 1978, pp. 283–284). A (f. 132r; Millás 1943–1950, p. 228): *ponimiento*. Q (Millás 1960, p. 251): *ponient*.
Comment: As opposed to *subimiento*.

PONIMIENTO DE LAS MAÑANAS (24:8)
ES: ocaso matutino; EN: morning setting.
A (f. 126v; Millás 1943–1950, p. 219): *ponimiento en las mannanas*.

PONIMIENTO DE LOS RAYOS DE PRIMA NOCHE (53:1)
ES: anochecer; EN: nightfall, end of evening twilight.

PONIMIENTO DE PRIMA NOCHE (24:3)
ES: ocaso vespertino; EN: evening setting.
A (f. 126v; Millás 1943–1950, p. 219): *ponimiento en los començamientos de las noches*.

PRESURA (24:13) / PRESSURA (27:52)
ES: rapidez; EN: swiftness.
B (Bossong 1978, p. 285): *pressura*.
Comment: As opposed to *tardamiento*.

PRESURADO (27:18) / PRESSURADO (27:18)
See *apresurado*.
ES: rápido; EN: swift, fast.
Comment: As opposed to *tardio*.

PRIETO (COLOR –) (35:3)
ES: color oscuro, negro; EN: black, dark color.
L (Rodríguez Montalvo 1981, p. 303).
Comment: This term appears in discussions of colors of eclipses.

PRIMA NOCHE (DE –) (24:3)
ES: vespertino; EN: evening.

PRIMER (5:4) / PRIMERO (2:7)
ES: primer, primero; EN: first.
A (f. 97r; Millás 1943–1950, p. 158) and L (Rodríguez Montalvo 1981, p. 303): *primero*.

PRIMO (4:7)
See *primer*.

PRINÇIPIO (1:1)
ES: principio; EN: beginning.

PROPIO (LUGAR –) (49:7)
See also *natural*.
ES: longitud sidérea; EN: sidereal longitude.
Comment: This refers to celestial longitude reckoned from the Head of Aries (Ari 0°), a fixed point on the 8th sphere, in contrast to celestial longitude reckoned from the vernal point, a fixed point on the 9th sphere.

PROPORÇION (17:16)
ES: proporción; EN: proportion.
B (Bossong 1978, pp. 286–287) and Q (Millás 1960, p. 242): *proporcion*.

PROPORÇIONALES (MINUTOS –) (17:10)
ES: minutos proporcionales; EN: minutes of proportion.
Comment: In A this term appears together with *grados* (f. 125v; Millás 1943–1950, p. 217).

PUNTOS DE LA EGUALDAD (38:6)
See also *egualdad*.
ES: equinoccios; EN: equinoxes.
Q (Millás 1960, p. 243): *punto de la egualdat*.

QUADRADO (32:35)
ES: cuadrado; EN: square.
Comment: This term refers to the result of multiplying a number by itself rather than to the geometrical figure. The latter meaning is found in B (Bossong 1978, p. 289) and L (Rodríguez Montalvo 1981, p. 304).

QUADRADURA (27:57)
ES: cuadratura; EN: quadrature.
B (Bossong 1978, pp. 289–290); L (Rodríguez Montalvo 1981, p. 304).

Comment: This term is only used in two chapters: in chapter 27 it refers to the relative positions of the Sun and the Moon when they are 90° apart in longitude, and in chapter 52 it is used in the context of the astrological theory of aspects.

QUANTIA (8:7)

ES: cuantía; EN: amount.

B (Bossong 1978, pp. 291–292); Q (Millás 1960, p. 246).

QUANTIDAD (33:18)

ES: cantidad; EN: quantity.

A (f. 133r; Millás 1943–1950, p. 233): *quantidades*. B (Bossong 1978, pp. 291–292) and Q (Millás 1960, p. 240): *quantidat*. L (Rodríguez Montalvo 1981, p. 252): *quantidat*.

QUARTO (0:5, 2:3)

ES: cuarto; EN: fourth.

A (f. 100r; Millás 1943–1950, p. 164); B (Bossong 1978, pp. 292–293); L (Rodríguez Montalvo 1981, p. 305); Q (Millás 1960, p. 246).

Comment: This term is used both as an ordinal number (0:3) and a fraction (2:3).

QUENTA (2:2)

ES: cuenta, cálculo; EN: calculation.

QUERDA (37:1)

See also *corda*.

ES: cuerda; EN: chord.

B (Bossong 1978, pp. 178–181). Q (Millás 1960, p. 239): *cuerda*.

QUINTO (2:11)

ES: quinto; EN: fifth.

RABE PRIMO (4:7)

A (f. 94r): *Rabeg I*.

Comment: This is the month *Rabīʿ I* in the Arabic calendar.

RABE SECUNDO (4:8)

A (f. 94r): *Rabeg II*.

Comment: This is the month *Rabīʿ II* in the Arabic calendar.

RAIAB (4:11)

A (f. 94r).

Comment: This is the month *Rajab* in the Arabic calendar.

RAIS (1:13) / RAIZ (1:14) / RAYS (1:13) / RAYZ (0:16)

ES: raíz; EN: radix.

B (Bossong 1978, pp. 295–296).

Comment: In the text, this term refers to the epoch at which a table begins, rather than to the value of a given quantity or mean motion at that time.

RAIZ (32:37)
ES: raíz cuadrada; EN: square root.
RAMADAN (4:13)
A (f. 94r).
Comment: This is the month *Ramaḍān* in the Arabic calendar.
RATIFICAR (0:6)
See *rectificar*.
RAYO (4:2)
ES: rayo; EN: ray.
B (Bossong 1978, p. 295); L (Rodríguez Montalvo 1981, p. 306).
RECTIFICAMIENTO (0:2) / RETIFICAMIENTO (0:15)
See *rectificar*.
B (Bossong 1978, pp. 298–299).
RECTIFICAR (0:11) / RATIFICAR (0:6) / RETIFICAR (0:9)
ES: observar, reconsiderar, recalcular; EN: to observe, to reconsider, to recompute.
B (Bossong 1978, p. 300); Q (Millás 1960, p. 239).
REGLA (3:2)
ES: regla; EN: rule.
A (f. 104v; Millás 1943–1950, p. 173); B (Bossong 1978, pp. 301–302); L (Rodríguez Montalvo 1981, p. 307).
REPARTIMIENTO (2:2)
ES: fracción; EN: fraction.
Comment: This term is only used once and it is given as a synonym of *fraçion*.
REPARTIR (5:1)
ES: distribuir; EN: to distribute.
RETROGRADAMIENTO (27:39)
ES: retrogradación; EN: retrogradation.
B (Bossong 1978, p. 304): *retrogramiento*.
RETROGRADAR (23:7) / RETROGADAR (23:10)
ES: retrogradar ; EN: to retrograde.
RETROGRADAÇION (23:1)
ES: retrogradación ; EN: retrogradation.
RETROGRADO (PLANETA –) (23:7)
ES: (movimiento) retrógrado (de un planeta); EN: retrograde (motion of a planet).
Comment: As opposed to *derecho*.
REVOLUÇION (4:4) / REUOLUÇION (13:25)
ES: revolución ; EN: revolution.
B (Bossong 1978, pp. 304–305). A (f. 101r; Millás 1943–1950, p. 166): *reuolucion*.
Comment: This term here means "cycle".

SAETA (37:1)

ES: sagita, seno verso; EN: sagitta, versine.

Q (Millás 1960, p. 240).

Comment: This term refers to a trigonometrical function (defined in 37:1 in chapter 4). In A the corresponding term is *sinu retornado* (f. 127r; Millás 1943–1950, p. 229), derived from the Arabic *al-jaib al-mankūs* (the versed sine).

SAFAR (4:6)

A (f. 94r): *Çaffar*.

Comment: This is the month *Ṣafar* in the Arabic calendar.

SAGITARIO (36:7)

ES: Sagitario; EN: Sagittarius.

A (f. 97r; Millás 1943–1950, p. 158); B (Bossong 1978, p. 29); L (Rodríguez Montalvo 1981, p. 309). A (f. 98r): *Sagitarius*. Q (Millás 1960, p. 256): *Sagitorio*.

SALIDA (24:14)

ES: orto; EN: rising.

Comment: This term only appears twice, in the same chapter, in the context of the visibility of the planets.

SASON (0:4) / SAZON (0:4)

ES: tiempo, época; EN: time, epoch.

B (Bossong 1978, p. 307), L (Rodríguez Montalvo 1981, p. 310), and Q (Millás 1960, p. 239): *sazon*.

SATURNO (22:2)

ES: Saturno; EN: Saturn.

SCURPION (29:13)

See *escurpion*.

SEGUNDO (2:8, 32:53) / SECUNDO (4:8) / SIGUNDO (0:5, 15:5)

ES: segundo; EN: second.

A (f. 98r; Millás 1943–1950, p. 160); B (Bossong 1978, p. 308); Q (Millás 1960, p. 252).

Comment: This term is used as an ordinal number, a unit of arc, and a unit of time.

SEÑAL (6:6)

A (f. 94r) and L (Rodríguez Montalvo 1981, p. 311): *señal*.

Comment: In context, this term refers to the weekday number associated with the first day of the year, the month, etc. In B (Bossong 1978, pp. 309–310) it means "point".

SEÑERA (LUNA –) (29:10)

ES: única, destacada; EN: unique, outstanding.

B (Bossong 1978, p. 310), L (Rodríguez Montalvo 1981, p. 311), and Q (Millás 1960, p. 239): *sennero*.

SEPTENTRION (1:23)

ES: norte; EN: north.

B (Bossong 1978, p. 311); L (Rodríguez Montalvo 1981, p. 312); Q (Millás 1960, p. 240).

SEPTENTRIONAL (20:1)
ES: septentrional; EN: northern.
A (f. 125v; Millás 1943–1950, p. 217); B (Bossong 1978, p. 311); Q (Millás 1960, p. 241).
Comment: As opposed to *meredional*.

SEPTIMO (2:13)
ES: séptimo; EN: seventh.

SEXTIL (RAYO –) (52:2)
ES: (rayo) sextil; EN: sextile (ray).
B (Bossong 1978, pp. 311–312).
Comment: This term is used in the context of the astrological theory of aspects.

SEXTO (2:12)
ES: sexto; EN: sixth.

SIGNO (15:5)
ES: signo; EN: sign.
A (f. 97r; Millás 1943–1950, p. 158); B (Bossong 1978, pp. 313–314); L (Rodríguez Montalvo 1981, p. 312); Q (Millás 1960, p. 241).
Comment: Throughout the text, one sign is defined as 30°.

—— (32:13)
ES: seno; EN: sine.
A (f. 127r; Millás 1943–1950, p. 229): *sinu egual*. B (Bossong 1978, pp. 313–314) and Q (Millás 1960, p. 241): *sino*.
Comment: This term refers to a trigonometrical function (defined in 37:1 in chapter 4).

SIGUNDO (15:5)
See *segundo*.

SIMPLE (AÑO –) (2:8)
See *comun*.
ES: (año) común; EN: common (year).
Comment: As opposed to *bisiesto*.

SINIESTRO (52:3)
ES: izquierdo; EN: left.
Comment: As opposed to *diestro*.

SOBEJANÇA (32:56)
ES: exceso; EN: excess.

SOBEJANOS (MINUTOS –) (33:18)
See *sobejança*.
ES: (minutos) de exceso; EN: (minutes) of excess.
B (Bossong 1978, p. 314).

SOBERANO (MES –) (30:11)
ES: soberano; EN: sovereign.
Comment: In context, this term, only appearing once, means the thirteenth month.

SOBIMIENTO (22:34)
See *subimiento*.

SOBREHAZ (34:16) / SOBREHAS (33:20)
ES: superficie; EN: surface.
B (Bossong 1978, p. 318): *sobrefaz*. Q (Millás 1960, p. 250): *sobrefaz*, *sobrefas*, and *sobrefraz*.

SODIACO (21:1)
See *zodiaco*.

SOL (0:11)
ES: Sol; EN: Sun.

SOMBRA (51:1)
ES: sombra; EN: shadow.
A (f. 130v; Millás 1943–1950, p. 230); B (Bossong 1978, pp. 319–320); L (Rodríguez Montalvo 1981, p. 314); Q (Millás 1960, p. 255).
Comment: See also *espandida* and *menguada*.

SUBIENTE (20:4)
ES: ascendente; EN: ascending.
Comment: As opposed to *deçendiente*.

SUBIMIENTO (24:1) / SOBIMIENTO (22:34)
ES: orto; EN: rising.
B (Bossong 1978, pp. 314–317).
Q (Millás 1960, p. 240): *sobimiento*; (p. 241): *subimiento*; (p. 251): *sobient* and *subient*.
Comment: As opposed to *ponimiento*.

—— (22:34)
ES: ascenso; EN: ascent.
Comment: As opposed to *abaxamiento*.

SUBIMIENTO DEL ALVOR (53:1)
See also *alvor*.
ES: amanecer; EN: daybreak.

SUBIMIENTO DE LAS MAÑANAS (24:3)
ES: orto matutino; EN: morning rising.
A (f. 126v; Millás 1943–1950, p. 219): *sobimiento* (or *subimiento*) *en las mannanas*.
Comment: As opposed to *ponimiento de las mañanas*.

SUBIMIENTO DE PRIMA NOCHE (24:4)
ES: orto vespertino; EN: evening rising.
A (f. 126v; Millás 1943–1950, p. 219): *subimiento en los començamientos de las noches*.

Comment: As opposed to *ponimiento de prima noche*.

SUBIMIENTO DEL CERCO YGUAL (32:23)

ES: ascensión recta; EN: right ascension.

SUJAMIENTO (22:18)

See *desviamiento*.

ES: desviación; EN: slant.

Comment: This term only appears once.

TABLA (0:16)

ES: tabla; EN: table.

A (f. 94r); B (Bossong 1978, p. 322); Q (Millás 1960, p. 243).

Comment: Despite Rico, the form *taula* does not occur in these canons. Moreover, it does not appear in B, L, or Q.

TAHMARMHE (5:7)

A (f. 96v): *Tirme*.

Comment: This is the month *Tīr Māh* in the Persian calendar.

TARDAMIENTO (24:13)

ES: lentitud; EN: slowness.

Comment: As opposed to *presura*.

TARDIO (MOVIMIENTO –) (0:3)

ES: (movimiento) lento; EN: slow (motion).

B (Bossong 1978, p. 324).

Comment: As opposed to *apresurado*.

TAURO (0:12)

ES: Tauro; EN: Taurus.

B (Bossong 1978, p. 29). A (f. 97v; Millás 1943–1950, p. 159): *Taurus*.

TEMPORALES (HORAS –) (28:4)

ES: (horas) desiguales; EN: seasonal (hours).

B (Bossong 1978, pp. 324–325); Q (Millás 1960, p. 240).

Comment: As opposed to *horas yguales*. On a given day of the year, a diurnal seasonal hour is one twelfth of the time from sunrise to sunset and a nocturnal seasonal hour is one twelfth of the time from sunset to sunrise.

TERCER (2:9)

ES: tercer, tercero; EN: third.

A (f. 99r; Millás 1943–1950, p. 162), L (Rodríguez Montalvo 1981, p. 316), and Q (Millás 1960, p. 252): *tercero*.

TINIEBLA (34:12)

ES: tinieblas; EN: darkness.

L (Rodríguez Montalvo 1981, p. 252): *tiniebra*.

Comment: The term appears in discussions of eclipses.

TORNADO (49:1)

See *alongado*.

ES: receso; EN: recess.

Comment: Together with *alongado*, these terms refer to the motion of "access and recess", that is, trepidation.

TORNAMIENTO (49:4) /TORNEAMIENTO (49:5)

See also *movimiento del llegamiento y del tornamiento*.

ES: receso; EN: recess.

B (Bossong 1978, pp. 123–124).

Comment: Together with *allegamiento* (or *llegamiento*), these terms refer to the motion of "access and recess", that is, trepidation.

TORNAR (11:2)

ES: volver, convertir; EN: to return, to convert.

B (Bossong 1978, p. 329).

TRESENO (4:4) / TREZENO (30:11)

ES: décimotercero; EN: thirteenth.

TRINO (52:4)

ES: trino; EN: trine.

Comment: This term is used in the context of the astrological theory of aspects: see Bossong 1978, p. 331. In L this astrological aspect is called *catamiento trino* (Rodríguez Montalvo 1981, p. 319).

TROPICO (0:12)

ES: solsticio; EN: solstice.

Comment: Here this term means solstice, as is clear from the following expressions: *tropico estiual* (38:6) and *tropico yvernal* (38:7). It is not found in B or in Q; rather, the term *mudamiento* is used for solstice.

— (53:2)

ES: vernal; EN: vernal.

Comment: This term is found in the expression *ygualdad tropica*, meaning "vernal equinox".

VARIANÇA (27:23)

ES: variación; EN: variation.

Comment: This term appears only once, and is equivalent to *desvariança* that appears five times.

VENUS (18:22)

ES: Venus; EN: Venus.

VERANO (32:22)

ES: primavera; EN: spring.

VERDURA (33:4)

ES: (coloración) verde; EN: green.

L (Rodríguez Montalvo 1981, p. 321).

Comment: This term appears in discussions of colors of eclipses.

VISIBLE (LUGAR –) (31:16)
ES: visible; EN: visible.
B (Bossong 1978, p. 333): *uisible*.

VISIESTO (2:3)
See *bisiesto*.

VISTA (2:2)
ES: visión; EN: vision.

XABEM (4:12)
A (f. 94r): *Xaben*.
Comment: This is the month *Sha*ʿ*bān* in the Arabic calendar.

XAHUEL (1:17) / XAVEL (4:14)
A (f. 94r): *Xauel*.
Comment: This is the month *Shawwāl* in the Arabic calendar.

YGUALAR (14:2)
See *egualar*.
ES: igualar; EN: to equate.

YGUALDAD (0:12)
See *egualdad*.
ES: equinoccio; EN: equinox.

YGUALDAD TROPICA (53:2)
See also *tropico*.
Comment: This term appears in the expression *punto de la ygualdad tropica*, meaning vernal equinox.

YGUALDAD DE VERANO (49:2)
ES: equinoccio de primavera; EN: vernal equinox.

YGUALES (HORAS –, DIAS –) (28:3, 29:9)
See *egual*.
ES: (horas) iguales; EN: equal (hours).
Comment: As opposed to *horas temporales*.

YNSTRUMENTO (0:8)
ES: instrumento; EN: instrument.
B (Bossong 1978, pp. 221–222): *estrumiente, estrumente, estrumento*, and *estrument*. Q (Millás 1960, p. 239): *estrumente*.

YNVOCAÇION (27:7) / YNCAÇION (30:18) / YNOVOÇAÇION (27:6) / YUCAÇION (33:5) / YUNCAÇION (27:3)
See *equaçion*.
ES: ecuación; EN: equation.
Comment: The corresponding term in A (f. 125v; Millás 1943–1950, p. 217) and B (Bossong 1978, pp. 202–203) is *eguacion*.

YVERNAL (TROPICO –) (38:7)
See *tropico*.
ES: solsticio de invierno; EN: winter solstice.

B (Bossong 1978, p. 336): *yuernal*. Q (Millás 1960, p. 241): *ivernal* ; and (p. 243): *yvernal*.
Comment: In Q these terms are both associated with *mudamiento*.

ZABA (32:59)
Comment: In this context where it refers to the Moon, *zaba* probably means hourly velocity in elongation. The term *zaba* may derive from the Arabic *sabq* used by al-Battānī (Nallino 1903–1907, 1:94, 2:336).

ZODIACO (21:5) / SODIACO (21:1)
ES: zodiaco; EN: zodiac.
B (Bossong 1978, p. 336): *zodiaco* and *zodiacho*. Q (Millás 1960, p. 240): *zodiaco*.

CHAPTER 4

ASTRONOMICAL COMMENTARY

Zijes are composed of a set of canons and a set of tables. These tables can be grouped as follows: calendar conversions; mean motions; solar, lunar, and planetary equations; trigonometry and spherical astronomy; planetary latitudes, visibilities, and retrogradations; lunar visibility; precession or trepidation; solar and lunar eclipses; astrological tables; geographical coordinates; and star lists. Some zijes may not have all these items and others may have additional items. In particular, planetary velocity tables appear in only a few, all of which derive from the Iberian Peninsula and the Maghrib.

The canons to the Castilian Alfonsine Tables refer to many tables which together form a zij, but the tables themselves are not extant. The canons do not refer either to a list of cities with their geographical coordinates or to a list of the fixed stars with their celestial coordinates. The tables described in the canons are similar to those in other zijes and, for this reason, we present some of them in this chapter to serve as examples of what the missing tables may have looked like.

One of the difficulties in deciding the character of the tables is that few parameters or numerical values appear in the canons. Such parameters are also clues to establish the relationships between zijes. Moreover, the persistence of some parameters furnishes evidence of the lines of transmission of scientific activity from one milieu to another. In the comments that follow comparative material is drawn from both earlier and later periods, because the astronomers under the patronage of Alfonso X worked in the mainstream of the Andalusian tradition that began in Muslim Spain and continued in North Africa, Christian Spain, and spread throughout Europe.

As a rule, these canons describe the use of the tables rather than their construction, but give no worked examples as are often found elsewhere. Throughout the canons the technical competence of the authors is evident, and they produced a text that accurately represented the scientific tradition in which they were embedded. It is therefore not surprising that the results of their efforts were widely diffused.

In the canons to the Castilian Alfonsine Tables the only earlier astronomers mentioned are Ptolemy and Azarquiel, and no cross references are given to other works produced at the court of Alfonso X.

Chapter 0

With the exception of this chapter and chapter 27, these canons have never been translated into English. Here we only offer a translation of chapter 0 (the prologue) because of its historical interest, generally following Rosen's translation without indicating where we have modified it (see Rosen 1976).

> [1] Judah, son of Moses, son of Mosca, and Rabbi Isaac ibn Sid say: [2] The science of astrology is a subject that cannot be investigated without observations. Yet, the observations made by the experts in this discipline cannot be completed by a single man; indeed, they cannot be completed in the lifetime of one man. On the contrary, when a result is attained, it is attained through the work of many men, laboring one after another for a long time. [3] The reason for this is that among the celestial motions are some that are so slow that they only complete a circuit after thousands of years. [4] It is therefore necessary to continue making observations because, by proceeding in this way, phenomena will become apparent at one time that were not apparent at another time. [5] We are now in our time in the first decade of the fourth century of the second millennium of the era of Caesar. [6] Two hundred years have passed since Azarquiel's observations, and discrepancies have appeared in some of the positions which he adopted that are so obvious and manifest to the senses that no excuse can be offered for [retaining] them. [7] At this time there appeared the happy reign, assisted by God, the rule of the very exalted and most noble lord, King Don

> Al[f]onso, may God preserve him! [8] Because he loved learning and appreciated it, he had instruments made, such as the armillary sphere and other devices, that are described by Ptolemy in the *Almagest*. [9] And he bade us observe in the city of Toledo, one of the principal cities in Spain—may God preserve it!— [10] in which Azarquiel made his observations. He [Don Alfonso] ordered [us] to rectify and correct the divergences and disagreements that had appeared in some positions of some of the planets and in other motions. [11] We obeyed his order as he demanded of us, reconstructing the instruments as best we could. We worked at making observations for a certain length of time and proceeded to observe the Sun throughout a complete year. [12] Before and after that time we observed it [the Sun] whenever it entered the equinoxes and the solstices and the other octants [*quartos*: usually 'quadrants'] of the heaven, namely, the mid-points of Taurus, Scorpio, Leo, and Aquarius. [13] Moreover, we observed some conjunctions of the planets with [*lit.* when they met] one another, as well as their conjunctions with some of the fixed stars. [14] We also observed many eclipses of the Sun and Moon. [15] We made [some] other observations in which we had doubts but repeated them many times to resolve the doubt; we did not abandon searching and investigating anything until the way to correct that which was in need of correction became evident to us. [16] Then, after completing our examination, we considered as verified what is certain or nearly so, and we based these tables on radices that were derived from those observations. [17] We added some chapters to them which seemed to us to be needed in this work. [18] We call this book, the book of the *Alfonsine Tables*, because it was made and compiled at the behest [of Don Alfonso], and we have divided it into 54 chapters, which follow.

0:1. The names of the authors of the Castilian Alfonsine Tables appear at the very beginning of this preliminary chapter: Judah ben Moses ha-Cohen and Isaac ben Sid. In general, the prologues and introductions to the almost 30 books concerning science (mostly astronomy), written by scholars in the service of Alfonso give the little information we have about his collaborators.

So, we are informed that Isaac ben Sid, also called Rabi Çag of Toledo, was the most productive scientific collaborator of Alfonso. He was also responsible for various treatises on the construction and use of instruments and clocks: *Libro de las armellas, Libro del astrolabio redondo, Libro del ataçir, Libro de la lamina universal, Libro del relogio de la piedra de la sombra, Libro del relogio del agua, Libro del relogio del argent vivo, Libro del relogio del palacio de las horas,* and *Libro del quadrante sennero.* Moreover, he was probably responsible for the first chapters of the *Libro de la alcora.* In 1277 he revised, or wrote, another treatise on instruments, *Libro del quadrante con que rectifican.* A brief note in Arabic in his own handwriting, written in Hebrew characters, has been preserved in Florence, Medicea Laurenziana, MS Or. 152 and, according to Villuendas, he was the copyist of this manuscript which contains, among other items, a treatise by Ibn Muʿādh of Jaén (d. 1093) (Villuendas 1978; Vernet 1978).

Judah ben Moses ha-Cohen was a physician who, during the reign of Ferdinand III, Alfonso's father, completed a translation into Latin of Azarquiel's *Treatise of the Azafea* in 1231. Under the patronage of Alfonso X, he took part in the translation of several treatises into Castilian: *Lapidario, Libro conplido, Libro de las estrellas de la ochaua espera, Libro de la alcora, Libro de las cruzes,* and possibly the *Liber Picatrix.* For this task he collaborated with some Christians: Garci Pérez, Guillén Arremón Daspa, and Johan Daspa. Of special interest is his participation in the translation of the *Lapidario* in 1243–1250 at the request of Alfonso before he became king. The prologue to the *Lapidario* provides us with information on Alfonso's early interest in learned matters (Rodríguez Montalvo 1981, pp. 17–19). For the passage below we generally follow Procter's translation without indicating where we have modified it (Procter 1945, p. 19):

> [In 1243 Alfonso] obtained [the Arabic manuscript] in Toledo from a Jew who held it hidden, who neither wished to make use of it himself nor that any other should profit therefrom. And when this book came into his possession, he had another Jew, who was his physician, read it. He was called Judah Mosca *el menor*: he was learned in the art of astrology (*astronomia*) and knew and understood well both Arabic and Latin. And when through the Jew his physician he understood the value and great profit which was in the book, he commanded him to translate it from Arabic into the Castilian language, so that everyone might better understand it and profit more

> from it. And one Garci Pérez his clerk aided in this translation (...) finished in [1250].

Based on Alfonso's royal titles in this text, Procter concluded that it was written after 1260. In this prologue we are also told that a Muslim scholar called Abolays preserved precious knowledge by translating a rare work from "Chaldean" into Arabic, and this is the work being translated into Castilian. This story was probably intended to promote an image of the king by drawing a parallel between his recovery of precious knowledge and Abolays's.

In 1276 Judah also participated in writing the final version of the *Libro de las estrellas de la ochaua espera,* together with two Italians, John of Messina and John of Cremona.

Although two authors are mentioned, there are some places in the text where the first person singular is used (see, e.g., 49:8). However, the first person plural is normally used. According to both Judah ben Asher II and Abraham Zacut, the only author of the "zij of Alfonso" was Isaac ben Sid (see chapter 5).

0:2. This sentence probably alludes to Hippocrates's first aphorism as reported in Seneca, *De brevitate vitae,* I.2 (see also Rosenthal 1966): "Life is short, the art is long".

0:3. Some celestial motions are so slow that to complete a revolution they need thousands of years. Presumably, the reference is to the slow motions of the fixed stars and of the planetary apogees.

0:5. "At the present time" (*agora*) here refers to the first decade (*la primera desena*) of the 14th century of the era of Caesar (whose epoch is Jan. 1, –37), that is, the period between 1262 (= 1300 – 38) and 1272 (= 1310 – 38). Therefore, the canons were completed no later than 1272. For the era of Caesar, also called the Spanish era, see Samsó 1992, pp. 38, 89.

0:6. Azarquiel (Ibn al-Zarqālluh), the first astronomer mentioned in the text, worked in Toledo and Córdoba about 200 years before the time mentioned in 0:5. In fact, his best known observations were made in 1075. The authors of these Castilian canons tell us that their motivation for constructing new tables was that they found "significant and noticeable divergences from observations" (*diversidades magnifiestas y paresçidas a los sentidos*) when they used Azarquiel's results (see 0:12, below).

0:8. Ptolemy is mentioned in a sentence stating that Alfonso ordered his astronomers to reproduce the instruments described in the *Almagest*.

0:9–10. The king ordered "us" (the authors of these canons) to replace and correct the positions and motions of some planets that Azarquiel had computed for Toledo. This seems to be a reference to a set of tables by Azarquiel, but it is not clear which set of tables is at issue. The candidates are the Almanac of Azarquiel (Millás 1943–1950) and what are now called the Toledan Tables whose canons, only preserved in Latin, are ascribed to Azarquiel (F. S. Pedersen 1987 and 2002).

0:11–12. Judah ben Moses ha-Cohen and Isaac ben Sid constructed instruments as commanded by the king (see 0:8), and observed the course of the Sun during one year. In particular, they say that they observed the positions of the Sun at the equinoxes and solstices, and at the octant points half-way between them, presumably to eliminate the divergences noted in 0:6, above.

This observational procedure is appropriate for determining the parameters of the solar model. For the use of observations at the octant points to determine the solar eccentricity and apogee, see Pseudo-Thābit's *On the Solar Year* (Neugebauer 1962b, and Morelon 1987). According to Bernard of Verdun (late 13th century), "Az-Zarqāl ... made four observations while the Sun was in the middle of the quadrants bounded by the solsticial and equinoctial points..." (*Tractatus super totam astrologiam,* quoted in Toomer 1987, p. 516). Azarquiel's value for the tropical longitude of the solar apogee was 85;49° for 1074–1075 (= 467 AH) according to Ibn al-Hā'im and Bernard of Verdun; and his sidereal longitude of the solar apogee was about 78;30° for 468 AH according to the Alfonsine translation of his treatise on the construction of the equatorium (Samsó 1994, essay X, pp. 8–9). Moreover, Samsó computes that, with the tables of Ibn al-Bannā' (d. 1321), the sidereal longitude of the solar apogee for 467 AH was 78;21,56° and the value for trepidation was 7;28°; adding them yields a tropical longitude for the solar apogee of 85;49,56°, and this result is very close to the value ascribed to Azarquiel for that date. So it is not clear if Azarquiel's observations played a role in determining the solar apogee. Isaac Israeli (*ca.* 1310) reports that Azarquiel observed the autumnal equinox of 1075 AD (*Yesod ʿOlam*, ed. Goldberg and Rosenkranz, vol. 2, f. 29b; cf. Mercier 1996, p. 313). As far as we can determine, there are no extant reports of observations of solstices, equinoxes, or of the Sun at the octant points by any of Alfonso's astronomers.

We have translated *rectificar* in these two sentences as "to observe", and the same translation applies to sentences 0:13–14 and 0:16. However, the verb *rectificar*, and its variant *retificar*, as well as the noun *rectificamiento* that occurs frequently in this chapter and occasionally elsewhere, may sometimes be better translated "to reconsider" or "to recompute". The same semantic range seems to apply to the Hebrew word *tiqqen* as used by Isaac Israeli of Toledo in his *Yesod ʿOlam* for, in reporting the eclipses of Isaac ben Sid, it seems to mean "to observe" (see below) but elsewhere "to compute" (e.g., treatise 5, chap. 15; ed. Goldberg and Rosenkranz, vol. 2, f. 41b).

0:13–14. They also observed planetary conjunctions and solar and lunar eclipses. It is indeed justified to translate *rectificar* by "to observe" in this case, for Isaac ben Sid, one of the authors of this text, made four observations of eclipses in Toledo, as reported by Isaac Israeli of Toledo in his *Yesod ʿOlam* (treatise 4, chap. 7; ed. Goldberg and Rosenkranz, vol. 2, ff. 11b and 12a). On the other hand, we are unaware of any reports concerning planetary conjunctions observed by Alfonso's astronomers. There follow the relevant passages on eclipses as reported by Isaac Israeli and comments on them.

> Eclipse 1: "Further, to verify this matter, even though there is no doubt in it for anyone who acknowledges the truth, I directed my attention to three lunar eclipses that R. Isaac ha-Ḥazzan ben Sid observed [*tiqqen*: usually 'established', or 'corrected'] and estimated [*amad*: can also mean 'to appraise' or 'to judge'] in the city of Toledo at the command of king Alfonso, according to what reached me in his own handwriting, how he observed and examined (*ʿiyyen*) them. The first eclipse: he wrote that he examined and saw (*ra'a*) that the Moon was eclipsed at opposition in the month of Tebet in the year 5026 AM [*Anno Mundi*] ... and he found that the time of eclipse-middle, i.e., true opposition, took place 3 and a third hours after the moment of midnight, of weekday 5 [i.e., Thursday] in Toledo, the 15th day of Tebet, ...".

The date given for this lunar eclipse, Tebet 15, 5026 AM, day 5 (= Thursday), corresponds to December 24, 1265 (JDN 2183457). The time given for eclipse-middle, $3^1/_3$h after midnight, corresponds to 15;20h after noon. "In his own handwriting"—it is not clear if Isaac ben Sid's notes on his observations were in Castilian or Hebrew.

> Eclipse 2: "It is what he wrote, and he said that he examined and saw (*ra'a*) that the Moon was eclipsed in the middle of the month of Tammuz in the year 26 [i.e., 5026] ... and he found that eclipse-middle, i.e., the moment of true opposition, took place after midnight of the night of the sabbath in Toledo, the 15th of Tammuz, by about 3 hours and 135 *ḥelaqim* [i.e., $^{135}/_{1080}$ of an hour]."

The given date for this lunar eclipse, Tammuz 15, 5026 AM, corresponds to June 19, 1266 (JDN 2183634). The time of eclipse-middle is given as $3^{135}/_{1080}$h after midnight, i.e., 15;7h after noon.

> Eclipse 3: "This is what he said. The Moon was eclipsed in the middle of the month of Tebet in year 27 [i.e., 5027], and he examined and found that eclipse-middle took place 6 hours and 666 *ḥelaqim* [i.e., $^{666}/_{1080}$ of an hour] after midday of weekday 2 [i.e., Monday] in Toledo, the 13th of Tebet."

The given date is equivalent to Dec. 13, 1266 (JDN 2183811). Tebet 13 in that year was a Monday. The time of eclipse-middle is given as $6^{666}/_{1080}$h, which corresponds to about 6;37h after noon or 18;37h after midnight. According to Oppolzer's canon of eclipses (No. 3829), there was a lunar eclipse on that date at 18;52 UT.

> Eclipse 4: "R. Isaac ben Sid said that the Sun was eclipsed at the end of the month of Ab of year 5023 AM ... and he examined and found that the moment of eclipse-middle took place at Toledo at about 2 hours after noon on weekday 1 [i.e., Sunday], the 28th of Ab."

For this solar eclipse, the date given, Ab 28, 5023 AM, corresponds to Sunday, Aug. 5, 1263 (JDN 2182585). There was a solar eclipse on that date which was not total anywhere in Spain.

Table 0A shows the times of the eclipses observed by Isaac ben Sid compared with modern values for the meridian of Greenwich, as given in Schroeter 1923, or Oppolzer 1887 in the case of the lunar eclipse for December 1266), and with values recomputed by means of a program for the Alfonsine Tables and the Toledan Tables (both for the meridian of Toledo): cf. Steele 2000, pp. 134–135.

Table 0A: Eclipse times

Eclipse No.	Date	Time of Eclipse-middle Text (h)	Modern (h)	Alf. (h)	Tol. (h)
4.	Aug. 5, 1263	14	13;53	13;44	14;19
1.	Dec. 24, 1265	3;20	3;46	3;14	3; 1
2.	Jun. 19, 1266	3; 7	3;36	3; 4	3; 1
3.	Dec. 13, 1266	18;37	18;52	18;45	18;45

0:16. The principles (*rayzes*), possibly including parameters, were carefully and repeatedly derived from these observations.

0:18. The title of this book in 54 chapters is explicitly given as *Libro de las tablas alfonsies.*

Chapter 1

1:1. This chapter is devoted to the different eras used in "these tables". Note that the astronomical day begins at noon.

1:3. The beginning of the world refers to the era of Creation, or the era of Adam. According to al-Bīrūnī (trans. Sachau 1879, p. 18), the Jews reckon the interval between Adam and Alexander as 3448 years, whereas the Christians take it to be 5180 years; al-Bīrūnī adds (p. 141) that the Jews prefer the era of Adam (or of the Creation) whereas the Christians prefer the era of the Flood. For the era of Adam, see Suter 1914, p. 109, and Neugebauer 1962a, p. 82. According to al-Khwārizmī, the interval from the era of Adam to the era of the Flood was 2280 years; see also Neugebauer 1962a, p. 84, for the contrasting views concerning the era of Adam according to Christian and Muslim traditions. In another book produced under the patronage of Alfonso and called *Siete Partidas*, the general prologue mentions several eras including the era of Adam (Craddock 1974, pp. 367–368).

1:4. The era of Alexander refers to the Seleucid era beginning on Oct. 1, –311 (JDN 1607739).

1:5. The era of Caesar is also called the Spanish era and began on Jan. 1, –37 (JDN 1707544).

1:6. The era of the Arabs, or the Arabic era, began on the Hijra which, for astronomical purposes, is taken to be noon, July 15, 622 (JDN 1948439), a Thursday, as is implied in 6:5 as well as in the Toledan Tables (cf. Toomer 1968, p. 18) and in al-Battānī's zij (Nallino 1903-1907, 2:199). Moreover, according to Ibn al-Muthannā (10th century), al-Khwārizmī also said that the Arabic era began on a Thursday (Goldstein 1967, p. 17). But, in the extant version of al-Khwārizmī's zij, the Hijra is taken to be noon, July 14, 622 (JDN 1948438), a Wednesday (cf. Neugebauer 1962a, pp. 10-11).

1:7–8. The era of Yazdegird, or the Persian era, began on June 16, 632 (JDN 1952063).

1:11-16. The era of Alfonso began at noon, Sunday, Dec. 31, 1251. Thus, the first day of this era, Jan. 1, 1252 (= 1290 – 38), is considered to be the epoch of these tables (JDN 2178351). This is indeed the year in which the reign of Alfonso X began, as mentioned in the text.

Note, however, that the *editio princeps* of the Alfonsine Tables (Ratdolt 1483) uses a different epoch, June 1, 1252 (JDN 2178503), 152 days later than that in the Castilian Alfonsine Tables. This is seen by inspection of the table on f. c1r–v, where the difference between the Incarnation (JDN 1721424) and the Alfonsine era used in this case is given as 2,6,57,59 days (= 457079d, expressed in sexagesimal form). The date used as epoch in the *editio princeps* is that of the beginning of the reign of Alfonso. This date is attested as such in a chronological note in the general prologue to the Alfonsine book called the *Siete Partidas*, where we are also given the time difference (in years and days) between June 1, 1252 and the beginning of various eras (Adam, Flood, Nabuchodonosor, Philippus, Seleucid, Caesar, Incarnation, Diocletian, Arabs, and Yazdegird): see Craddock 1974. The same list of eras is given in al-Khwārizmī's zij: the era of Nabuchodonosor does not refer to Nebuchadnezzar, the king of Babylon mentioned in the Bible (e.g. 2 Kgs 24:1), but is to be understood as the era of Nabonassar, used in the *Almagest*, whose epoch is Feb. 26, –746 (Neugebauer 1962a, p. 82). Further remarks on the differences between eras may be found below in Section 6.2.1.

1:17. Indeed, Jan. 1, 1252 = Shawwāl 16, 649 AH as stated in the text. It is also claimed that this date is equivalent to Isfandar 18, 620 AY (Anno Yazdegird), but this is a mistake for Isfandar 24, 620 AY, presumably because the 5 epagomenal days that come after the 8th month have been neglected (see sentences 5:3 and 5:11, below, where the 8th month is said to have 35 days).

1:20–22. The longitude of Toledo is reckoned from the western meridian, taken as the prime meridian. In a quite confused manner, the authors explain that the distance from Toledo to the western meridian is 28°; the distance from Toledo to Arin (Ujjain) is 62° (= 90° – 28°); and the distance from Toledo to the eastern meridian is 152° (= 180° – 28°). The use of this prime meridian is attested in Andalusian and Maghribī sources going back to Maslama al-Majrīṭī (see Comes 1992–1994). The very same value (28°) for the longitude of Toledo appears in *al-Zīj al-Muqtabis* of Ibn al-Kammād (early 12th century) (Madrid, Biblioteca Nacional, MS 10023, f. 54v), as well as in other texts written before the Castilian canons of the Alfonsine Tables: see the zijes of Abū l-Ḥasan ʿAlī al-Marrākushī (d. 1262), Muḥyī al-Dīn al-Maghribī (d. *ca.* 1290) (Kennedy and Kennedy 1987, p. 357), and Ibn Isḥāq of Tunis (beginning of the 13th century) (Mestres 2000, p. 26).

1:23. The latitude of Toledo is given as 39;54°, a value already attested in the Toledan Tables (Millás 1943–1950, pp. 49, 64; Toomer 1968, p. 36; F. S. Pedersen 2002, p. 431).

Chapter 2

2:1. This chapter deals with the Julian calendar.

2:3. The Roman years refer to the Julian calendar, where the length of the year is given as $365^1/_4$ days. There are 3 years of 365 days and a leap year of 366 days, and the leap day is added at the end of February (2:4). Hence, in the Julian calendar a complete cycle consists of 4 years.

2:7. In this text, the year begins in January, whereas in other texts the year begins in different months, especially in March, as in the Tables of Toulouse (Poulle 1994, p. 59).

Chapter 3

3:1. This chapter deals with the leap years in the era of Alfonso.

3:4. The rule given for determining the leap year according to the era of Caesar derives from the fact that 38 ≡ 2 (mod. 4).

Chapter 4

4:1. This chapter is devoted to the Julian calendar.

4:4. 1 Arabic year = $354^1/_5\ ^1/_6$ days = $354^{11}/_{30}$ days. Hence, in the Arabic calendar a complete cycle consists of 30 years.

The leap day is added at the end of Dhu l-ḥijja, and the 11 leap years are: 2, 5, 7, 10, 13, 16, 18, 21, 24, 26, 29.

Chapter 5

5:1. This chapter is devoted to the Persian calendar.

5:3. 1 Persian year = 365 days. Hence, in the Persian calendar a complete cycle consists of 1 year. In the text it is said that each month has 30 days except for month 8 that has 35 days. In fact, there are 12 months of 30 days each, and 5 epagomenal days inserted between month 8 and month 9 (see al-Bīrūnī, trans. Sachau 1879, p. 53).

Chapter 6

6:1. This chapter is devoted to the determination of weekdays.

6:5. In order to know the weekday of the beginning of any year for any era, the text indicates that after converting into days the time elapsed between the date under consideration and the beginning of the corresponding era, one must add 2 days if the Alfonsine era is concerned, 3 days in the case of the Persian era, and 5 days in the case of the Arabic era. This is indeed correct, for the epoch of the Alfonsine era (Jan. 1, 1252) was a Monday (weekday 2), that of the Persian era (June 16, 632) was a Tuesday (weekday 3), and that of the Arabic era (July 15, 622) a Thursday (weekday 5).

6:9. In the Arabic calendar the cycle for weekdays is 210 years (= 30 · 7); in the Christian calendar it is 28 years (= 4 · 7); and in the Persian calendar this cycle is only 7 years.

Chapter 7

7:1. This chapter deals with the determination of weekdays in various calendars by means of a table.

7:1. This is the first time a specific table is mentioned. The text refers to a table to find the weekday of the beginning of each year and each month.

7:2. This table must contain sub-tables for the Arabic, Christian, and Persian calendars.

7:7–11. This table must contain sub-tables for the years in cycles of 28 years for the Christian calendar, and cycles of 7 years for the Persian calendar, as well as for the months of the year, both for common years and for leap years.

All the details given seem to describe a multi-calendaric table very much like those in al-Khwārizmī's zij (cf. Suter 1914, p. 110, table 2 for the Arabic calendar; and pp. 111–112, tables 2a and 2b for the Persian calendar); in al-Battānī's zij (Nallino 1903–1907, 2:7, for the Arabic calendar); and in the Toledan Tables (cf. Toomer 1968, pp. 24–25, tables 9 and 10 for the Arabic calendar; and p. 26, table 11a for the Christian calendar; F. S. Pedersen 2002, pp. 929–940).

Chapter 8

8:1. This chapter explains how to convert, by calculation, a date given in one era into another era.

8:2. The day is the common unit used for computations involving years of different lengths.

8:4. The proposed example consists in finding the number of Julian years in 300 Arabic years. The length of the Arabic year is given here as $354\,{}^{3}/_{10}\,{}^{2}/_{30}$ days, whereas in 4:4 it was given equivalently as $354\,{}^{1}/_{5}\,{}^{1}/_{6}$ days.

8:5. The result of the multiplication is given as 6,368,600 days, whereas it should be 106,310 days. Note that $106{,}310 \cdot 60' =$ 6,378,600, which in turn is a mistake for 6,368,600 (in Castilian it is easy to read, or copy, *setenta y ocho* as *sesenta y ocho*).

8:6. The mistaken order of magnitude in the intermediary computation did not prevent the authors from giving the correct answer: 291 Jul. years 23d.

Chapter 9

9:1. The text refers to another specific table to convert dates from one era to another. Although the eras are not specified, it is likely that the Arabic, Alfonsine, and Persian eras were intended, as was the case in chapter 7.

9:2. This table must contain sub-tables for "collected years" for the three eras.

9:3. The term *offiçio* here probably means "rank" indicating the order of magnitude.

9:4. This table must contain sub-tables for the months in a year for the three eras.

9:6–10. All the calculations involving the number of days are performed in base 60. Moreover, in a cross-reference to this chapter given in chapter 12, the text uses the term *dias compuestos*, meaning that the number of days is expressed in sexagesimal form (12:2).

9:11. This table must contain sub-tables for *anni expansi* for the three eras. All the details given seem to describe a multi-calendaric table very much like the one for the same purpose in the Toledan Tables (cf. Toomer, 1968, pp. 14–16, tables 1, 2 and 3 for the Arabic, Christian, and Persian calendars; F. S. Pedersen 2002, pp. 877–894).

Chapter 10

10:1. This chapter gives the differences in days between the epochs of different eras.

10:2. Alfonsine era – Persian era = 619 Per. years 353d
Indeed, $2178351 - 1952063 = 619 \cdot 365 + 353$.

10:3. Alfonsine era – Persian era = 619 Jul. years $198^1/_4$d
Indeed, $2178351 - 1952063 = 619 \cdot 365.25 + 198.25$.

10:4. Alfonsine era – Persian era = 638 "old" years 200d
If *años antiguos* here refer to Arabic years, then the number of extra days should be 202 rather than 200.

10:5. Alfonsine era – Persian era = 226288d
Indeed, 2178351 – 1952063 = 226288.

10:6. The number (1,2,51,48) is given in sexagesimal form, for $1 \cdot 60^3 + 2 \cdot 60^2 + 51 \cdot 60 + 48 = 226288$d. Therefore, the term *compostura de los dias* means "number of days expressed in sexagesimal form".

10:7. Alfonsine era – Arabic era = 648 Ar. years 282d
Indeed, 2178351 – 1948439 = 229812. In 648 Arabic years there are 229630 days, because they consist of 21 complete cycles (223251d) plus 11 ordinary years of 354 days each and 7 leap years of 355 days each.

10:8. Alfonsine era – Arabic era = 629 Per. years 327d
Indeed, $2178351 - 1948439 = 629 \cdot 365 + 327$.

10:9. Alfonsine era – Arabic era = 629 Jul. years $169^3/_4$d
Indeed, $2178351 - 1948439 = 629 \cdot 365.25 + 169.75$.

10:10. Alfonsine era – Arabic era = 229912d
Indeed, 2178351 – 1948439 = 229912.

10:11. The number (1,3,51,52) is given in sexagesimal form, for $1 \cdot 60^3 + 3 \cdot 60^2 + 51 \cdot 60 + 52 = 229912$d.

10:12. Persian era – Arabic era = 10 Ar. years 80d = 9 Per. years 339d
Indeed,
$$\begin{aligned} 1952063 - 1948439 &= 3624 \\ &= 6 \cdot 354 + 4 \cdot 355 + 80 \\ &= 2124 + 1420 + 80 \\ &= 9 \cdot 365 + 339. \end{aligned}$$

10:13. Persian era – Arabic era = 9 Jul. years $336^3/_4$d
Indeed, $1952063 - 1948439 = 9 \cdot 365.25 + 336.75$.

10:14. Persian era – Arabic era = 3624d
Indeed, 1952063 – 1948439 = 3624.

10:15. The number (0,1,0,24) is given in sexagesimal form, for $0 \cdot 60^3 + 1 \cdot 60^2 + 0 \cdot 60 + 24 = 3624$d. Note that *çifra* is the term used for zero.

According to Toomer (1968, p. 18), some manuscripts of the Toledan Tables give scattered information on the intervals between different eras in sexagesimal form (but see F. S. Pedersen 2002, pp. 896–901). Moreover, in al-Bīrūnī's *Chronology of Ancient Nations* (trans. Sachau 1879, pp. 133, 404), there is a table of the intervals in days, in sexagesimal form, between various eras. This indicates that the sexagesimalisation of days is not an innovation by Alfonso's astronomers. For Ibn Yūnus's list, see Delambre 1819, p. 96; for al-Khwārizmī's list, see Suter 1914, p. 109, and Neugebauer 1962a, pp. 82–83. See also Section 6.2.1, below.

Chapter 11

11:1. This chapter explains how to convert, by calculation, a date given in one era into another era.

Cross-references to chapter 8 and chapter 10 appear in 11:4 and 11:10, respectively.

Chapter 12

12:1. This chapter explains how to determine, by means of a calculation, the differences between the eras.

12:2. The text refers to the table already mentioned in chapter 9, and contains cross-references to chapters 9 and 10.

Chapter 13

13:2. This chapter addresses the same problem as chapter 12, and complements it.

13:11. The text indicates that in the sub-table corresponding to the Arabic era, the *fraçiones* of a day are such that 1 day = 30 *fraçiones*. This convention is coherent with the fact that 1 Arabic year is taken as $354^{11}/_{30}$ days.

13:34. We are given here the time difference between the epoch of the Alfonsine era and the religious definition of the epoch of the Arabic era as 648 Ar. years 281 days and 18 hours (erroneously called *fraçiones*). In 10:7 the time difference, 648 Ar. years and 282 days, was given with respect to the epoch of the astronomical Arabic era, 6 hours earlier. Note that the day for religious purposes in the Arabic calendar begins at sunset, whereas the astronomical day begins at noon.

Chapter 14

14:1. The title of this chapter is "How to obtain the mean motions of the planets and other motions reckoned from the epoch of the Alfonsine era" (in Toledo), but the chapter deals with the problem of determining the positions of the planets for places other than Toledo. On the other hand, chapter 15 bears the title: "How to obtain the positions of the planets at the cities to the east or to the west of the city of Toledo", whereas the content of chapter 15 corresponds to the title given to chapter 14. Thus, the titles seem to have been interchanged; as far as we know, no one has noticed this peculiarity, and Rico says nothing about it in his edition of this text.

14:2. The text specifies that the tables are computed for noon at Toledo.

14:3–4. Here *longura* means geographical longitude; no other term for geographical longitude occurs in the text.

14:5–10. The text correctly explains how to compute the time interval between Toledo and some other city (by dividing the difference of their longitudes by 15), how to compute the differences in mean motion, mean argument, and mean center during this time interval, and how to obtain the corresponding values for a city other than Toledo.

Chapter 15

15:1. See comments to 14:1.

15:3. In the tables for the mean motions there seem to be specific subtables for each planet and for each of the following quantities: mean motion in longitude, argument of anomaly, and center. The argument for all of them is the time difference between the date under consideration and the epoch, given in Julian years, months, days, and hours (15:2).

15:4. In each of these cases there are sub-tables for collected years (15:4), expanded (*expansi*) years (15:6), months in a leap year (15:9), months in a common year (15:10), days (15:12), hours (15:13), and minutes (15:13).

15:5. The entries in the table are given in signs, degrees, minutes, and seconds.

15:18. This table uses signs of 30°, in contrast to the Parisian version of the Alfonsine Tables which uses "physical" signs of 60° as described in John of Saxony's canons (Ratdolt 1483, f. a2r; cf. Poulle 1984, p. 30).

The tables for the mean motions described here follow the pattern traditionally found in Arabic zijes (as well as in the Toledan Tables).

The text indicates (15:4) that the collected years in the table are presented at intervals of 20 years, where years from 1 to 20 are the expanded years. In the mid-13th century there was not yet a settled tradition for presenting tables for mean motions according to the Latin calendar, and different cycles were used in tables adapted from Arabic zijes. For instance, Petrus Alphonsi (*ca.* 1116) arranged his tables of mean motion for cycles of 30 years in his adaptation of Maslama's version of al-Khwārizmī's zij (Neugebauer 1962a, p. 134); whereas in the adaptations of the Toledan Tables for Marseille and Novara cycles of 28 years were used and in the adaptation of the Toledan Tables for Toulouse cycles of 24 years were used (see Paris, Bibliothèque nationale de France, MSS lat. 14704 and 16658; cf. Poulle 1994, and F. S. Pedersen 2002, pp. 1197–1207, 1210). Although the Castilian Alfonsine Tables led to the prevalence of a pattern based on groups of 20 collected years, this period had been used previously. According to Mercier (1987, p. 101), a version of the zij of al-Khwārizmī preserved in a 12th century manuscript (Paris, Bibliothèque nationale de France, MS lat. 16208) has radices adapted to the Latin calendar in steps of 20 years; Abraham Ibn ʿEzra (d. 1167) also used a cycle of 20 years for arranging his tables of mean motions (Millás 1947, pp. 14, 37, and 67). Moreover, the use of steps of 20 years is already found in al-Battānī's zij (see some of his tables for Roman years: Nallino 1903–1907, 2:73, 77, 102, 103, 107, and 188); this zij was well known to Alfonso's astronomers, for they made a translation of it.

Chapter 16

16:1. This short chapter on the solar equation explains how to compute the true longitude of the Sun.

16:3. A table for the solar equation is mentioned in which the argument is the mean solar anomaly, here called *argumento* (16:2), which is defined as the difference between the mean solar longitude (*medio curso del sol*) and the longitude of the solar apogee (*aux*).

16:6. The true longitude is obtained from the mean longitude by adding, or subtracting (according to the standard rule given in 16:5), the solar equation found in the table. For the underlying solar model, see Fig. 4.1.

No numerical data are given but we are confident that the maximum solar equation was 2;10° in the Castilian Alfonsine Tables. The evidence is found in several sources.

1. A table with a maximum equation of 2;10° seems to underlie the 12 solar positions for 1277 in the *Libro del quadrante con que rectifican* (recomputed by Samsó 1984, pp. 82–86), as well as to underlie the table for the solar positions in *Libro del astrolabio llano*, datable to 1278 (Chabás 2003).

2. In his *Expositio* of 1321 (edited in Poulle 1980, pp. 250–268), John of Murs reports two observations of autumnal equinox, one by Ptolemy on September 25, 132 at 2 p.m. (*Almagest* III.7; Toomer p. 168) and the other by Alfonso on September 16, 1252, without specifying the hour, and he assumes that the Sun returned to the same mean position. He correctly cites the mean position of the Sun at the time of the first observation given in the *Almagest*: 182;10° (Poulle 1980, p. 255), which results from adding the mean center of the Sun (116;40°) and Ptolemy's longitude for the solar apogee (65;30°). It follows that at the time of the second observation the mean Sun was again at 182;10°. Now, at autumnal equinox, the true Sun has to be at 180° and, if one assumes as does John of Murs, that the solar apogee at the time of the second observation is about 90°, it follows that 2;10° is the maximum solar equation (cf. North 1996, p. 461). Therefore, we infer that the maximum solar equation of 2;10° was in John of Murs's source which he calls "Alfonso" (see Goldstein and Chabás 2001).

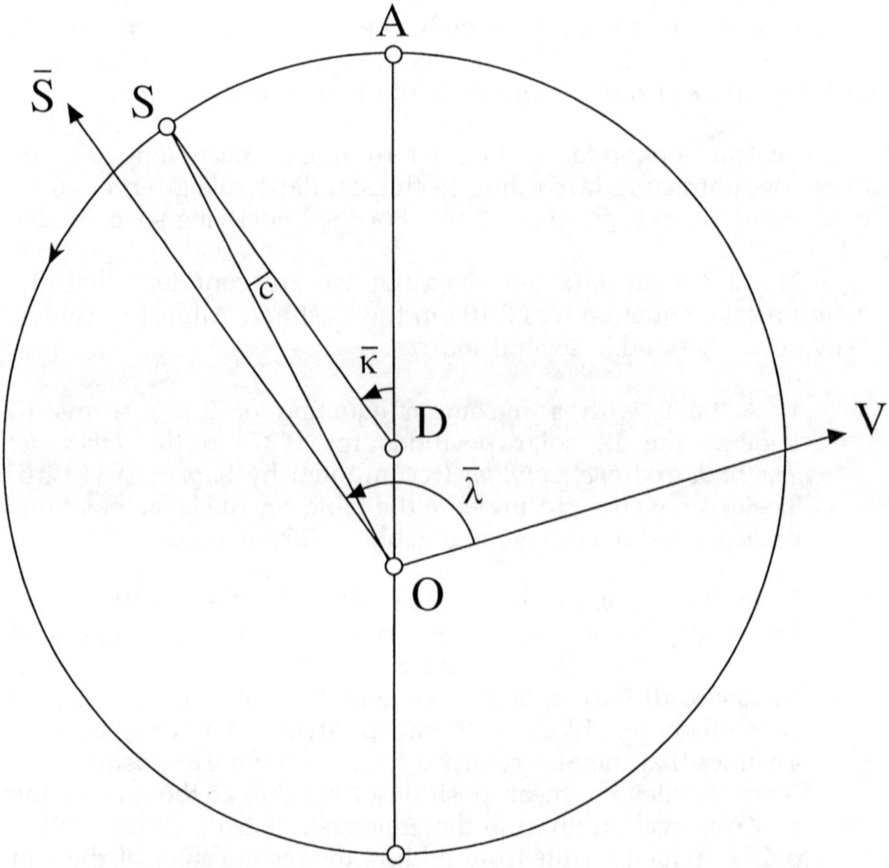

Fig. 4.1. The model for solar motion: O is the observer; D is the center of circle ASP (whose radius is 60) on which the Sun moves; OD is the eccentricity; A is the solar apogee; P is the solar perigee; OV is the direction to Aries 0°; O$\bar{S}$ is the direction to the mean Sun whose mean longitude is $\bar{\lambda}$; DS is parallel to O$\bar{S}$; and $\bar{\kappa}$, called the argument of center, is the difference between the solar mean longitude and the longitude of the solar apogee ($\bar{\kappa} = \angle VO\bar{S} - \angle VOA$). The equation, c, is $\angle \bar{S}OS$ which is equal to $\angle OSD$, and it is the difference between the mean solar longitude ($\angle VO\bar{S}$) and true solar longitude ($\angle VOS$); c is computed as a function of $\bar{\kappa}$.

Table 16A: Solar equation

Distance from apogee (°)	Solar equation (°)
0, 1	0; 2,10
0, 2	0; 4,19
...	
1,32	2;10, 0
1,33	2;10, 0
1,34	2;10, 0
...	
2,59	0; 2,24
3, 0	0; 0, 0

3. In the tables of John Vimond, discussed in Section 6.3.1 (below), that are independent of John of Murs, the maximum solar equation is 2;10° (see Chabás and Goldstein 2003).

Moreover, a maximum solar equation of 2;10° is a plausible parameter, for it lies within the range of other attested values: in the zijes of al-Khwārizmī and Ibn Mu[c]ādh it is 2;14°, in the Toledan Tables it is 1;59,10°, and al-Bīrūnī ascribes to the Sindhind tradition the use of 2;10,46,40° (Samsó 1987b, p. 176; Samsó 1992, p. 157).

See Table 16A for an excerpt of the table for the solar equation, with a maximum of 2;10,0° corresponding to the arguments 92°–94° (here expressed in physical signs as 1,32° to 1,34°) for the solar distance from the apogee, found in various manuscripts in the Alfonsine corpus, e.g., Madrid, Biblioteca Nacional, MS 4238, ff. 46v–47v: *Prima tabula equacionis solis*.

Chapter 17

17:1. This chapter on the lunar equation explains how to compute the true longitude of the Moon. Unlike the solar model described in chapter 16, the lunar model differs from the standard Ptolemaic model.

17:3–4. We are told to compute the expression

$$0;0,5,30 \cdot \sin(\bar{\lambda}_M - \lambda_A)$$

where $\bar{\lambda}_M$ is the mean longitude of the Moon and λ_A is the longitude of the solar apogee.

17:8. The result of adding this correction to, or subtracting it from, the mean motion of the Moon (according to a rule specified in 17:6–7) is called "equated mean motion" (*medio curso ygualado*), which is the starting point for determining the true longitude.

This correction is reminiscent of a lively tradition represented in other zijes in al-Andalus and the Maghrib prior to the writing of the Alfonsine Tables. In Ibn al-Kammād's *al-Zīj al-Muqtabis* (Madrid, Biblioteca Nacional, MS 10023, f. 8rb–8va), we are told that in order to obtain the true position of the Moon in computing an eclipse, we must subtract the solar apogee from the center, take the sine of the difference and multiply it by two-fifths of a minute (24 seconds), probably an error for two-fifths of a degree (24 minutes). The result is to be subtracted from, or added to, the equated position of the Moon to get the *locus lune uisibilis a puncto circuli signorum* [...] *in horis eclipsis luminarium*. According to Mestres (2000, p. 50), the same correction is found in chapter 15 in the zij of Ibn Isḥāq. A similar passage is found in the zij of Ibn al-Bannā' of Marrākesh (1256–1321) (Vernet 1952, pp. 88–89) where there is even a table for this correction, called "Table of Diversion" (*al-ṣarf*: the translation of this term is uncertain), whose entries, c, were computed by the following rule (where the argument, a, is the difference in longitude between the mean Moon and the solar apogee):

$$c = 0;24 \sin a.$$

Puig (2000, pp. 72–73, 79) reports a similar procedure in the zij of Ibn al-Hā'im (*ca.* 1205). The astronomical significance of this correction has not been explained satisfactorily.

17:9. The double elongation (*longura doblada*) is defined as twice the difference between equated mean longitude of the Moon and the mean longitude of the Sun.

17:10–20. The double elongation serves as the argument for the equation of center and for the minute of proportion. The expression

Table 17A: Lunar equation

Argument (°)	Equation of center (°)	Min. prop. (min)	*Div. diam.* (°)	Equation of anomaly (°)
1	0; 9	0	0; 3	0; 4,46
2	0;18	0	0; 5	0; 9,31
...				
95				4;56, 0
...				
103			2;40	
...				
109			2;40	
...				
114	13; 9			
115	13; 9			
...				
179	0;23	60	0; 4	0; 5,40
180	0; 0	60	0; 0	0; 0, 0

equaçion de la diversidad del diametro del piçiclo (17:15) refers the angular difference between the mean and true lunar epicyclic apogees, corresponding to the Latin *equatio diversitatis diametri*. The procedure for finding the true longitude of the Moon then follows in the usual way (Neugebauer 1969, pp. 197–198), but no numerical data are provided.

See Table 17A for an excerpt of the table for the lunar equation found in various manuscripts in the Alfonsine corpus, e.g., Madrid, Biblioteca Nacional, MS 10002, ff. 14r–15r. The maximum equation of center is 13;9° corresponding to 114°–115° of double elongation, and the maximum equation of anomaly is 4;56,0° for 95°. The maximum *diversitas diametri* is 2;40° for arguments 103°–109°.

Chapter 18

18:1. This chapter concerns the planetary equations.

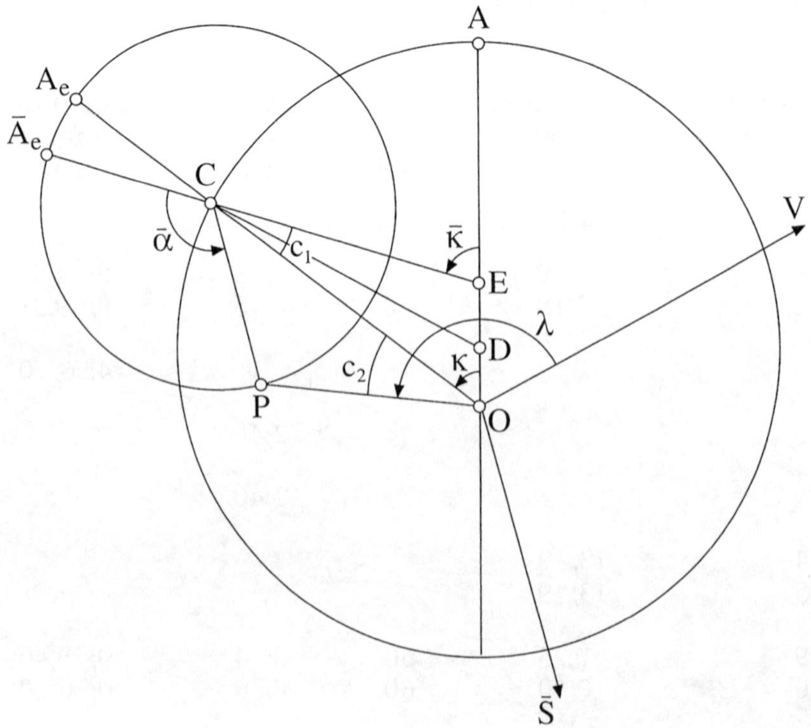

Fig. 4.2. The model for the motion of an outer planet: O is the observer; D is the center of the deferent circle, AC, whose radius is 60; E is the equant point; OD (= DE) is the eccentricity; A is the apogee of the deferent; C is the center of the epicycle; P is the position of the planet on the epicycle; OV is the direction to Aries 0°; and $O\bar{S}$ is the direction to the mean Sun (for an outer planet, this direction is parallel to the direction from the center of the epicycle to the planet, CP). $\bar{A}_e$ is the mean apogee of the epicycle and A_e is the true apogee of the epicycle. The mean argument of center, $\bar{\kappa}$, is measured about E from the apogee A; the true argument of center, κ, is measured about O from the apogee A; and the equation of center, c_1, is equal to $\bar{\kappa} - \kappa$. The mean argument of anomaly, $\bar{\alpha}$, is measured about C from $\bar{A}_e$, and the equated argument of anomaly is $\bar{\alpha} + c_1$ ($\angle A_eCP$). The equation of the equated anomaly is c_2. The true longitude of the planet, λ, is measured about O from the direction OV, and it is the sum of the longitude of the apogee, the true argument of center, and the equation of the equated anomaly. Note that arrows indicate the positive sense of rotation for the various angles.

18:2–21. This passage describes the procedure to compute by means of tables the true position (*lugar çierto*) of the three superior planets (*los planetas tres de suso*). For Ptolemy's model that underlies these tables, see Fig. 4.2.

18:5. The center, i.e., the difference between the mean position of a planet and its apogee (18:4), serves as the argument for entering the table of the equation of center for each planet, and the algebraic sum of the center and its equation is the equated center (18:8).

18:13. The equated argument of anomaly, i.e., the algebraic sum of the mean argument and the equation of center (18:11–12), is also used as argument for entering the table of equations for each planet.

18:14. The expression *longura mas luenga* refers to the angle to be subtracted at apogee from the corresponding mean value for the equation of anomaly, and the expression *longura mas çerca* refers to the angle to be added at perigee to the corresponding mean value for the equation of anomaly. The expressions in Latin are: *longitudo longior* and *longitudo propior*.

18:22. This passage is devoted to the equations of Venus and Mercury, and indicates that their mean motions are equal to the mean motion of the Sun. In all other respects the procedure is the same as that for the superior planets. This procedure applies to planetary tables modeled after those in the *Almagest*. Although no parameters are given, it is most likely that the maximum equation of center for Venus is the same as that for the Sun, namely 2;10°, because the equality of their apogees as well as their maximum equations was commonly assumed, as is the case, e.g., in the tables of al-Battānī and the Toledan Tables (1;59,10° for the Sun and 1;59° for Venus: see Nallino 1903–1907, 1:114, 2:81 and 128–129; Toomer 1968, pp. 45, 56, 65). This practice is explained by al-Bīrūnī who attributes it to Ḥabash (Goldstein 1977b, p. 167):

> Ḥabash made tables for [the planets] and their motions whose results do not differ much from those I have mentioned except for the apogee of Venus. Its [Venus's] equation with reference to the center of its equant sphere is equal, according to Ptolemy, to the solar equation, and it was in the *Zīj al-Shāh* that the corrected solar longitude is the corrected argument of Venus. This is impossible unless their apogees and equations are equal, and this is how they are in it [the

> *Zīj al-Shāh*]. This opinion was transferred [by Ḥabash] to the principles of Ptolemy, so he set the apogee of Venus which is [also] the apogee of the Sun according to the moderns, and its equation for the argument, equal to those of the Sun. But Ptolemy derived the equation and the apogee of the Sun from the solstice [and equinoxes], and nothing requires that they apply to the apogee and equation of Venus except the opinion found in the *Zīj al-Shāh*. Then al-Battānī followed him [Ḥabash?] in this, and I will not add any more here but will discuss it in [my] book: *Clarification of the Intellects Concerning the Zīj of Muḥammad al-Battānī*.

On the other hand, in the zij of Ibn al-Kammād the maximum equation of the Sun is 1;52,44° whereas the maximum equation of center for Venus is 1;59°, but the apogee is 76;45,21° for both of them (Chabás and Goldstein 1994, p. 33).

See Table 18A for an excerpt of the table for the planetary equation of Saturn found in, e.g., Madrid, Biblioteca Nacional, MS 4238, ff. 49v–51v. In each column we display the first two and the last two rows as well as the range of the extremal values. Tables of the same form are presented for all five planets.

Chapter 19

19:1. This short chapter on the lunar nodes defines the position of the ascending node (*cabeça del dragon*) as the difference between its mean position and 360° or 12 signs (19:2).

Chapter 20

20:1. This chapter concerns solar declination.

20:2. The solar longitude (*lugar natural del sol*) serves as the argument for the table of solar declination (*declinaçion*). For the meaning of *natural*, see chapter 49. No parameter is given in the canon, but in the Castilian Alfonsine corpus there is a table for this purpose, given to seconds, and based on the value of 23;32,30° for the obliquity of the ecliptic. As Comes (1992, p. 152) has noted, this table is included in chapter 3 the *Libro del relogio de la piedra de la sombra*, a treatise written by one of the authors of these canons, Isaac ben Sid (see Rico 1863–1867, vol. 4, p. 6). The same parameter, 23;32,30°, is the maxi-

Table 18A: Equation of Saturn

Argument (°)	**Equation of center** (°)	**Min. prop.** (min)	***Long. long.*** (°)	**Equation of anomaly** (°)	***Long. prop.*** (°)
1	0; 7	60	0; 0	0; 6	0; 0
2	0;14	60	0; 1	0;12	0; 1
...					
87		1			
88		1			
...					
90	6;31				
91				6;12	
...					
93				6;12	
94	6;31				
...					
100			0;21		
...					
102					0;25
...					
106			0;21		
...					
111					0;25
...					
179	0; 8	60	0; 1	0; 8	0; 1
180	0; 0	60	0; 0	0; 0	0; 0

mum value in Ibn Isḥāq's solar declination table; cf. Mestres 1999, pp. 54, 275. In the *Libro del quadrante con que rectifican*, Isaac ben Sid reports a variant of this parameter for the obliquity: 23;32,29° (see Dorce 1999, pp. 169–70, and the references given there).

In the heading of chapter 32 the obliquity is given as 23;33°, a value that appears frequently in early Islamic astronomical texts (e.g., the Almanac of Azarquiel; see Millás 1943–1950, p. 174).

20:3. A second expression for declination, *arredramiento del çerco egualador* (distance from the equator), is used here.

Table 20A: Declination

Solar long. (°)	**Declination** (°)
1	0;24, 0
...	
30	11;31,36
...	
60	20;15, 0
...	
90	23;33,30

See Table 20A for an excerpt of the table for the solar declination, with a maximum of 23;33,30°, found in the Toledan Tables (Toomer 1968, p. 30; F. S. Pedersen 2002, pp. 961–966) and in various manuscripts in the Alfonsine corpus, e.g., Oxford, Bodleian, MS Can. Misc. 27, f. 82r.

Chapter 21

21:1. This chapter is devoted to lunar latitude.

21:3. The argument of the Moon is defined as the difference between its longitude and that of the lunar node, here called *geuzaar* (Arabic: *jawzahar*). The corrresponding term in the canons to the Toledan Tables is *geuzahar* (F. S. Pedersen 2002, p. 444).

21:4. The argument of the Moon serves as the argument in the table for the latitude (*latitud*) of the Moon.

21:5. A second expression for latitude, *arredramiento del zodiaco* (distance from the ecliptic), is used here. In the title (21:1) it appears as *arredramiento del sodiaco*.

See Table 21A for an excerpt of the table for the lunar latitude, with a maximum of 5;0°, found in various manuscripts in the Alfonsine corpus, e.g., Oxford, Bodleian, MS Can. Misc. 27, f. 91r.

Table 21A: Lunar latitude

Argument (°)	Latitude (°)
1	0; 5,13
...	
30	2;29,52
...	
60	4;19,47
...	
90	5; 0, 0

Chapter 22

22:1. This chapter is devoted to planetary latitude.

For the superior planets, Ptolemy's procedure in the *Almagest* is correctly described, but the argument of latitude is not properly defined, for it is simply identified with the center in all cases (22:2). For the definition of "center", see comments to 18:5.

For the inferior planets the latitude is described in terms of three components called *latitud primera, latitud segunda,* and *latitud terçera* (for these three components and the way they are computed, see Neugebauer 1975, pp. 223–224). The text gives step-by-step instructions for computing the latitude at a given time, for determining its proper direction to the north or to the south, and then discusses whether the latitude is increasing or decreasing at that time. As was the case for the superior planets, here too the argument of latitude is not properly defined (22:11 and 22:26). For the inferior planets a column for the third component of latitude is described (22:12 and 22:28). As far as we know, the only astronomer before 1400 to tabulate the third component was John Vimond (see Section 6.3.1, below).

The tables described in this chapter, both for the superior and the inferior planets, are in the style of *Almagest* XIII.5 and the zij of al-Battānī, as opposed to those in the *Handy Tables* and those in the zij of al-Khwārizmī.

The terminology used for "inclination" and "slant" is *declinaçion* and *desviamiento* (with variants for the latter: *declinamiento* and *sujamiento*, each occurring only once), and it is the same as in the Alfonsine version in Castilian of the tables in the Almanac of Azarquiel, uniquely preserved in Paris, Bibliothèque de l'Arsenal, MS 8322 (Millás 1943–1950, pp. 217–218).

22:30 and 22:32. The text gives a rule for determining the slant (*desviamiento*) of Mercury that does not correspond to the rules in the *Almagest*. According to the text, the table for the latitude of Mercury is divided in two parts: values less than 93° or more than 267°, and values between 93° and 267°. This range is not the same as in the *Almagest* and the coefficients are ignored (see Neugebauer 1975, p. 224).

22:34–43. This passage concerns what was called "ascent" and "descent" (*sobimiento* and *abaxamiento*) of planets. These terms are defined as follows: when the planet is northern and its latitude is increasing, the planet is said to be ascending; when the planet is northern and its latitude is decreasing, the planet is said to be descending; when the planet is southern and the absolute value of its latitude is increasing, the planet is said to be descending; and when the planet is southern and the absolute value of its latitude is decreasing, the planet is said to be ascending. These definitions fully agree with those given in chapter 47 of the canons to al-Battānī's zij, which can thus be considered as a likely source.

No parameters are given for the superior planets. For Venus the maximum latitude is given as 8;36° (22:42): this parameter occurs in the canons of Ibn al-Kammād's zij (see Chabás and Goldstein 1994, p. 32) and in the canons to the tables of Judah ben Asher II (d. 1391) of Burgos (Vatican, MS Heb. 384, f. 303a–b). We assume 8;36° is a corruption of the value 8;56° found in the *Handy Tables* (Neugebauer 1975, p. 1014; see also Kennedy 1956, p. 173) and in al-Battānī's zij (Nallino 1903–1907, 1:116). For Mercury the maximum latitude is given as 4;38° (22:43), a parameter not found elsewhere except in the canons of Judah ben Asher II, and it seems to be a corruption of the value 4;18° in the *Handy Tables* and in al-Battānī's zij. Despite the similarities in terminology and presentation with the Almanac of Azarquiel (Millás 1943–1950, p. 218), that work cannot be its source, for the maximum *declinacion* for Venus is 6° (rounded from 6;22°, the value in the *Almagest*), and the maximum *declinacion* for Mercury is 4° (rounded from 4;5°, the value in the *Almagest*): see Table 22A.

Table 22A: Maximum latitudes for Venus and Mercury

	Venus (°)	Mercury (°)
Almagest (1)	6;22	4; 5
Handy Tables (2)	8;56*a*	4;18*b*
Mumtaḥan zij (3)	8;56	4;18
al-Farghānī (4)	9	
al-Battānī (5)	6;22	4; 5
al-Battānī (6)	8;56	4;18
Almanac of Azarquiel (7)	6	4
Ibn al-Kammād (8)	8;35	3;52
Ibn al-Kammād (9)	8;36	4;18
Castilian Alfonsine Tables (10)	8;36	4;38
Tables of Barcelona (11)	8;55	3;52

a. 8;51° in Stahlman 1959, p. 332.
b. 3;52° in Stahlman 1959, p. 334.

(1) Toomer 1984, 633f.
(2) Neugebauer 1975, p. 1014.
(3) Kennedy 1956, p. 173.
(4) Golius 1669, p. 74.
(5) Nallino 1903–1907, 2:141.
(6) Nallino 1903–1907, 1:116.
(7) Millás 1943–1950, p. 218.
(8) Chabás and Goldstein 1994, p. 32.
(9) Chabás and Goldstein 1994, p. 32.
(10) Chapter 22.
(11) Millás 1962, p. 227.

Chapter 23

23:1. This chapter concerns direct and retrograde motion.

23:5. The text refers to a table with the 4 stations (*estados*) of each planet. This is probably the same as the table found in *al-Muqtabis* of Ibn al-Kammād (Chabás and Goldstein 1994, p. 33) with only four values for each planet: the positions of the first and the second stationary points for arguments of 0° and 180°.

23:14. The word used for apogee is *aux* (Arabic: *awj*), and that for perigee is *hadit* (Arabic: *ḥaḍīḍ*: see Goldstein 1967, p. 29:16; for the variant *elhadhid*, see Suter 1914, p. 14, note 28). The term, *aux*, appears in the same form in medieval Latin where it occurs frequently in astronomical texts in contrast to *hadit* which only occurs rarely.

See Table 23A for the table in *al-Muqtabis* of Ibn al-Kammād for the planetary stations (Madrid, Biblioteca Nacional, MS 10023, f. 46r).

Chapter 24

24:1. This chapter and the one following it address the same topic, planetary visibility. Chapter 24 reports the headings of the tables for the planetary phases of both superior and inferior planets. It agrees very closely with the corresponding chapter in the Toledan Tables (F. S. Pedersen 2002, pp. 474–475).

24:3. For the three superior planets it is said that the morning risings (*subimientos de las mañanas*) occur for arguments of anomaly from 0° to 180°, and the evening settings (*ponimiento de prima noche*) from 180° to 360°.

24:4–8. For Venus it is said that the evening risings (*subimiento de prima noche*) occur for arguments of anomaly from 1° to 137°, the evening settings from 137° to 180°, the morning risings from 180° to 223°, and the morning settings (*ponimiento de las mañanas*) from there to 360°.

24:9–12. For Mercury it is said that the evening risings occur for arguments of anomaly from 1° to 112°, the evening settings from 112° to 180°, the morning risings from 180° to 248°, and the morning settings from 248° to 360°.

The extremal values of all the above-mentioned intervals agree with those given in the canons to the Toledan Tables (F. S. Pedersen 2002, pp. 476–479), in the zij of al-Battānī (Nallino 1903–1907, 2:142–143) which, in turn, were taken from the table for the planetary phases in Ptolemy's *Handy Tables* (Stahlman 1959, pp. 340–346). The extremal values are identified as arguments of anomaly in chapter 48 of al-Battānī's canons (Nallino 1903–1907, 1:117). As noted by Toomer (1968, p. 75), the headings of the columns in the *Handy Tables*

Table 23A: Planetary stations

	Saturn (°)	Jupiter (°)	Mars (°)	Venus (°)	Mercury (°)
I	3s 22;44	4s 4; 5	5s 7;28	5s 15;51	4s 27;14
II	8s 7;16	7s 25;55	6s 22;32	6s 14; 9	7s 2;46
III	3s 25;30	4s 7;11	5s 19;15	5s 18;21	4s 24;42
IV	8s 4;30	7s 22;49	6s 10;45	6s 11;39	7s 5;18

I: first station at apogee.
II: second station at apogee.
III: first station at perigee.
IV: second station at perigee.

as well as in the zij of al-Battānī and the Toledan Tables are associated with the wrong columns of entries.

See Table 24A for an excerpt of the table for the planetary phases found, e.g., in the Toledan Tables (Toomer 1968, pp. 73–76) and in Madrid, Biblioteca Nacional, MS 10053, f. 14v. In this table the numbers in the headings appear as they do in the manuscripts and have not been corrected.

Chapter 25

25:1. This chapter is the continuation of chapter 24, referring to the appearance (*paresçimiento*) and disappearance (*ascondimiento*) of the planets, and gives an explanation for using the corresponding table. The procedure is to compute the true elongation of the planet (25:2–5) and then to compare it with the entry for each sign in the table, given in degrees and minutes (25:6).

25:13–14. The texts refers to the argument of anomaly (*argumento*) for some of the extremal values used in the case of Venus and Mercury.

This procedure seems to correspond exactly with a table for the planetary phases found in the *Handy Tables* and the zij of al-Battānī, with 2 columns for each of the superior planets and 4 columns for each of the inferior planets, and where the entries repre-

Table 24A: Planetary phases

	Saturn		Jupiter		Mars	
	M.R. (°)	**E.S. (°)**	**M.R. (°)**	**E.S. (°)**	**M.R. (°)**	**E.S. (°)**
Aries	29;28	13;46	19;33	9;28	29; 0	14;12
Taurus	26;26	14; 7	18;21	9;38	27;11	15; 8
...						
Aquarius	21;16	14;40	14;44	10;16	22;14	16; 7
Pisces	26;46	14; 0	18;11	9;38	27;11	15; 8

Venus

	E.R. 1–137 (°)	**M.S. 224–360 (°)**	**M.R. 180–223 (°)**	**E.S. 138–180 (°)**
Aries	15;31	4;25	3;36	2;27
Taurus	13;48	4;29	4; 9	3;30
...				
Aquarius	12;47	8;29	1;36	1;14
Pisces	15;20	7;43	2;43	1;31

Mercury

	E.R. 1–152 (°)	M.S. 249–360 (°)	M.R. 180–248 (°)	E.S. 153–180 (°)
Aries	24;10	12;24	22;43	12; 9
Taurus	21;15	12;18	24;23	12;12
...				
Aquarius	20;15	14; 7	14;25	11;32
Pisces	24;38	12;14	18;22	11;47

E.R.: evening rising.
M.S.: morning setting.
M.R.: morning rising.
E.S.: evening setting.

sent the true elongation of the corresponding planet given for the beginning of each zodiacal sign.

The same table is also found in Toledan material: the Toledan Tables (Toomer 1968, pp. 73–76; F. S. Pedersen 2002, pp. 1530–1537), and the Almanac of Azarquiel (Millás 1943–1950, p. 219). Note that the terms used for rising and setting in that manuscript are *sobimiento* and *ponimiento*.

This table for planetary visibility (without a title) is also found in a 13th-century Latin manuscript (Madrid, Biblioteca Nacional, MS 10053, f. 14r) that contains many tables clearly related to the Castilian Alfonsine Tables (as we shall see), and a translation into Latin of Azarquiel's treatise on the *saphea* ascribed to Judah ben Moses. For the authorship of this work, also attributed to Guillelmus Anglicus, see Millás (1943–1950, pp. 453-454) and Hilty (1955, pp. 16–18). In his lengthy description of the manuscript, Millás (1942, pp. 180–202) did not mention this table.

Chapter 26

26:1. This chapter seems to refer to the "transit" of one planet over another, and the wording agrees very closely with that in the canons to the Toledan Tables (F. S. Pedersen 2002, pp. 480–481). The transit (*alcançamiento*) of one planet over another occurs when their difference in distance vanishes.

26:3–5. The text defines the distance (*longura*) of the planet from the highest point of its heaven as the difference between 180° and the equated argument of anomaly of the planet (or between 360° and the equated argument of anomaly, if it exceeds 180°).

This obscure discussion is probably related to the astrological problem of transits in thickness, discussed by Ibn Muᶜādh, among others: see Samsó 1997, p. 91; and Kennedy 1958.

Chapter 27

The text of chapter 27 was transcribed and translated in Goldstein *et al.* (1994): that transcription differs from the one given in chapter 2 of this book for, instead of relying on Rico, we now report the readings in the manuscript itself.

27:2. The "table for the daily unequal motion of the planets" is mentioned in this sentence and its use is explained in this chapter: it was the most popular medieval table for planetary and lunar velocities. This table has been edited and analyzed in Goldstein *et al.* (1994), based almost entirely on copies associated with John of Lignères (see Section 6.3.3, below). In addition to the 21 manuscripts and editions cited in the above-mentioned article, 21 more copies in Latin have been identified. In 1994 only one copy independent of John of Lignères had been located (Madrid, MS 10053), a 13th-century Latin manuscript already mentioned in the comments on chapter 25. For a list of the 46 manuscripts and editions that we have identified, see below. None of the manuscripts or editions explains how the table was constructed, and none indicates the name of the author. Although copies of this table are often found among 13th-century copies of the Toledan Tables (well before the time of John of Lignères), we are informed by F. S. Pedersen that no detailed canon for them has been located. There is, however, a short chapter of 3 lines in these canons that gives instructions for deriving the daily velocity of a planet by computing the difference between its positions on two successive days (Cb 224, in F. S. Pedersen 2002, pp. 482–483). Despite brief glosses on the use of this table found in three manuscripts (A, Na, and Fj: for an explanation of the sigla, see below), chapter 27 of the Castilian canons is the earliest detailed description of it.

27:6–7. Both the equated argument of center and the equated argument of anomaly serve as the arguments in this table. The table corresponding to this description consists of 3 columns for each planet where one column lists the argument at intervals of 6°, and two other columns give the components of velocity that depend on the center and the anomaly, respectively. The headings for these two columns in Latin are: *motus centri* or *motus puncti;* and *motus argumenti* or *motus portionis.*

27:9. The daily velocity (called here "daily unequal motion") is found by adding appropriate entries.

27:27. Planets are designated "heavy" (*pesado*) and "light" (*liviano*), for "slow" and "swift", with respect to each other where the planet farther from the Earth is "heavy" and the planet closer to the Earth is "light": see Goldstein *et al.* 1994, p. 87.

There follows a new edition of the planetary velocity table, based largely on manuscripts unavailable to us in 1994. We have found the planetary velocity table in a total of 41 manuscripts (37 in Latin, 2 in Hebrew, and 2 in Arabic) and 5 printed editions in Latin, and it is likely that more manuscript copies will be discovered in the future. Most of the Latin manuscripts of this table can be assigned to one of two principal families: family α that consists primarily of 13th-century copies, and family β that is associated with John of Lignères.

Family α:

A	Madrid, Biblioteca Nacional, MS 10053, ff. 9v–10r
Fa	Florence, Biblioteca Medicea Laurenziana, MS Ashburnhamianus 211, ff. 235v–236r
Fj	Florence, Biblioteca Nazionale Centrale, MS conv. soppr. J.V.6, f. 57r–v
Gr	Groningen, University Library, MS 102, ff. 129v–130r
Lr	London, Royal Astronomical Society, MS Add. 1 vol. 2, ff. 37v–38r
Na	Naples, Biblioteca Nazionale, MS VIII-C-49, f. 66r–v
Pa	Paris, Bibliothèque de l'Arsenal, MS 877, f. 56r–v
Vb	Vatican, Biblioteca Apostolica, MS Barberinianus lat. 296, f. 21r
α′	Common readings in all MSS of family α except for Na.

Family β:

Basel, Universitätsbibliothek, MS F.II.7, f. 69r
Bonn, Universitätsbibliothek, MS S 498, f. 52r–v
Budapest, National Library, MS 62, f. 22r
Cologne, Historischer Archiv der Stadt, MS W*178, f. 9r
Cracow, Jagiellonian Library, MS 550, f. 41r
Cracow, Jagiellonian Library, MS 551, f. 79v
Cracow, Jagiellonian Library, MS 553, f. 106r
Cracow, Jagiellonian Library, MS 564, p. 128
Cracow, Jagiellonian Library, MS 578, p. 168
Cracow, Jagiellonian Library, MS 609, f. 71v
Cracow, Jagiellonian Library, MS 610, f. 337
Erfurt, Biblioteca Amploniana, MS F 37, f. 114v
Erfurt, Biblioteca Amploniana, MS F 377, f. 46v
Florence, Biblioteca Medicea Laurenziana, MS San Marco 185, f. 108v
Madrid, Biblioteca Nacional, MS 4238, f. 62r
Madrid, Biblioteca Nacional, MS 10002, f. 38r
Moscow, The Russian State Library, MS F.68 (450), f. 43v (acc. to Rosińska)
Munich, Bayerische Staatsbibliothek, MS Clm 26667, f. 29v
Oxford, Bodleian Library, MS Canon. Misc. 27, f. 90r
Oxford, Bodleian Library, MS Canon. Misc. 499, f. 142r/132r
Oxford, Bodleian Library, MS Laud. Misc. 594, f. 155v
Paris, Bibliothèque nationale de France, MS lat. 7282, f. 117v
Paris, Bibliothèque nationale de France, MS lat. 7286C, f. 55r
Paris, Bibliothèque nationale de France, MS lat. 7295A, f. 162v
Rome, Biblioteca Casanatense, MS 653, f. 33v
Vatican, Biblioteca Apostolica, MS Ottobonianus 1826, f. 121r
Vienna, Nationalbibliothek, MS Vin. 5268, f. 26v
Tabule Astronomice Alfonsi Regis, J. L. Santritter, 1492, ff. 140, 148, 156, 164, 174
Tabule Astronomice Divi Alfonsi Regis, P. Liechtenstein, 1518/1521, ff. 69, 73, 77, 81, 85
Alphonsi … Tabule, L. Gaurico, 1524, ff. 44v–46v
Tabulae Astronomicae, quas vulgo... resolutas vocant, J. Schöner, 1536, 1551
Divi Alphonsi ... tabulae, P. Du Hamel, 1545, 1553, pp. 86–90

In our previous article we only knew of one manuscript belonging to family α (MS A). Subsequently, we identified a second member of this family (MS Na) and F. S. Pedersen kindly provided us with photographs of six additional copies of this table, together with information about the manuscripts in which it appears. Family β was already well represented in our previous article (21 items, including the printed editions) and it has grown by eleven (2 printed editions and 9 manuscripts: Bonn, Budapest, Cracow MS 609, Florence, Madrid MS 4238, Oxford MS 594, Rome, Vatican, and Vienna). Indeed, every early edition of the Alfonsine Tables, but for the *editio princeps*, included this planetary velocity table, a fact which was first reported in the scholarly literature in 1994. This may serve as a warning that later editions of a text should not be assumed to be copies of the first edition.

Here are some examples of entries in family α that differ systematically from family β:

Saturn, col. 3, 30°: 5;1 in α vs. 5;0 in β.
Jupiter, col. 5, 78°: 4;15 in α vs. 4;45 in β.
Jupiter, col. 5, 84°: 3;10 in α vs. 3;34 in β.
Jupiter, col. 5, 90°: 2;5 in α vs. 2;22 in β.
Mars, col. 6, 180°: 38;30 in α vs. 38;40 in β.
Venus, col. 9, 144°: 60;48 in α vs. 60;54 in β.
Venus, col. 9, 168°: –45;0 in α vs. –65;0 in β.
Venus, col. 9, 174°: –17;0 in α vs. –87;0 in β.
Mercury, col. 11, 168°: –83;0 or –93;0 in α vs. –103;0 in β.

Some of these variants would have no significant effect on computing a planet's velocity (e.g., 5;1° vs. 5;0° for Saturn), whereas others would have a significant effect (e.g., –17;0° vs. –87;0° for Venus). Admittedly, we have not found a computation based on this table in any medieval manuscript. Moreover, the column for the daily progress of the Moon (col. 14) is not found in family β but it is found in family α with one exception (Vb). On the other hand, column 14 appears by itself, in Roman numerals, in an early 13th-century manuscript (MS Ct: Cambridge, Trinity College, MS O.8.34, f. 1v).

In addition to these two families, there are two Latin manuscripts which do not fit easily in either of them. London, British Library, MS Arundel 377, ff. 77v–78r (La), is an exceptional 13th-century manuscript because it shares some characteristics with both families as well as having variants of its own. For instance, Jupiter, col. 5, 90°, is a case where La agrees with all manuscripts in family α, that is, the entry is 2;5 rather than 2;22 as in family β. The situation is reversed, for example, in the case of Venus, col. 9, 174°, where La agrees with all manuscripts in family β, that is, the entry is –87;0 rather than –17;0 as in family α. But for Mercury, col. 11, 18°, all manuscripts in families α and β have 50;58 whereas La has 50;55. Another unusual feature of La is that the table on f. 77v has a row for argument 0° for the motions of the Sun, the Moon, Venus, and Mercury; this feature does not appear on f. 78r for the outer planets. The table in Mp (Montpellier, Bibliothèque Interuniversitaire, MS H 323, f. 295v) is lacking the entries for Mercury and the Moon; most of the readings in this manuscript agree with family α but a few agree with family β.

The Hebrew and Arabic manuscripts are:

Jg Juan Gil: ex-London, Jews College, MS 135, f. 98b
Ya Judah ben Asher II: Vatican, Biblioteca Apostolica, MS Heb. 384, f. 363a

Az1 Ibn ʿAzzūz: Rabat, Bibliothèque Générale, MS D 2461, pp. 410–426
Az2 Ibn ʿAzzūz: Rabat, Bibliothèque Ḥasaniyya, MS 8772, pp. 82–98 (no columns for Mercury)
Az Common readings in Az1 and Az2

Variant readings found in the text of Juan Gil are only displayed occasionally because the entries in his table are given at 5° intervals and only for columns (2), (4), (6), (9), and (11). Judah ben Asher II has entries at 6° intervals with two columns for each planet: there is one column for each planet that is usable for comparative purposes: columns (2), (4), (6), (8), and (10). The entries in other column for each planet are simply the differences between the entries in the first column and 0;59,8° for Saturn, Jupiter, and Mars; 1;35,59° for Venus; and 4;5,32° for Mercury (with some mistakes in the computations of the differences).

Juan Gil of Burgos (*ca.* 1350) constructed a set of astronomical tables that are uniquely preserved in a Hebrew translation, and he depended on the Andalusian astronomer, Ibn al-Kammād, for many of them. He was known as an astrologer and wrote a set of annotations in Castilian on Ptolemy's *Tetrabiblos* (Beaujouan 1969, p. 11; cf. Chabás and Goldstein 2000, p. 152). The astronomical tables of Judah ben Asher II are uniquely preserved in the Vatican manuscript cited above (see Goldstein 1998, pp. 179–181). He was the great-grandson of Rabbi Asher ben Yeḥiel (d. 1328) who moved from Cologne to become the chief rabbi of Toledo in 1305. Ibn ʿAzzūz al-Qusanṭīnī (d. 1354) was an astronomer active in Fez, Morocco, who worked in the Andalusian tradition. His treatise, *al-Zīj al-Muwāfiq*, is extant in two copies (listed above) and the planetary velocity table appears in both of them (see Samsó 1997, pp. 95, 104–105: in some cases our readings differ from Samsó's).

The two Arabic manuscripts form a separate family. In many cases the readings in them agree with family α rather than with β:

> Mars, col. 6, 120°: 34;10 in α and Az vs. 34;0 in β.
> Mars, col. 7, 18°: 10;55 in α and Az vs. 10;58 in β.
> Venus, col. 8, 60°: 59;5 in α and Az vs. 59;10 in β.

But there are cases where Az differs from both α and β, and some where Az agrees with β against α:

> Saturn, col. 3, 30°: 5;0 in α vs. 5;1 in Az and β.
> Jupiter, col. 5, 6°: 8;50 in α and β vs. 9;0 in Az.
> Mars, col. 7, 180°: –53;50 in α and β vs. –53;30 in Az.
> Venus, col. 9, 174°: –17;0 in α vs. –87;0 in Az and β.
> Mercury, col. 11, 162°: –84;0 in α vs. –94;0 in Az and β.

Similarly, Ya is distinct from the other families.

The edition that follows is based on all manuscripts except those in family β which seem to represent a later stage in the tradition. The base manuscript is Na with variants noted from A, Ct, Fa, Fj, Gr, Jg, La, Lr, Mp, Pa, Vb, Ya, Az1, and Az2. For column 14, the base manuscript is Na with variants noted from A, Ct, Fa, Fj, Gr, Lr, Pa. There are three copies of column 14 in Lr: Lr1 (38r), Lr2 (56v), and Lr3 (64r). The Arabic and Hebrew manuscripts do not have the columns for lunar motion that appear here. In Jg there is a column for daily lunar velocity at 5° intervals whose entries range from 12;5° (for argument 0°) to 14;16° (for argument 180°).

Columns (2), (4), (6), (8), and (10) are for the center (Ar. *markaz*) for each planet. Columns (3), (5), (7), (9), and (11) are for the anomaly (Ar. *ḥiṣṣa*) of each planet. The entries in columns (2) to (11) are given in minutes and seconds per day. Column (12) is for the hourly lunar motion and it is also given in minutes and seconds. Column (14) gives multiples of the mean daily progress of the Moon in degrees and minutes.

Saturn

(1)	(2)	(3)	
6	1;44	5;43	Az: 37
12	1;45	5;36	Mp: 46; Az: 30
18	1;46	5;24	Mp: 44
24	1;46	5;12	A: 22
30	1;47	5; 1	La,Az: 0
36	1;48	4;46	Az2: 56
42	1;48	4;32	Ya: 49
48	1;49	4;18	Ya: 50; α′: 3;18 La: 4;16 Az: 3;16
54	1;50	3;50	Ya: 51; Az: 3;48
60	1;51	3;20	Ya: 52; Az1: 3;24
66	1;52	2;52	Ya: 53; Mp: 2;51
72	1;53	2;22	Ya: 55; Az: 2;52
78	1;54	1;50	La,Mp,α′,Az: 55 Ya: 56
84	1;56	1;18	A: 57 Ya: 58; La,Az: 1;15
90	1;58	0;36	Ya: 2; 0
96	2; 0	0; 0	Ya: 1
102	2; 1	–0;36	Ya: 3
108	2; 3	–1;20	A: 2 Ya: 4; Az: 1;14
114	2; 4	–2; 0	Ya: 6; Az: 1;54
120	2; 5	–3;40	La,Mp,α′,Az: 6 Jg Ya: 7; La,Mp,α′: 2;40 Az: 2;30
126	2; 6	–3;10	La,Mp,α′,Az: 7 Ya: 8
132	2; 7	–4;50	La,Mp,α′,Az: 8 Ya: 9; La,Mp,α′: 3;50 Az: 3;30
138	2; 8	–4;30	La,Mp,α′,Az: 9 Ya: 10
144	2; 9	–5; 0	La,Mp,α′,Az: 10 Ya 11; Az: 5;10
150	2;10	–5;20	Jg,La,Mp,α′,Az: 11 Ya: 12; Az: 5;30
156	2;11	–5;50	La,Mp,α′,Az: 12 Ya: 13; Az: 5;55
162	2;12	–6;15	Ya: 14; Az: 6;20
168	2;12	–6;40	La,Mp,α′,Az: 13 Ya: 14
174	2;13	–7; 0	La,Mp,α′,Az,Ya: 14
180	2;14	–7;15	Az: 10

In MSS A, La, Lr, Pa, Vb, and Az, the entry in col. 2 for argument 120° is 2;6 (rather than 2;5 as in Na). The following entries in Na were shifted one line downwards. In Ya, col. 2, a sequence of entries are all shifted up one place: the entries for arguments 42° to 150° in Ya correspond to the entries for 48° to 156° in α′.

Jupiter

(1)	(4)	(5)	
6	4;32	8;50	Az1: 30; Az: 9;0
12	4;33	8;42	
18	4;34	8;30	
24	4;36	8;18	Az: 13
30	4:37	8; 7	Jg: 38; Az: 8;6
36	4;39	7;50	
42	4;40	7;44	La,Mp,α′,Az: 34
48	4;42	7;34	La,Mp,Az: 14
54	4;44	6;50	
60	4;46	6;20	
66	4;48	5;45	
72	4;50	5; 5	
78	4;53	4;15	
84	4;55	3;10	
90	4;58	2; 5	Az: 1
96	5; 0	1;10	
102	5; 3	0; 0	
108	5; 6	–1;15	
114	5;10	–2;30	
120	5;13	–3;45	
126	5;16	–5; 0	
132	5;19	–6;15	
138	5;21	–7;25	
144	5;23	–8;32	
150	5;25	–9;40	Ya: 21
156	5;27	–10;40	Ya: 23
162	5;28	–11;35	Ya: 25
168	5;29	–12;25	
174	5;30	–12;50	
180	5;30	–13; 0	Az: 10

Mars

(1)	(6)	(7)	
6	25;43	11; 5	
12	25;50	11; 0	
18	26; 0	10;55	Ya: 18
24	26;15	10;48	
30	26;30	10;42	
36	26;45	10;36	
42	27; 0	10;30	Ya: 26; Az: 10;36
48	27;15	10;20	La,Mp,α′: 24; Az: 23
54	27;30	10;11	Az: 15
60	27;50	10; 0	Az: 3
66	28;25	9;48	
72	29; 0	9;30	
78	29;40	9;10	
84	30;20	8;50	A: 55
90	31; 0	8;25	
96	31;35	7;55	
102	32;10	7;10	Ya: 25
108	32;15	6;20	La,Mp,Ya: 55 Az: 50
114	33;30	5;20	
120	34;10	4; 0	Jg: 0 (?)
126	34;50	2;10	
132	35;35	0; 0	Az: 30
138	36; 5	–2;18	Az: 6; A: 2;10 Az: 3;58
144	36;35	–7;45	
150	37; 0	–13; 0	A: 15
156	37;20	–25;20	Az: 20;20
162	37;40	–29;30	
168	38; 0	–39;29	
174	38;20	–49;20	
180	38;30	–53;50	Na mg.: 40; Az: 53;30

Venus

(1)	(8)	(9)	
6	57; 8	15;30	
12	57;10	15;28	
18	57;13	15;26	
24	57;14	15;23	Az: 16
30	57;21	15;18	Mp: 12; Jg: 15;20
36	57;27	15;12	Mp: 22
42	57;34	15; 4	La: 24
48	57;42	14;56	Az: 44
54	57;50	14;46	
60	58; 0	14;36	A: 57;0 Ya: 58;8; Jg,La,Mp,α′,Az: 14;34
66	58;12	14;20	
72	58;24	14; 2	Ya: 22 Az: 23
78	58;36	13;40	
84	58;50	13;19	Mp: 2 Az: 15
90	59; 5	12;48	Ya: 20; Az1: 12;40
96	59;20	12;18	Az1: 8
102	59;34	11;30	Az: 31
108	59;47	10;20	
114	59;59	8;50	
120	60;11	7; 5	
126	60;23	2; 0	La,Az: 5
132	60;36	0; 0	Az: 53; Az: 2;0
138	60;41	–5; 0	Az1: 3;5 Az2: 4;5
144	60;48	–7;20	A: 61;48 Az: 60;47; La: 2 Az2: 6
150	60;54	–14;10	A,Ya: 61
156	61; 0	–25;50	Lr: 24 Az: 20
162	61; 5	–42;30	
168	61; 9	–45; 0	La,Mp: 65
174	61;13	–17; 0	La,Mp,Az: 87
180	61;15	–96; 0	A: 36 Mp: 9

La has entries for argument 0° for columns 8 to 11.

Mercury

(1)	(10)	(11)	
6	56;10	51;10	Az1: 15
12	56;15	51; 5	Lr: 50
18	56;21	50;58	La: 55 Az1: 35
24	56;28	50;30	Ya: 21
30	56;36	49;50	Ya: 37
36	56;46	49; 0	
42	56;57	47;30	Az1: 56; Az1: 47;50
48	57; 8	45;30	La: 7; Az1: 45;50
54	57;22	43; 0	La,Ya,Az1: 20
60	57;34	40; 0	
66	57;46	37; 0	
72	58; 0	34; 0	La: 5
78	58;14	31; 0	Az1: 34
84	58;28	28; 0	Az1: 58
90	58;44	25; 0	Az1: 24
96	59; 0	19; 0	
102	59;16	12; 0	
108	59;34	5; 0	La: 0
114	59;42	–3; 0	Az1: 2
120	60;10	–11; 0	Ya: 11
126	60;28	–19; 0	
132	60;44	–31; 0	
138	61; 0	–44; 0	Ya: 9
144	61;14	–58; 0	Ya: 44
150	61;26	–71; 0	
156	61;36	–83; 0	
162	61;45	–84; 0	La: 44 Vb: 83 Az1: 94
168	61;53	–93; 0	α′: 83 La,Az1: 103
174	62; 0	–108; 0	Ya: 61; La: 110;0
180	62; 5	–120; 0	Ya: 61;5 Az1: 62;0; Fa,Jg,La,Pa,Vb,Az1: 112;0 Lr: 102;0

The entries in col. 11 in La are shifted up one row (a problem caused by introducing a row for 0°); an extra entry (110;0°) was added for argument 174° but the maximum (112;0°) was maintained for argument 180°. The values in Jg, but for argument 180°, are too discordant to be useful as variant readings.

Moon

(1)	(12)	
6	29; 0	
12	29; 4	
18	29; 8	
24	29;13	
30	29;22	
36	29;33	
42	29;46	
48	30; 2	
54	30;19	
60	30;37	Fa: 27
66	30;58	
72	31;18	
78	31;39	
84	32; 3	
90	32;26	
96	32;49	Fa: 33
102	33;15	
108	33;44	
114	34;17	
120	34;44	α′: 49
126	35;17	
132	35;44	
138	36;11	
144	36;38	
150	37; 1	Fa: 33
156	37;19	Fa: 44
162	37;33	Fa: —
168	37;44	Fa: —
174	37;52	
180	37;56	α′: 46

There is no col. 12 in La.

Lunar Progress: the entries in col. 13 are days, and those in col. 14 are given in degrees.

(13)	(14)	
1	13;52	α′: 53*
2	27;46	
3	41;39	
4	55;32	
5	99;28	Ct,Lr2,Lr3: 69;25* Fa: 69;28
6	83;17	Lr2,Lr3: 27
7	97;10	Lr3: 79
8	111; 3	A: 13
9	124;56	
10	138;49	
11	152;42	Lr2,Lr3: 45
12	166;38	Ct,Lr2,Lr3: 35*
13	180;28	Lr1: 108
14	194;21	
15	208;14	A: 308 Lr2,Lr3: 204
16	222; 7	
17	236; 0	A: 226
18	249;52	
19	263;45	Lr2,Lr3: 47
20	277;38	
21	291;31	
22	315;24	Ct,Lr2,Lr3: 305*
23	319;17	
24	333;10	
25	347; 3	
26	0;56	
27	14;48	
28	28;41	
29	30;32	A: blank Ct: 42;34*
30	—	A: 30;32

Variants marked with an asterisk (*) are properly computed as multiples of about 13;52,55°, rounded to minutes. We have no explanation for the choice of this parameter.

Chapter 28

28:1. The title agrees with that of chapter 97 in *al-Zīj al-Shāmil fī Tahdhīb al-Kāmil* of Ibn al-Raqqām (*ca.* 1250–1315): "On Determining the Ascendent and the Tenth (house) by Means of the Hours Passed of the Day and Night" (Kennedy 1997, p. 49), but the text differs.

28:5. The text mentions a *tabla de los subimientos de la villa,* that is, a table for oblique ascensions. Most sets of medieval astronomical tables contain at least one such table (e.g., al-Battānī's zij: Nallino 1903–1907, 2:65–71).

See Table 28A for an excerpt of the table for the oblique ascension "ad civitatem toleti" (latitude 39;54°) as found in Oxford, Bodleian, MS Can. Misc. 27, ff. 99v–100r, where the column labeled "O.A." is for oblique ascension, and the column labeled "S.H." is for the length of a diurnal seasonal hour, in degrees, as a function of the solar longitude. (On a given day of the year, a diurnal seasonal hour is one-twelfth of the time from sunrise to sunset and a nocturnal seasonal hour is one-twelfth of the time from sunset to sunrise.) The same table for Toledo is found in the Toledan Tables (Toomer 1968, p. 36; F. S. Pedersen 2002, pp. 997–1003).

28:9–10. The text mentions a *tabla de los subimientos del cerco derecho,* that is, a table for rising times at *sphaera recta*. Again, most sets of medieval astronomical tables contain such a table. The argument of the table here is called *grados yguales*. This expression is very unusual, but we find it elsewhere in our text (chapters 39, 41, 48, and 52) as well as in a table for the length of daylight in Madrid, Biblioteca Nacional, MS 10053, f. 13r–v, where *gradus equales* appears in the heading. Normally, the argument of a table in Latin is called *linee numeri*.

28:11. The text mentions a *tabla de la equaçion de las casas de la villa,* that is, a table for domification (a table for determining the astrological houses). For the various methods of domification, see North 1986. Moreover, the authors indicate that "in our tables" there is a specific table of this kind for the city of Toledo (28:13). This table gives, for a particular locality, the longitudes of the cusps (i.e., the boundaries of the astrological houses on the ecliptic) as a function of the longitude of the ascendant, that is, the point on the ecliptic that crosses the eastern horizon at a given time. Note that the text mentions the *çentros destas casas* where *çentros* is a proper translation of the Arabic for the cusps: according to Kennedy (1996, p. 537), "The houses are twelve divisions of the ecliptic which, unlike zodiacal

Table 28A: Oblique ascensions and diurnal seasonal hours

	Aries		Taurus		Gemini	
	O.A. (°)	S.H. (°)	O.A. (°)	S.H. (°)	O.A. (°)	S.H. (°)
1	0;33	15; 3	18;43	16;41	40;39	18; 2
...						
30	18; 4	16;38	39;50	18; 0	68;37	18;34

	Cancer		Leo		Virgo	
	O.A. (°)	S.H. (°)	O.A. (°)	S.H. (°)	O.A. (°)	S.H. (°)
1	69;44	18;34	105;30	17;57	143;34	16;35
...						
30	104;14	18; 0	142;18	16;38	180; 0	15; 0

	Libra		Scorpio		Sagittarius	
	O.A. (°)	S.H. (°)	O.A. (°)	S.H. (°)	O.A. (°)	S.H. (°)
1	181;13	14;57	218;59	13;19	257; 1	11;58
...						
30	217;42	13;22	255;46	13; 0	291;23	11;26

	Capricorn		Aquarius		Pisces	
	O.A. (°)	S.H. (°)	O.A. (°)	S.H. (°)	O.A. (°)	S.H. (°)
1	292;28	11;26	321; 0	12; 3	342;34	13;25
...						
30	320;10	12; 0	341;56	13;22	360; 0	15; 0

signs, vary (...). The initial points of the divisions are called *cusps* (Arabic *marākiz*, [which literally means] 'centers'; singular *markaz*)."

Tables for domification are common in the astronomical literature, but often they were computed with inconsistent values of the obliquity for different localities (see, e.g., the Toledan Tables: Toomer 1968, pp. 140–146; F. S. Pedersen 2002, pp. 1075–1121). The only complete table for Toledo known to us is in the *Almanaque Perpetuo* of Ferrand Martines, preserved in Vienna, Nationalbibliothek, MS Vin. 2488, ff. 77v–83r (Chabás 1996, 288–294).

See Table 28B for an excerpt for Aries of the tables for domification for Toledo, as found in the *Almanaque Perpetuo* of Ferrand Martines where the entry in the first column is the longitude of the ascendant. There are similar sub-tables for the other zodiacal signs.

28:12. On the expressions *comienço natural* and *començamiento propio* related to the *movimiento del llegamiento y del tornamiento*, see the comments on chapter 49.

Chapter 29

29:1. The title, "conjunction of the days with their nights", should read "equation of the days with their nights", which is the usual medieval expression for the equation of time.

29:2. Equator, horizon, and meridian are called *çerco equinoçial*, *çerco del orisonte*, and *çerco de mediodia*, respectively.

Table 28B: Domification for Toledo

Aries	2 Tau (°)	3 Gem (°)	4 Cnc (°)	5 Cnc (°)	6 Leo (°)
1	3; 3	2;50	0;30	28;30	28;34
...					
				(Leo)	(Vir)
30	26;57	22; 4	16;40	18;53	24; 8

29:11–15. The text gives a sound explanation of the equation of time and the behavior of this function throughout the year (cf., e.g., Goldstein 1974, pp. 102–104). As noted in the text, the equation of time affects significantly the position of the Moon for, in half an hour, the Moon can move as much as 0;18° (29:10–12).

29:16. The maximum value for the equation of time is given as 7;48°. Note that the values for its two components in 29:14 (3;15°) and 29:15 (4;32°) add up to 7;47°. These values are compatible with that for the maximum given in the same sentence expressed in units of time: $^{1}/_{2} + ^{1}/_{48}$h = 0;31,15h (corresponding to 7;49°). In any case, this maximum differs from that given in the zij of al-Battānī and in the Toledan Tables which both have 7;54°. In other tables in the Alfonsine corpus, the maximum is 7;57°.

In Table 29A we present an excerpt of the table for the equation of time in Madrid, Biblioteca Nacional, MS 4238 (f. 61v) and MS 10002 (f. 25r); Segovia, Catedral, MS 110, f. 46r–v; and Oxford, MS Can. Misc. 27, f. 79r. The extremal values in this table are:

Min:	0; 0°	Aqu 18–25°
max:	5;21°	Tau 25–27°
min:	2;49°	Leo 5°
Max:	7;57°	Sco 8–9°

According to F. S. Pedersen (2002, p. 984), this table, usually found among the tables ascribed to John of Lignères, perfectly agrees with the description given by Peter of St. Omer in his *Tractatus de semissis* of 1293–1294. In Fig. 4.3 we display a graph of this table.

Table 29A: Equation of time

	Cap (°)	Aqu (°)	...	Sgr (°)
1	3;46	0;38	...	7; 5
2	3;37	0;34	...	7; 0
...				
30	0;42	0; 6	...	3;54

Two other values are given in this sentence: Cap 6;57° for the mean position of the Sun, and Cap 7;29° for the true position of the Sun. The next sentence seems to indicate that these longitudes are in fact radices (29:17).

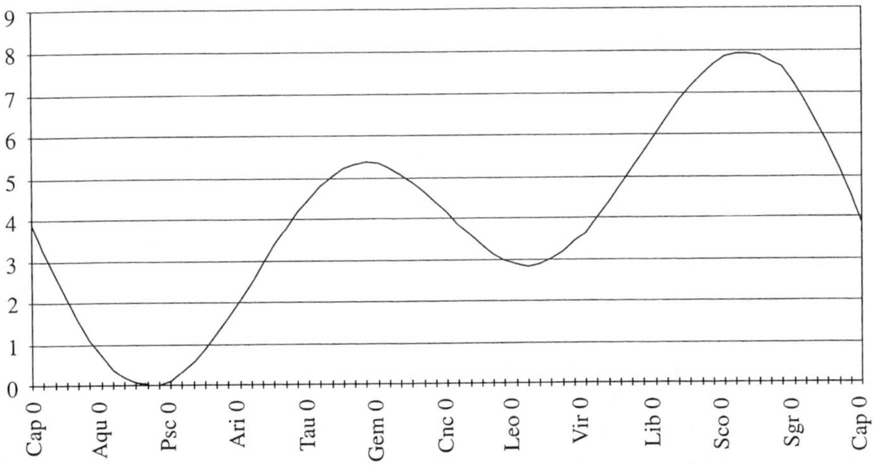

Fig. 4.3. The annual variation in the equation of time: a graphic representation of the table described by Peter of Saint Omer, usually attributed to John of Lignères.

The first value is indeed the radix for the mean Sun at epoch (noon of Jan. 1, 1252), computed with the Toledan Tables. The time interval between the epoch (JDN 2178351) and the Hijra (JDN 1948439: see our comments to 1:6, above) is 229912 days. Using the parameter in the Toledan Tables for the mean motion of the Sun (0;59,8,11,28,27°/d), one obtains 163;15,43° for the progress of the Sun between these two dates. Since the radix for the Sun at the Hijra given in the Toledan Tables is 3s 23;41,11°, one gets 276;56,54° = Cap 6;57° for the mean longitude of the Sun at the beginning of the "era of Alfonso" in agreement with the text. Note that this is one of the few chapters to include numerical data.

The two values given, Cap 6;57° and Cap 7;29°, are sidereal longitudes, and their difference, the solar equation, is 0;32°. With the value for the solar apogee in al-Battānī's zij, chapter 28, of 82;17° (Nallino 1903–1907, 1:44) and his equation table (whose maximum value is also in the corresponding table in the Toledan Tables), the solar equation corresponding to Cap 6;57° is 0;31,7°, i.e., 0;1° less than the value in the text. On the other hand, according to the Parisian Alfonsine Tables the radix for the Sun at this epoch (noon, Jan. 1, 1252) is Cap 16;48°, the solar equation is 0;42°, and the solar apogee is 88;40°.

29:18–24. This describes correctly the procedure for calculating the equation of time which, in modern terms, is equivalent to the formula

$$E = \alpha_2 - \alpha_1 - (\bar{\lambda}_2 - \bar{\lambda}_1)$$

where E is the equation of time, α_2 and α_1 are the right ascensions of the true solar longitudes corresponding to some day in the year and January 1, respectively, and $\bar{\lambda}_2$ and $\bar{\lambda}_1$ are the mean solar longitudes on those days, respectively.

Chapter 30

30:1. This chapter concerns conjunctions and oppositions.

30:2. Two distinct tables are mentioned: one for mean conjunction, another for mean opposition. Each of them has four columns: days and hours, mean motion of the luminaries, lunar argument of anomaly, and lunar argument of latitude. This is the standard format already found, for instance, in the zij of al-Khwārizmī (Suter 1914, pp. 183–186), in the zij of al-Battānī (Nallino 1903–1907, 2:84–87), and in the Toledan Tables (Toomer 1968, pp. 78–81; F. S. Pedersen 2002, pp. 1327–1350).

30:2–4. Both tables have sub-tables for collected years, expanded years, and months in a leap year or in a common year.

30:11. For a year with 13 conjunctions or 13 oppositions, the 13th entry in the table for the months of that year is called *mes soberano*.

30:14. For determining true syzygy, one first has to find the solar equation using a table for this purpose called *tabla de la ynvocaçion del sol*. Note the change in terminology, for in chapter 16 the text refers to the *equaçion del sol*.

30:17. Analogously, one has to find the lunar equation using a table called *tabla de la ynvocaçion de la luna*.

30:20–23. The method presented here for determining the time from mean to true syzygy is that already described in *Almagest* VI.4: add to the elongation of the true Moon from the true Sun one twelfth of it and divide the result by the hourly lunar velocity. In al-Battānī's zij true hourly lunar velocity is specified whereas here we are only told to divide by the hourly lunar velocity (but see 31:17, below).

30:25. Here we are told that the tables for conjunctions and oppositions use the meridian of Toledo.

Chapter 31

31:1. This chapter concerns parallax.

31:3. The table for parallax (*diversidad del catamiento*) mentioned in this sentence, and described in the passage following it, is of the same type as that found in the *Handy Tables*: it is composed of 12 sub-tables, one for each zodiacal sign, giving the components in longitude and latitude for each integer hour before and after noon, when the Moon is at mean distance. The table is to be used when the Moon is in conjunction with the Sun (*en las horas del eclipsy*, as mentioned in 31:1), and the entries are actually the difference between lunar and solar parallax. This type of table is also found in al-Battānī's zij (Nallino 1903–1907, 2:95–101) and in the Toledan Tables (Toomer 1968, pp. 97–112; F. S. Pedersen 2002, pp. 1351–1408).

31:4. The entries for both components of parallax are given in minutes of arc.

31:11. The text refers to a table called *tabla del endereçamiento* used to correct the components of parallax when the Moon is not at mean distance.

31:17–18. The method to determine the time from mean to true syzygy is again presented, here specifying true hourly lunar velocity.

Madrid, Biblioteca Nacional, MS 10053, f. 10v, has a table of the kind mentioned in 31:3, but it has neither a title nor headings. Only the sub-tables for Cancer and Leo are reproduced in Table 31A, where the components of parallax are given in arcminutes. The entries are identical to those found in a parallax table for geographical latitude 41;44° and longest daylight 15 hours (Toomer 1968, p. 104).

Table 31A: Parallax

Cancer	**Parallax**		**Leo**	**Parallax**	
Time (h)	**(min)**	**(min)**	**Time (h)**	**(min)**	**(min)**
–7;30	39	38	–7;18	46	28
–6	41	32	–6	48	23
–5	40	28	–5	45	21
–4	36	24	–4	41	19
–3	29	21	–3	33	17
–2	20	19	–2	24	17
–1	10	18	–1	15	17
Noon	0	17	Noon	5	19
1	10	18	1	6	22
2	20	19	2	15	26
3	29	21	3	17	29
4	36	24	4	29	33
5	40	29	5	32	38
6	41	32	6	32	41
7;30	49	38	7;18	32	42

Chapter 32

32:1. The title refers to parallax for any arbitrary time (in contrast to the table of parallax described in the previous chapter that was only valid at the time of conjunction for a solar eclipse) and for all places, specifically those with geographical latitudes greater than 23;33°. For this value of the obliquity of the ecliptic see the comments on chapter 20. For an illustration of lunar parallax as a function of the Moon's zenith distance see Fig. 4.4.

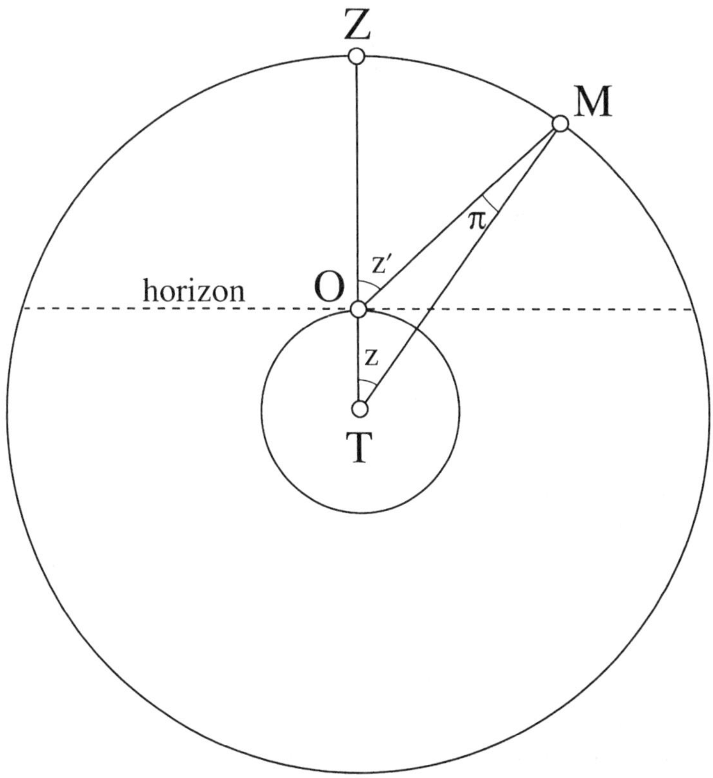

Fig. 4.4. Lunar parallax: T is the center of the Earth, O is the observer, Z is the zenith for that observer, and M is the position of the Moon. The true zenith distance of the Moon is z (= ∠ZTM) and the apparent zenith distance of the Moon is z′ (= ∠ZOM). The lunar parallax, π, is the difference between angles z′ and z. TM is the true distance from the Earth to the Moon, and it is measured in units of Earth radii (OT).

32:2–20. This is a long description of the way to find the altitude of the Moon for a given time, a given longitude, and a given place: cf. al-Battānī, chapter 39 (Nallino 1903–1907, 1:79, 256).

32:5. The text refers to the *comienço natural,* i.e., the zero point for counting tropical longitudes or the vernal equinox (see 49:2, below).

32:21–43. This is a description of the way to find the angle between the altitude circle and the ecliptic at the position of the Moon: cf. al-Battānī's zij, chapter 39 (Nallino 1903–1907, 1:79–81).

32:44–56. The table for parallax in altitude described here is the 9-column parallax table found in *Almagest* V.18 and in al-Battānī's zij (Nallino 1903–1907, 2:93–94). This table is also found in the *Libro de las armellas* just after chapter 40 on parallax (Madrid, Universidad Complutense, MS 156, f. 147r; cf. Rico 1863–1867, vol. 2, p. 57; Cárdenas 1974, p. 1123), and in Madrid, Biblioteca Nacional, MS 10053, f. 9r. The description of the procedure to compute the lunar parallax by means of this table follows very closely that presented in *Almagest* V.19. For a survey of parallax theory in Islamic astronomy, see Kennedy 1956b.

32:57–61. This is a description of a computation of the components of parallax in longitude and latitude; the parallax in longitude is then converted from an arc into time. With the parallax in longitude, measured in hours, the time of true conjunction of the Sun and the Moon can be corrected to the time of apparent conjunction.

32:59. The expression *zaba de la luna* probably means hourly velocity in elongation. The term *zaba* may derive from the Arabic *sabaq* used by al-Battānī with this meaning in chapter 42 of his zij: see Nallino 1903–1907, 1:94, 2:336.

32:62–65. This is a description of lunar latitude, adjusted for parallax.

32:63. As in 21:3, the term *geuzaar* (Arabic: *jawzahar*) refers to the lunar node.

32:66. Here we are told that this entire computation is only of interest when there is a solar eclipse.

See Table 32A for an excerpt of the table for parallax in altitude found in Madrid, Biblioteca Nacional, MS 10053, f. 9r, where

Table 32A: Parallax in altitude

1 (°)	2 (°)	3 (°)	4 (°)	5 (°)	6 (°)	7 (°)	8 (°)	9 (°)
2	0; 7	1;54	0;23	0; 3, 0	0;50	0;14	0;11	0;15
4	0;13	3;48	0;45	0; 7, 0	1;40	0;28	0;22	0;30
...								
88	2;51	53;33	10;17	1;18,20	24;50	59;47	59;45	59;50
90	2;51	53;34	10;17	1;19, 0	25; 0	60; 0	60; 0	60; 0

the headings, here labeled (1)–(9), correspond to those in *Almagest* V.18 (see Toomer 1984, p. 265).

Chapter 33

33:1: This chapter concerns solar eclipses.

33:2. The limits for a solar eclipse are given as the intervals in the argument of lunar latitude from 1° to 12° and from 168° to 180°. In the canons to the Toledan Tables (F. S. Pedersen 2002, pp. 456–457), the same limits are given in the same way and differ from those in earlier texts that we have examined (e.g., Ptolemy, al-Battānī, and al-Khwārizmī). It is noteworthy that these limits only apply when the lunar latitude is northern; nothing is said about these limits when the lunar latitude is southern.

In John of Lignères's *Priores astrologi*, chapter 33 (Saby 1987, 1:226), these limits are reported as from 0° to 12° and from 5s 18° to 6s 0°. In some manuscripts of this text, e.g., Liège, Bibliothèque de l'Université, MS 354C, f. 181rb, this chapter is numbered 32 (Chabás 1997).

According to *Almagest* VI.5, there are different limits in the argument of lunar latitude for solar and for lunar eclipses. For the Sun there is no symmetry: for northern lunar latitudes the limits are 17;41° from the nodes, whereas for southern lunar latitudes the limits are 8;22° from the nodes. On the other hand, Ptolemy's limit for the argument of lunar latitude for lunar eclipses is symmetrical about the nodes and in all cases it is 12;12°. What distinguishes solar from lunar eclipses in this context is that parallax only affects solar eclipses. Hence the silence on the eclipse limits when the lunar latitude is southern does not imply that symmetry around the nodes was assumed.

33:4. Read from 1° to 24° (instead of 1° to 3°) as in 33:10 and in the canons to the Toledan Tables: the lunar parallax in latitude is always southern at the time of a solar eclipse for places north of geographical latitude 24° (F. S. Pedersen 2002, pp. 458–459). As is the case here, the corresponding remark in the canons to the Toledan Tables comes in the chapter on solar eclipses, not in the chapter on parallax.

33:10. The factor $11^1/_2$ converts parallax in latitude, π_β, into a correction to ω, the argument of latitude. This factor appears in the Toledan Tables, and it can be explained from the following formula:

$$\Delta\omega = \pi_\beta \,/\, \sin i,$$

where i, the inclination of the lunar orb to the ecliptic, is 5°, and thus $\Delta\omega = 11;28^\circ\ \pi_\beta$ (cf. Toomer 1984, p. 311 n. 72).

33:13–14. The text refers to two tables for solar eclipses at greatest distance and least distance, respectively; each of them displays two columns of entries: digits of eclipse and minutes of immersion (*caymiento*). The argument for these tables is the argument of lunar latitude at the middle of the eclipse (33:12). This means that the arrangement of these tables is the same as in *Almagest* VI.8, the zij of al-Khwārizmī (Suter 1914, p. 193), and some of the tables associated with the Toledan Tables (Toomer 1968, pp. 86–87); but differs from that in the zij of al-Battānī (Nallino 1903–1907, 2:91) and the Almanac of Azarquiel (Millás 1943–1950, p. 234).

33:18. The text refers to an interpolation table to be used for computing eclipses when the Moon is neither at greatest nor at least distance, as in the Toledan Tables (Toomer 1968, pp. 116–117).

33:20. The text refers to a table for the digits of the diameter and such a table was included in the Toledan Tables (Toomer 1968, p. 113).

This chapter is closely related to the chapter "On finding the time of a solar eclipse" in the canons to the Toledan Tables (F. S. Pedersen 2002, pp. 456–461), and one finds whole sentences that are identical:

Canons to the Alfonsine Tables:	Canons to the Toledan Tables:
(33:5) E despues sabe el lugar de la conjunçion eguada y el movimiento de la ladesa ques yguada con la yucaçion de la luna y con la longura y su dosena parte.	(Cb 187) Scire etiam te oportet locum coniunctionis aequatum, et argumentum latitudinis aequatum per longitudinem et eius duodecimam partem (...)
(33:10) E despues toma los menudos de la diversydad del catamiento en la ladesa y es todavia meredional en Toledo y en cada çibdad cuya anchura fuere mas de 24 grados multiplicados en 11 grados y medio. y guarda lo que se juntare de grados y de menudos.	(Cb189) Postea vero multiplica minuta diversitatis aspectus in latitudine—quae semper est meridiana in omnibus regionibus quarum latitudo fuerit maior XXIIII gradibus—in XI et dimidio, et quod exinde provenerit minue de (...)

Both texts describe an iterative procedure for finding the time of the apparent conjunction (*conjunçion visible*/*coiunctio visibilis*), that is, the true conjunction adjusted for parallax (cf. *Almagest* VI.10; Toomer 1984, pp. 310–311).

Note that the expressions *movimiento de la ladesa* (33:4) and *movimiento de la latitud* (33:9) are used for "argument of latitude" instead of the expected *argumento de latitud*.

It is thus most likely that both texts have a common source which we presume to be the lost Arabic version of the canons to the Toledan Tables.

Chapter 34

34:1: This chapter concerns lunar eclipses.

34:3–4. The text refers to two tables for lunar eclipses at greatest distance and least distance, respectively; each of them has three columns of entries: digits of eclipse, minutes of immersion (*caymiento*), and half totality (*medio de la mora*), corresponding to the arrangement in the Toledan Tables (Toomer 1968, pp. 91–93).

34:7. The text refers again (as in 33:18) to an interpolation table to be used for computing eclipses when the Moon is neither at greatest nor at least distance.

The rules for computing the circumstances of a lunar eclipse agree quite closely with those in the canons to the Toledan Tables (see F. S. Pedersen 2002, pp. 466–471).

Chapter 35

35:1. In this chapter we are given two different criteria for assigning a color to an eclipse, without specifying whether they are valid for both lunar and solar eclipses or just for one type of eclipse.

The tradition for assigning colors to eclipses has its ultimate roots in the Babylonian omen literature going back to the series of tablets called *Enūma Anu Enlil* that was compiled in the 2nd millennium; this series continued to be copied for as long as the Babylonian scribes remained active, i.e., until about the 1st century AD (Rochberg 1988, espec. pp. 27, 105). Ptolemy referred to the colors of eclipses in the *Tetrabiblos* II.9 (Robbins 1940, pp. 191–193) and in the *Optics* II.93 (Smith 1996, p. 108; Petri 1968, p. 97), without giving any quantitative criteria. Indian astronomers often discussed colors of eclipses: they made the color depend on the magnitude of the eclipse (Brahmagupta, in the case of the lunar eclipses; see Pingree 1976, p. 166), its phases (*Āryabhaṭīya*, Gola 46; see Clark 1930, p. 81), or the altitude of the eclipsed body. The zij of al-Khwārizmī probably dealt with colors of eclipses, for al-Bīrūnī (d. 1048), *India*, chapter 59, refers explicitly to it (trans. Sachau 1910, 2:114). This is confirmed by Ibn al-Muthannā (10th century) in his *Commentary on the Astronomical Tables of al-Khwārizmī*, where he states that, in the case of a lunar eclipse, the color varies during the eclipse (Goldstein 1967, pp. 234–235).

35:2–7. The first set of criteria uses lunar latitude as the independent variable: see Table 35A.

Kennedy (1971, pp. 71–72) has noted that al-Bīrūnī, in *al-Qānūn al-Mas*ᶜ*ūdī*, VIII.7 (section 2), and VIII.11, deals with the colors of lunar and solar eclipses. Al-Bīrūnī gives a sequence of colors for lunar eclipses corresponding to six intervals of 0;10° in lunar latitude beginning at 0°: deep-black, greenish, reddish, yellowish, dusty, and "similar to dust" (perhaps to be read as "gray": cf. Petri 1968, p. 97), a graduation that corresponds very closely to that given in 35:3–7. This line of transmission probably reached Ibn al-Kammād in whose *al-Muqtabis*, uniquely preserved in a Latin version in Madrid, Biblioteca

Table 35A: Colors of eclipses

Lunar latitude (minutes)	**Color**
1 to 10	very black (*bien prieto*)
10 to 20	greenish
20 to 30	reddish
30 to 40	yellowish
40 to 50	bright
50 to [60]	brighter

Nacional, MS 10023, there is such a table (Chabás and Goldstein 1994, pp. 18–19). Some time later, in an anonymous handbook, based on the tables of Ibn Isḥāq al-Tunīsī, we also find tables for the colors of both solar and lunar eclipses (Mestres 1996, p. 419).

35:8–12. The second set of criteria uses the lunar anomaly as independent variable: see Table 35B.

This list is very similar to the list for the colors of lunar eclipses according to al-Ṣūfī (10th century) as reported in his *Kitāb al-ᶜAmal bi'l-Asṭurlāb*, chapter 151, where the argument of latitude is the independent variable (Kennedy and Destombes 1966, p. 9).

Table 35B: Colors of eclipses

Lunar anomaly	**Color**
6 signs	black (*prieto*)
5 or 4 signs	greenish
3 or 9 signs	yellowish
2 or 10 signs	bright (*abierto*)
1 or 11 signs	bright "towards clarity"

This evidence, taken altogether, seems to indicate that, in this regard, the astronomers of Alfonso X followed Hindu traditions transmitted by al-Ṣūfī and al-Khwārizmī.

In the canons to the Toledan Tables there is a discussion of colors of eclipses, but no table, that corresponds to the two lists, shown above. See Version Cc, Section 154, in F. S. Pedersen 2002, pp. 652–653:

> When you want to know the colours of an eclipse, consider the latitude of the moon. If it is from one minute to 10, the eclipse will be entirely black (*nigerrima*); if it is up to 20 minutes, it will be black containing greenness (*nigra habens in se viriditatem*); and if it is up to 30, it will be black with redness (*nigra cum rubedine*); and if it is up to 40, it will be black with pallor (*nigra cum pallore*); and if it is up to 50, it will be of grey colour (*grisii coloris*); and if it exceeds 50, it will be grey with whiteness (*grisia cum albedine*).
>
> There is also another way to do it which is related to the first way when we investigate the knowledge of the colour [of the moon]. According to this, you are to consider: if the place of the moon is 6 signs greater than its apogee, it will be entirely black; and if it is 5 or 6 signs greater than its apogee, it will be black with greenness; and if it is 4 or 8 [signs] greater than its apogee it will be black with redness; and if it is 3 or 9 signs greater than its apogee, it will be black with pallor; and if it is 2 or 10 signs greater than its apogee, it will be grey; but if it is 1 or 11 signs greater than its apogee, it will be grey with whiteness, if God wills.

Moreover, in the same tradition, Abraham Ibn ʿEzra devoted a short chapter of about 9 lines to the color of eclipses in his *Book on the Foundations of the Astronomical Tables* (Millás 1947, p. 167), but there are no numerical criteria in it. The text shows that Ibn ʿEzra was well acquainted with the works of al-Khwārizmī, al-Ṣūfī, and Ibn al-Muthannā, among many others. The Latin version is preserved in various manuscripts, and amongst them is Madrid, Biblioteca Nacional, MS 10053, a manuscript mentioned several times above in relation to some tables associated with the "original" Alfonsine Tables. The subject of the color of eclipses is also found in the *Ars inveniendi eclipsim solis et lunae* by Robert Grosseteste (d. 1253), and it gives the same two sets of criteria as in the Castilian canons to the

Alfonsine tables (Lértora 1989). Both Ibn ʿEzra and Grosseteste probably depended on the tradition of al-Khwārizmī. Finally, in the *editio princeps* of the Alfonsine Tables (Ratdolt 1483, f. 18v) there are two lists for the colors of eclipses.

In sum, we have a long literary tradition for assigning colors to eclipses by one or another arbitrary set of criteria, unrelated to any observational data. To be sure, there is a variation in the colors of lunar eclipses, but modern explanations depend on completely different criteria. On the other hand, the claim that there is a variation in the colors of solar eclipses has no basis in fact, and was probably introduced simply on analogy with lunar eclipses.

Chapter 36

36:1. This chapter concerns the visibility of the lunar crescent. For the theory of lunar crescent visibility in Ibn Yūnus (*ca.* 1000) and an analysis of a large number of tables constructed for this purpose, see King 1987 and 1988.

36:9–11. The criteria for visibility are:

$$9° < \lambda_m - \lambda_s < 15°, \text{ for Cap } 1° < \lambda_m < \text{Gem } 30°$$
$$10° < \lambda_m - \lambda_s < 24°, \text{ for Cnc } 1° < \lambda_m < \text{Sgr } 30°$$

where λ_m and λ_s are the longitudes of the Moon and the Sun. These limits for the elongation are identical to those given for the same purpose in Ibn al-Raqqām's zij, chapter 137 (Kennedy 1997, p. 53). There is no chapter on lunar crescent visibility in the canons to the Toledan Tables.

36:13. The text refers to a *tabla del desviamiento de la luna*, and to a procedure that is rather close to the one described by Maimonides (see Neugebauer 1956, pp. 136–146).

36:18. The "arc of setting" (*arco del ponimiento*) is defined as

$$\sigma(\lambda) = \rho(\lambda + 180) - 180$$

where $\sigma(\lambda)$ is the setting time of λ and $\rho(\lambda)$ is its rising time.

36:19–20. The "arc of vision" (*arco de la vista*) is obtained by adding (subtracting) to (from) the arc of setting the northern (southern) apparent latitude of the Moon times the tangent of the geographical latitude. The "arc of vision" is not to be confused with the expression *arcus visionis*, which refers to the arc on an altitude circle between the Sun (below the horizon) and the horizon when a planet or a star becomes visible following (or preceding) conjunction with the Sun.

36:21–26. The criteria for lunar visibility stated in the text can be summarized as follows (where η is the elongation from the corrected longitude of the Moon to the true position of the Sun):

arc of vision $< 10°$, the Moon is not visible

arc of vision $\geq 10°$, the Moon is visible if $\eta \geq 13°$

arc of vision $\geq 12°$, the Moon is visible if $\eta \geq 11°$

arc of vision $\geq 13°$, the Moon is visible if $\eta \geq 10°$

arc of vision $\geq 14°$, the Moon is visible.

These criteria are almost the same as those given by Maimonides (*Sanctification*, 17:15–21; Gandz et al. 1956, pp. 71–72), although Maimonides adds another criterion for a value of the arc of vision greater than 11° that seems to have been dropped from our text. The second condition is identical with that in the zij of Ibn al-Raqqām (Kennedy 1997, p. 53). As far as we have been able to determine, the criteria given by Maimonides are closer to the Castilian Alfonsine canons than those given in any other medieval text. Maimonides's source for these criteria has not been identified, and it may have been independently available to Alfonso's astronomers (cf. Neugebauer 1949, p. 360).

Chapter 37

37:1. The purpose of this chapter is to explain the construction of a trigonometric table and its use. The text generally follows the method presented by Ptolemy in *Almagest* I.10 (for a modern version of it, see Aaboe 1964, pp. 112–125). The basic terminology for trigonometric functions already appears in the title of the chapter: chord (*querda* or *corda*); sine (*signo*); cosine ([*signo del*] *complimiento*); and versine or sagitta (*saeta*). Note that the text also uses the term *signo* for "[zodiacal] sign" (e.g., 15:5). In the comments that follow, the initial letter of a trigonometric function is capitalized when normed to 60, rather than to 1 (see Fig. 4.5).

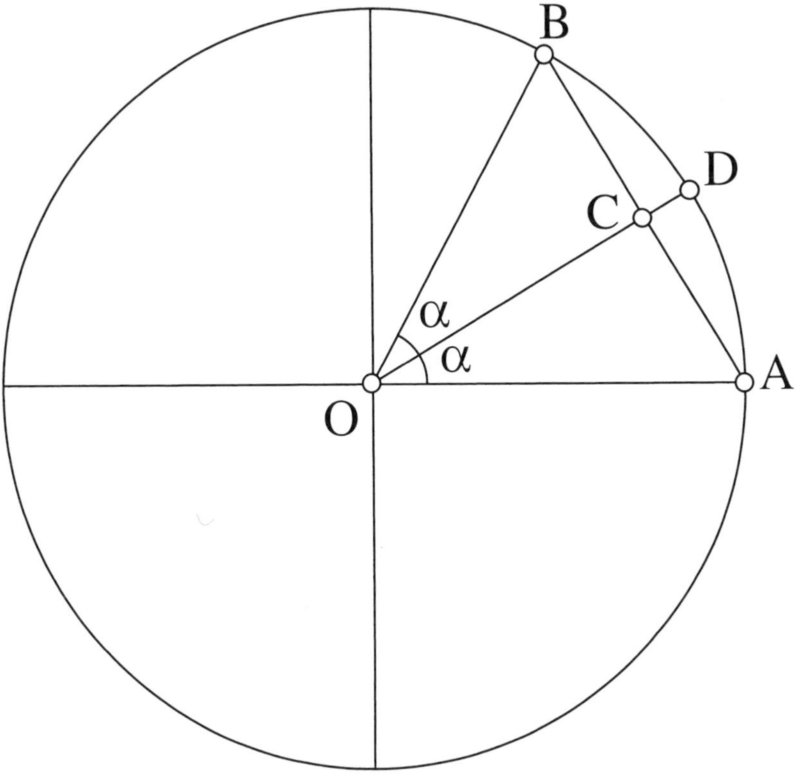

Fig. 4.5. Trigonometric functions: when OA = 60, AB = Crd 2α, AC = Sin α, OC = Cos α, and CD = Vers α = 60 – Cos α.

37:2–5. The ratio of the circumference to the diameter is given in the name of Ptolemy as 3;8,30 (equivalent to 3.14166...), and this is indeed the value for π reported by Ptolemy in *Almagest* VI.7 (Toomer 1984, p. 302).

37:6–7. Ptolemy is cited for the number of parts in the diameter (120) and in the circumference (360), as in *Almagest* I.10.

37:12–13. $(120^2 + 60^2)^{1/2}$ = 103;55,23 and this is indeed the value for Crd 120° in Ptolemy's table (*Almagest* I.11).

37:15. Crd 36° = 37;4,55 as in *Almagest* I.11.

37:16. Crd 144° = 114;7,36 (but 114;7,37 in *Almagest* I.11).

37:17. Crd 30° = 31;3,30 as in *Almagest* I.11.

37:21. The expression used here for "supplementary arc" is *arco con que se cumple a medio çerco*.

37:21–24. These sentences describe a special case of what is usually called Ptolemy's theorem (Toomer 1984, p. 53) which, in modern notation, corresponds to

$$\cos(\alpha + \beta) = \cos\alpha \cdot \cos\beta - \sin\alpha \cdot \sin\beta.$$

37:25. Crd 168° = 119;20,33 (but 119;20,34 in *Almagest* I.11).

37:26. Crd 12° = 12;32,36 and Crd 6° = 6;16,49 as in *Almagest* I.11.

37:27. Crd 3° = 3;8,28 as in *Almagest* I.11.

37:28. Crd $1^1/_2°$ = 1;34,15 as in *Almagest* I.11.

37:29. Crd $^3/_4°$ = 0;47,8.

37:31. Crd 1° = 1;2,5 is a scribal error for 1;2,50 (cf. *Almagest* I.11).

37:36,40. It is explicitly stated that the Castilian Alfonsine Tables included a table of chords (*cordas complidas*), sines, cosines, and versines. As the examples above indicate, the entries for the chords were given to three significant places. A table with columns for sines, cosines, and versines is found in the Almanac of Azarquiel (Millás 1943–1950, p. 229), and a table with columns for chords, cosines, and versines in the zij of Ibn al-Kammād (Madrid, Biblioteca Nacional, MS 10023, f. 46v). But no such table is found in the zij of al-Battānī or in the Toledan Tables.

37:44. The text refers to a figure, not extant in the manuscript, to illustrate the cosine of an angle.

37:45. The text describes an expression that is equivalent to the modern formula, Sin α = (Crd 2α)/2. From here to the end of the chapter, the text explains the use of a table in which the range of the argument seems to have been 0° to 180° (37:53,58).

The *Libro del relogio de la piedra de la sombra,* also authored by Isaac ben Sid, has a table of sines (not chords) for all integers from 1° to 90° (Madrid, Universidad Complutense, MS 156, f. 173r; cf. Rico 1863–1867, vol. 4, p. 10; and Cárdenas 1974, p. 1128). The entries, given to minutes, are displayed under columns headed *el sino,* and they are easily recomputed from Ptolemy's table of chords (*Almagest,* I.11) by means of the expression in 37:45. No other trigonometric function appears in this text. But in the Almanac of Azarquiel (Paris, Bibliothèque de l'Arsenal, MS 8322, f. 127r; cf. Millás 1943–1950, p. 229), there is a table of Sines, Cosines, and Versines at 3°-intervals: see Table 37A.

Chapter 38

38:1. In this chapter some trigonometric problems are considered.

38:2–4. To determine the observer's latitude, φ, by means of the observed meridian altitude of the Sun, h, and the solar declination, δ, found in a table (see chapter 20), an expression, equivalent to the modern

Table 37A: Trigonometric functions

(°)	Sine	Cosine	Versine
3	3; 8	59;45(*sic*)	0;35(*sic*)
6	6;16	59;40	0;24(*sic*)
...			
30	30; 0	51;58	8; 2
...			
60	51;58	30; 0	30; 0
...			
90	60; 0	0; 0	60; 0
93			63; 8
...			
180			120; 0

formula, is described:

$$\varphi = 90° - h + \delta,$$

where δ is sometimes positive and sometimes negative. This is a typical problem in spherical trigonometry, quite common in the astronomical literature. It is also found, for instance, in some other books written or translated by the Alfonsine astronomers: the *Libro de la açafeha,* chapter 9 (see Rico 1863–1867, vol. 3, p. 170; Puig 1987b); the *Libro de la lamina universal,* Book II, chapter 20 (see Rico 1863–1867, vol. 3, p. 78); and the canons of al-Battānī's zij, chapter 14 (Nallino 1903–1907, 1:29–30), as well as in the Castilian translation of these canons, made by the astronomers of Alfonso X (Bossong 1978, p. 71).

38:5. The observer's latitude can also be determined by observing the highest and lowest altitudes, h_1 and h_2, of a star near the celestial north pole. The instructions are equivalent to the modern expression:

$$\varphi = (h_1 + h_2)/2.$$

The same instructions are also found in Book II, chapter 31, of the *Libro del astrolabio llano* (Rico 1863–1867, vol. 2, p. 276); and in chapter 18 of the *Libro del quadrante con que rectifican* (Rico 1863–1867, vol. 3, p. 316).

38:6–9. The inverse problem to that in 38:2–4 is presented here: given the geographical latitude, φ, determine the meridian altitude of the Sun, h. The solution is, of course:

$$h = 90° - \varphi + \delta,$$

where δ may take on both positive and negative values.

This problem is discussed in the canons to the Toledan Tables (F. S. Pedersen 2002, pp. 408–409), and in chapter 15 of the canons to al-Battānī's zij (Nallino 1903–1907, 1:30). In the translation into Castilian made by the Alfonsine astronomers, we read (Bossong 1978, pp. 71–72):

> Quando quisieres saber que tanta sera la altura del Sol en la hora del medio dia en qual dia quier del anno,

> sabe la declinacion del grado do es el Sol. Et si fuere septe[n]trional mengua la de la ladeza del logar, y si fuere meridional: annade la sobre la ladeza del logar. Et lo fuere de la ladeza despues del annadimiento, o del menguamiento: mengua lo de .XC. et lo que fincare; essa sera la altura del Sol en la hora del medio dia.

This is similar, but not identical, to sentences 38:8–9. Moreover, Ibn Sid had already formulated the same problem in chapter 1 of his *Libro del quadrante sennero,* extant in Paris, Bibliothèque de l'Arsenal, MS 8322 (Millás 1960, pp. 241–242):

> Et si tu quisieres saber que tanta es su altura en el cerco de mediel dia, annade so arredramiento sobre la altura de la Cabeça de Aries que a en el logar do tu fueres, si fuere ell arredramiento septentrional, et mengua lo della si fuere meridional. Et lo que fuere della despues dell annadimiento o del menguamiento, esso sera su altura en el çerco de mediel dia.

The three fragments presented here deal with the very same problem and were written in the same language, probably by the same person, who reformulated the problem each time, rather than repeating himself.

Chapter 39

39:1. This chapter addresses a problem of transforming ecliptical coordinates to equatorial coordinates, specifically how to compute the declination of a celestial body from its longitude and latitude. In 39:1 declination is called "the distance from the equator" (*alongamiento del eguador del dia*), but in 39:3 and 41:2 an equivalent expression (*arredramiento del equinoçio)* is used. See Fig. 4.6 for an illustration of the three great circles that define the three coordinate systems used for fixing positions on the celestial sphere: the horizon, the ecliptic, and the equator. In each case, the second coordinate (called altitude, latitude, and declination, respectively) is the distance from a given point on the celestial sphere to the circle that defines that system along a perpendicular to that circle.

39:2–3. The authors refer to a list of stars compiled by them that included stellar declinations among other quantities.

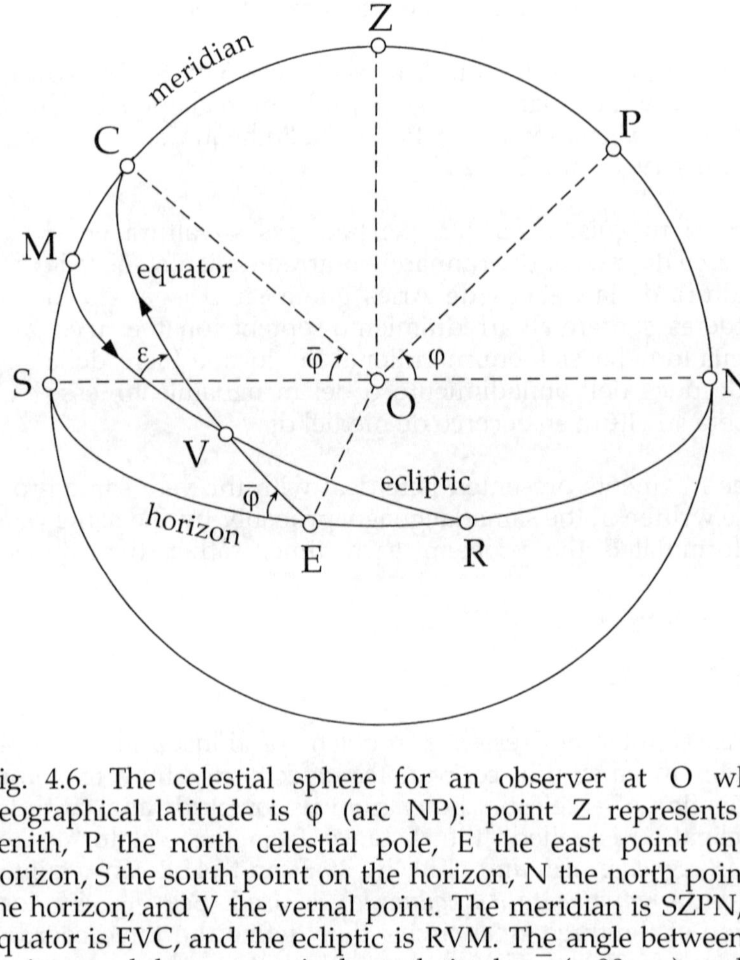

Fig. 4.6. The celestial sphere for an observer at O whose geographical latitude is φ (arc NP): point Z represents the zenith, P the north celestial pole, E the east point on the horizon, S the south point on the horizon, N the north point on the horizon, and V the vernal point. The meridian is SZPN, the equator is EVC, and the ecliptic is RVM. The angle between the horizon and the equator is the co-latitude, $\bar{\varphi}$ (= 90 – φ), and the angle between the equator and the ecliptic (called the obliquity of the ecliptic) is ε. Note that the positive sense on the ecliptic (marked by an arrow) is contrary to the daily rotation about the celestial pole (marked by an arrow on the equator).

39:5. Since this sentence is parallel to 39:9 below, it should be: "(...) e no oviere ladesa en (*read*: el) su arredramiento es tanto del equinoçio (*read*: del equinoçio es tanto) como toda la declinaçion y es meredional."

The aim of the whole chapter is to present and explain a procedure equivalent to the algebraic expression:

$$\sin \delta = \frac{\sin (\beta + \beta') \cdot \cos \varepsilon}{\cos \beta'}$$

where δ is the declination of the celestial body, β is its latitude, ε is the obliquity of the ecliptic, and β' is the latitude of a point on the equator that has the same longitude as the celestial body. Both β and β' can be positive or negative.

39:17. The computation of β' requires the use of a table of the *subimientos del çerco derecho*, that is, a table for the rising times at *sphaera recta*. The argument for this table is $\lambda + 90°$, and it is called *grados yguales* (cf. 28:9–10).

The terminology used in this chapter is quite confusing: β is called *ladesa* (39:4), and ε *toda la declinaçion* (39:5), but the declination δ is called *arredramiento* (39:3), and β' *declinaçion* (39:17).

As was the case in the previous chapter, the subject treated in chapter 39 is also addressed in the canons to al-Battānī's zij (chapter 18), translated into Castilian by Alfonso's astronomers, and in Ibn Sid's *Libro del quadrante sennero* (chapter 1). Almost the same terminology is used in all three Castilian treatises, although the last two use *declinaçion del grado do es* for β'. In the *Libro del quadrante sennero*, $\beta + \beta'$ is called generically *arco salido*, whereas in the canons of al-Battānī and in those for the Castilian Alfonsine Tables it is called *ladesa eguada* (39:18). As pointed out by Nallino (1903–1907, 1:32 n. 1), the expression *declinatio gradus in quo stella est* already appears in the Latin translation of al-Battānī's canons.

Chapter 40

40:1. This chapter explains how to compute the diurnal arcs above the horizon described by the Sun, the Moon, the planets, and the stars at any time of the year and at any geographical place.

40:2. The text calls for a table of the *subimientos de aquella villa*, that is, a table for the rising times at *sphaera obliqua* for the latitude of a given geographical place. Toledo is explicitly mentioned.

40:4. The length of daylight, $d(\lambda)$, when the Sun is at λ, and $\rho(\lambda)$ is its oblique ascension for a given geographical place, is properly defined as:

$$d(\lambda) = \rho(\lambda + 180°) - \rho(\lambda)$$

(Neugebauer 1975, p. 40). The length of night, $n(\lambda)$, is correctly defined in the text as:

$$n(\lambda) = 360° - d(\lambda).$$

40:11. The computation requires the use of a table of declinations, already mentioned in chapter 20.

The instructions given in this chapter to compute the arc described by a celestial body above the horizon is equivalent to the algebraic expression:

$$P = 90° - \arcsin\left(\frac{\sin\varphi \cdot \sin\delta}{\cos\varphi \cdot \cos\delta}\right),$$

that is,

$$\cos P = \tan\varphi \cdot \tan\delta,$$

where P is the arc sought, φ is geographical latitude of the observer, and δ is the declination of the celestial body observed.

As was the case in chapters 38 and 39, the subject treated here is also examined in the canons to al-Battānī's zij (chapters 13 and 19), translated into Castilian by the Alfonsine astronomers, and in Isaac ben Sid's *Libro del quadrante sennero* (chapter 3).

Chapter 41

41:1. This chapter concerns risings, settings, and culminations of the Moon, the planets, and the fixed stars.

41:2. The expression *lugar natural* refers to tropical longitude (see chapter 49), and *parte* here means direction, either north or south.

41:3–7. We are given rules for the meridian crossing of a star in contrast to the meridian crossing of the point on the ecliptic which has the same longitude as that of the star.

41:8–9. We are given a rule for using a table of the *subimientos del çerco derecho*, that is, a table for normed right ascension, α', beginning with Cap 0°, where $\alpha'(\lambda) = \alpha(\lambda) + 90°$. Therefore, if for instance, $\lambda = 270°$, then its right ascension $\alpha(270°) = 270°$ and its normed right ascension $\alpha'(270°) = 0°$. The *alongamiento primero* mentioned in 41:9 stands for the right ascension: $\alpha(\lambda) = \alpha'(\lambda) + 270°$.

41:10–13. The authors incorrectly give an expression for the ascensional difference, γ, here called *arco de la diversidad*. The correct procedure is described in chapter 46:7–14 (see also Nallino 1903–1907, 1:187–189; Neugebauer 1962a, pp. 47–48; and Goldstein 1967, pp. 202–205).

41:14–16. We are given the correct rules for finding the oblique ascension (*alongamiento eguado*), ρ, as the algebraic sum of α and γ. Then we are given an incorrect rule for finding the culminating point that ought to be

$$\rho(\lambda) = \alpha'(\lambda_M)$$

where λ is the longitude of the rising point on the ecliptic (point R in Fig. 4.6), and λ_M is the longitude of the point on the meridian (point M in Fig. 4.6) at the same time (*grado con que se para la estrella en medio çielo*).

41:17–21. These paragraphs concern the risings, settings, and culminating points of the stars, the Moon, and the planets, for a given locality. Sentence 21 explicitly mentions a table "that we compiled" for the fixed stars that displays for the latitude of Toledo their risings, settings, and culminations. As far as we know, no such table survives, but among the Toledan Tables is a star list, one column of which has the heading *gradus cum quo celum mediant* (Kunitzsch 1966, pp. 90–91; Toomer 1968, pp. 128–131). There is also a description of the rules for computing these three quantities in the zij of Ibn al-Raqqām, as reported by Kennedy (1997, p. 49).

Chapter 42

42:1. This chapter deals with the problem of finding the meridian altitude of a planet or a fixed star.

42:2. In order to determine the meridian altitude, one has to compute

$$h = 90° - \varphi + \delta,$$

as was the case for the meridian altitude of the Sun at noon (38:6–9).

The authors remark that "our" star list (*nuestra tabla*) included values for the meridian altitude at Toledo of each star, whereas in 39:2–3 they said that the list included declinations. Other quantities in this table are mentioned in 41:21. It may be of interest to note that the star list in the Parisian version of the Alfonsine Tables (Ratdolt 1483) does not display stellar declinations, but among the Toledan Tables is a star list that does include declinations (Kunitzsch 1966, pp. 87–94; F. S. Pedersen 2002, pp. 1494–1497).

Chapter 43

43:1. This chapter addresses the problem of determining the time of day from the solar altitude for a given locality.

43:2–5. The procedure described is equivalent to the expression:

$$\text{Vers } t = \text{Vers } d - \frac{\text{Sin } h_t \cdot \text{Vers } d}{\text{Sin } h_m}$$

where t is the hour angle, h_t is the solar altitude at time t, h_m is the meridian solar altitude on that day, and d is half the diurnal arc.

This formula is of Indian origin and appeared in the zij of al-Khwārizmī according to Ibn al-Muthannā (Goldstein 1967, pp. 83, 208–209; Charette 1998, p. 27). For a discussion of this problem and an explanation of its antecedents, see Nadir 1960. For Ibn Yūnus's treatment of timekeeping, see King 1973.

This problem is also analyzed in chapter 16 of the canons to al-Battānī's zij (Nallino 1903–1907, 1:30–31, 189–191) and is therefore included in the Alfonsine translation of it (Bossong 1978, pp. 72–73). The inverse problem, that is, determining the solar altitude at a given time of day, is explained in chapter 4 of the *Libro del relogio de la piedra de la sombra,* another Alfonsine treatise where the same formula is used (see Rico 1863–1867, vol. 4, p. 7).

Chapter 44

44:1. This chapter deals with the determination of the time of night for a given locality, given the altitude of a star.

44:2. The procedure described here is analogous to that in chapter 43, and the expression required is

$$\text{Vers } t = \text{Vers } d - \frac{\text{Sin } h_t \cdot \text{Vers } d}{\text{Sin } h_m}$$

where h_t here is the altitude of the star at time t, h_m is the meridian altitude of the star, and d is half the diurnal arc of the star.

This topic is also analyzed in chapter 16 of the canons to al-Battānī's zij (see Nallino 1903–1907, 1:33–34, 195–196) and is therefore included in the Alfonsine translation (Bossong 1978, pp. 80–81).

Chapter 45

45:1. This short chapter explains how to determine the number of equal hours in the time from sunrise to sunset (the length of daylight).

45:2. The text mentions a specific table for Toledo where the entries are the numbers of equal hours and the argument is the solar longitude. This table is found in Madrid, Biblioteca Nacional MS 10053, f. 13r–v. For other localities we are told to compute the diurnal arc in order to derive the length of daylight in equal hours.

See Table 45A for the table in Madrid, Biblioteca Nacional, MS 10053, f. 13r–v (*Tabula horarum diei equalium ad meridiem Toleti*). The maximum length of daylight (14;51h) occurs at Gem 30° (= Cnc 0°) and corresponds to a geographical latitude of 39;54°; both this maximum length of daylight and this latitude are attested for Toledo in many medieval manuscripts (see, e.g., Toomer 1968, p. 36).

Table 45A: Length of daylight (for Toledo)

	Ari (h)	**Tau (h)**	**Gem (h)**	**Cnc (h)**	**Leo (h)**	**Vir (h)**
1	12; 3	13;21	14;25	14;51	14;29*b*	13;16
2	12; 5	13;23	14;27	14;51	14;20	13;13
3	8	26	29	51	18	11
4	11	28	30	50	17	8
5	13	31	32	50	15	6
6	16	33	33	50	13	3
7	19	36	34	49	11	13; 1
8	21	38	36	49	9	12;58
9	24	40	37	48	7	56
10	26	43	38	48	5	53
11	29	45	40	47	3	50
12	32	47	41	46	14; 1	48
13	35	50	42	46	13;58	45
14	37	52	43	45	56	42
15	40	54	44	44	54	40
16	42	56	45	43	52	37
17	45	13;58	14;46	14;42	50	12;35
18	48	14; 1	14;46	14;41	47	12;32
19	50	3	47	40	45	29
20	53	5	48	38	43	27
21	56	7	48	37	40	24
22	12;58	9	49	36	30*c*	21
23	13; 0	11	49	35	36	19
24	3	13	50	33	33	16
25	6	15	50	32	31	13
26	8	17	50	30	28	11
27	11	18	51	29	26	8
28	13	20	51	27	23	5
29	13;16	14;22	14;51	14;25	21	12; 3
30	13;18	14;24	14;51	14;27*a*	18	12; 0

a. Instead of 14;24.
b. Instead of 14;22.
c. Instead of 13;38.

	Lib (h)	**Sco (h)**	**Sgr (h)**	**Cap (h)**	**Aqu (h)**	**Psc (h)**
1	11;57	10;39	9;35	9; 9	9;28*d*	10;40*i*
2	11;55	10;37	9;33	9; 9	9;40	10;47
3	52	34	31	9	42	49
4	49	32	30	10	43	52
5	47	29	28	10	44	54
6	44	27	27	10	47	47*j*
7	41	24	26	11	49	10;59
8	39	22	24	11	51	11; 2
9	36	20	23	12	53	4
10	33	17	22	12	53	7
11	31	15	20	13	57	10
12	28	13	19	14	59	12
13	25	10	18	14	2	15
14	23	8	17	15	4	18
15	20	6	16	9;16	9; 6*e*	20
16	18	4	15	9;17	9; 8*f*	23
17	15	10; 2	9;14	18	10	25
18	12	9;59	9;14	19	13	28
19	10	57	13	20	15	31
20	7	55	12	22	17	33
21	4	53	12	24	20	36
22	11; 2	51	11	24	22	39
23	10;59	49	11	27	24	41
24	57	47	10	27	27	44
25	54	45	10	28	29	47
26	52	43	10	30	32	49
27	49	42	9	31	34	52
28	47	40	9	33	37	55
29	10;44	9;38	9; 9	9;35	9;39*g*	11;57
30	10;42	9;36	9; 9	9;36	9;42*h*	12; 0

d. Instead of 9;38.
e. Instead of 10; 6.
f. Instead of 10; 8.
g. Instead of 10;39.
h. Instead of 10;42.
i. Instead of 10;44.
j. Instead of 10;57.

Chapter 46

46:1. This chapter concerns the length of a seasonal hour (*tiempo de una hora temporal*) and the oblique ascensions for any locality.

46:2–3. These sentences refer to a table for finding the length of both diurnal and nocturnal seasonal hours at Toledo for any solar longitude.

46:4. This sentence refers to a table of oblique ascensions, here called *subimientos de los signos*, for the latitude of Toledo.

46:6. We are given here a rule for determining the length of seasonal hours from the diurnal arc for a given locality.

46:7–14. The procedure described requires the computation of the ascensional difference, that is, the difference between the right ascension and the oblique ascension for a given place whose geographical latitude is φ. According to the instructions, the ascensional difference, γ, is to be computed by means of the following formula:

$$\mathrm{Sin}\,\gamma = 60 \cdot \frac{\mathrm{Sin}\,\varphi \cdot \mathrm{Sin}\,\delta}{\mathrm{Cos}\,\varphi \cdot \mathrm{Cos}\,\delta}$$

or, equivalently,

$$\mathrm{Sin}\,\gamma = 60 \cdot \tan\varphi \cdot \tan\delta.$$

This seems to be the correct version of the formula which is garbled in 41:13 (see Fig. 4.7). Note that the terminology used here is no longer vacillating as was the case in chapter 41.

Chapter 47

47:1. This short chapter explains how to convert equal hours (*horas yguales*) into seasonal hours (*horas temporales*), and *vice versa*.

47:2–4. The text indicates that one needs to know the length of a seasonal hour in degrees (where 1 day is 360°) for each day of the year (*los tiempos de una hora temporal de aquel dia*), thus suggesting that,

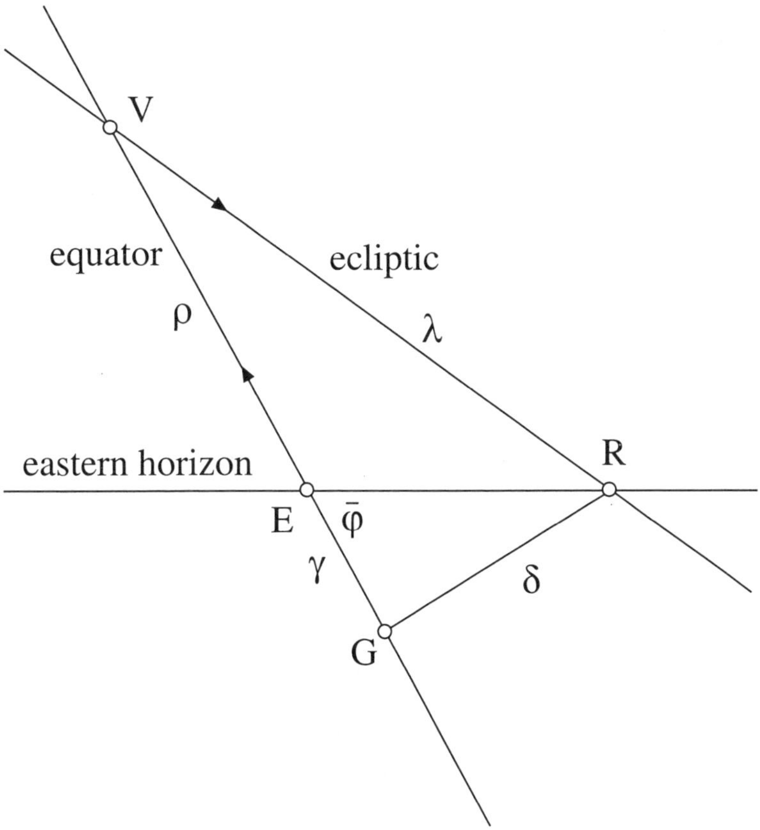

Fig. 4.7. A portion of the celestial sphere represented as a plane: V is the vernal point, GEV is the equator, VR is the ecliptic, and ER is the eastern horizon. In right spherical triangle RGE, arc GR (= δ) is the declination of arc VR (= λ); the angle between the horizon and the equator is $\bar{\varphi}$; and arc GE (= γ) is the ascensional difference. Arc EV (= ρ) is the oblique ascension of λ, and arc VG (= $\alpha = \rho + \gamma$) is the right ascension of λ.

although no specific table is mentioned, the Castilian Alfonsine Tables had a table for this purpose.

Chapter 48

48:1. This chapter concerns anniversaries of nativities, the world, and other events in an astrological context: an anniversary of a given moment in time takes place after an integer number of years have gone by, where the year is expressed as 365 days plus a number of hours and minutes or as 365 days plus a number of time-degrees such that 15° = 1 hour. In many zijes there is a table for the "excess of revolution", that is, the difference between the length of the year and 365 days for expanded and collected years, e.g., al-Khwārizmī (Suter 1914, p. 230); al-Battānī (Nallino 1903–1907, 2:188); and Ibn al-Kammād (Chabás and Goldstein 1994, p. 35). For a list of parameters see Kennedy 1956a, p. 147. This topic is discussed in chapter 53 of the canons to al-Battānī's zij (Nallino 1903–1907, 1:128–129; Bossong 1978, pp. 84–86), and in the canons to the Toledan Tables (F. S. Pedersen 2002, pp. 484–487).

48:4. The term *buc* is used to designate the hourly velocity of the Sun. It is also used in the Toledan Tables (F. S. Pedersen 2002, pp. 482–485) where it is applied to the daily velocity of planets as well. This word, ultimately of Sanskrit origin (*bhukti*: see Neugebauer 1962a, p. 57), is usually given in Arabic sources as *buht* (Nallino 1903–1907, 2:324).

48:7. This sentence is an instruction for finding at a given moment the ascendant, the 12 astrological houses (where the cusps are called *çentros*, as in 28:11), and the true positions of the planets. This information normally appears in a horoscope.

48:8. This passage indicates that the authors made a table for this purpose. This table is called *tabla de la hera de las revoluçiones* and its entries are given in hours and minutes.

48:11. This sentence refers to another version of the same table, called *la tabla de los subimientos de la revoluçion* (table for the ascensions of revolution) for collected and expanded years. The units for the entries are not indicated but we presume that they were in degrees and minutes.

48:14. The text refers to the anniversary of the world and a table for finding it. The anniversary of the world (*revoluciones de los años del mundo*) is defined as either the entry of the Sun in the vernal equinox (*ygualdad del verano*) or its entry in the point at the beginning of Aries. In contrast to modern Castilian, *verano* here means "spring". The distinction between tropical and sidereal years is the subject of chapter 49.

See Table 48A for excerpts of the tables for the "excess of revolution" in Madrid, Biblioteca Nacional, MSS 4238 (f. 35r–v) and 10002 (f. 4v). The first entry in both tables gives the difference between a tropical year and 365 days: in Table 48A this difference is expressed in degrees of which one day is 360°, and in Table 48B in hours. Subsequent entries in these tables are simply multiples of the first entry.

Chapter 49

49:1. This chapter deals with trepidation, called in the text "access and recess", with vacillating terminology. The term "trepidation" does not occur in medieval texts, but it is standard in modern discussions of "motion in access and recess".

49:2. The expression *los lugares naturales* refers to tropical longitudes, as opposed to sidereal longitudes, called *propios* (essential) in 49:7. On the terminology for sidereal and tropical coordinates, see Samsó 1997, pp. 107–108. On p. 108 n. 92, Samsó refers to chapter 43 in the anonymous canons to Ibn Isḥāq's zij (preserved in Hyderabad, Andra

Table 48A: Excess of revolution in time-degrees

Years	**Excess of revolution (°)**
1	87;19, 6
...	
20	306;22, 0
...	
100	91;50,24
...	
800	14;40, 6

Table 48B: Excess of revolution in hours

Years	**Excess of revolution (h)**
1	5;49,16
...	
20	20;25,20
...	
100	6; 6,40
...	
800	0;53,20

Pradesh, MS 298) for the distinction between sidereal (*dhātī* = essential) and tropical (*ṭabīʿī* = natural). Moreover, Ibn al-Bannā' says that in order to obtain the tropical (*al-ṭabīʿī*) longitude of a fixed star, one should add the variable precession to its sidereal (*al-dhātī*) longitude (Vernet 1952, p. 46 of the Arabic text). See also Ibn al-Kammād's star list (Goldstein and Chabás 1996, p. 326).

The vernal equinox serves as the origin for tropical longitude. The expression used to designate the 9th sphere, i.e., the highest sphere with no stars on it, is *çielo alto raso no estrellado*.

49:3–5. In order to calculate the amount of trepidation, two tables must be used. The first is a table for the mean motion of "the head of Aries" (sidereally fixed), arranged as the other mean motion tables; and the second is a table for the equation of "access and recess" whose argument is derived from the first table. No parameters are given but we are told that the entries in the second table are positive from 0° to 180° (access), and negative from 180° to 360° (recess).

49:7. The text explicitly says that after adding the equation (of access) to the sidereal positions, or subtracting the equation (of recess) from them, the resulting positions are tropical, i.e., in the 9th sphere with no stars; it then says that all the planetary positions obtained from the tables described previously are sidereal. The expression used to designate the 8th sphere, i.e., the starry sphere, is *çielo estrellado*. The term *forma* used here is a translation of Arabic *ṣūra* (form), a technical term for "constellation" (cf. Casanovas 1987, p. 81).

49:8. At arguments 0° and 180° the amount of trepidation is 0°. The description of the table for the equation of access and recess here is consistent with formulas for Azarquiel's third model for trepidation which had a great impact on subsequent Andalusian and Maghribī astronomers (Comes 1996):

$$\sin P = \sin \delta \,/\, \sin \varepsilon$$

and

$$\delta = r \sin i,$$

where P is the distance from the sidereally fixed Aries 0° to the vernal equinox, ε is the obliquity of the ecliptic, r is the radius of a small circle on which Aries 0° is located, and i is the argument on that circle, which is to be understood as derived from the first table, mentioned above (cf. Samsó 1992, pp. 235–236). The same results for P, very nearly, can be obtained by using the formula:

$$\sin P = \sin (P_{max}) \cdot \sin i$$

as may have been suggested by Ibn al-Hā'im (Comes 2001, pp. 347–348; cf. Chabás and Goldstein 1994, p. 27). This formula, in turn, can be approximated by

$$P \approx P_{max} \cdot \sin i$$

as remarked by Neugebauer (1962a, p. 184). The period for the argument, i, is given different values in Andalusian astronomical texts, and the value for P_{max} varies as well. For Azarquiel's third model, the period is 3874 Julian years and P_{max} is about 10;24° (Samsó 1994, Essay VIII, p. 22, and Essay X, p. 12). In Pseudo-Thābit, *On the motion of the eighth sphere*, the period is 4077 Julian years, and P_{max} is 10;45° (Neugebauer 1962b, p. 297); and in the Tables of Ibn al-Kammād P_{max} is 9;59° ≈ arcsin (10;24°/60) (Chabás and Goldstein 1994, p. 25; cf. Mancha 1998).

In this chapter only a periodic term is described, whereas in the Parisian version of the Alfonsine Tables there are two terms: a periodic term and a linear term depending directly on time (see

Mercier 1977, pp. 58–59; North 1996, pp. 462–464). The amount of trepidation is a measure of the difference between the tropical and sidereal years; here the sidereal year is fixed and the tropical year varies, whereas in the Parisian version of the Alfonsine Tables, the reverse is the case (see Goldstein 1994, pp. 191, 194). The periodic term in the Parisian Alfonsine Tables corresponds to the equation, $P = P_{max} \cdot \sin i$, where the period of i is 7,000 years, and the value of P_{max} is 9°: it is quite possible that the table in the Parisian version of the Alfonsine Tables was taken from the (lost) Castilian Alfonsine Tables, but with a variable sidereal year instead of a fixed sidereal year (see chapter 6, below).

Chapter 50

50:1. This short chapter concerns variable obliquity of the ecliptic (*declinaçion universal*) because it refers to a motion of the pole. The same terminology is used by Ibn al-Raqqām for the same purpose (cf. Kennedy 1997, p. 59).

50:2–3. The table described here agrees with the table that appears in Azarquiel's *Treatise on the motion of the fixed stars* which is only extant in Hebrew (see Millás 1943–1950, pp. 269, 332). This chapter gives no entries in the table of variable obliquity, but Azarquiel's table ranges from 23;33,0° to 23;53,0°. The descriptions given in chapters 49 and 50 conform to Azarquiel's third model for trepidation which, in contrast to Pseudo-Thābit's model, exhibits two independent but related kinematic models, one for access and recess and another for variable obliquity (Comes 1996, pp. 352–354). Azarquiel's model is based on the model described in the treatise, *On the motion of the eighth sphere*, attributed to Thābit (Neugebauer 1962b). One of Azarquiel's modifications was to introduce a table for variable obliquity, instead of the cumbersome calculations required by Pseudo-Thābit's model. Therefore, chapters 49 and 50 refer to Azarquiel's model, whereas the Toledan Tables depend on Pseudo-Thābit's model (Toomer 1968, pp. 118– 122). The Parisian Alfonsine Tables treat trepidation in a completely different way without variable obliquity.

It should be noted that in chapter 20 there is a description of a table for solar declination for a fixed value of the obliquity.

See Table 50A for the table for variable obliquity found in Azarquiel's *Treatise on the motion of the fixed stars* as reproduced by Millás (1943–1950, p. 332).

Table 50A: Variable obliquity

Argument (°)	Obliquity (°)
10	23;33, 9
20	23;33,36
30	23;34,20
40	23;35,20
50	23;36,34
60	23;38, 0
70	23;39,35
80	23;41,16
90	23;43, 0
100	23;44,44
110	23;46,25
120	23;48, 0
130	23;49,26
140	23;50,40
150	23;51,40
160	23;52,24
170	23;53,51
180	23;53, 0

Chapter 51

51:1. This chapter deals with the problem of finding the length of a shadow projected by a gnomon as a function of solar altitude, and vice versa. The title corresponds to that in Ibn al-Raqqām's zij, chapter 90: "On the Obtaining (*istinbāṭ*) of the Shadow by Means of Altitudes, and Vice Versa" (cf. Kennedy 1997, p. 49).

51:2–8. The procedure described here is found in many zijes with corresponding tables: see, for example, al-Khwārizmī (Suter 1914, p. 174), al-Battānī (Nallino 1903–1907, 2:60), and Ibn al-Kammād (Chabás and Goldstein 1994, p. 34).

Chapter 52

52:1. This chapter concerns the projection of the rays of the planets in the zodiac, and its title refers explicitly to al-Battānī. The corresponding passage is in chapter 54 of the canons to his zij (Nallino 1903–1907, 1:129–131 and 306 ff.; Bossong 1978, p. 86) and in the canons to the Toledan Tables (F. S. Pedersen 2002, pp. 474–475). For the general topic, see Kennedy and Krikorian-Preisler 1972. The title corresponds to that in Ibn al-Raqqām's zij, chapter 113: "On Determining the Positions of the Rays of the Stars in the Sector (*niṭāq*) of the Ecliptic Without Adding (?) the Horizon" (cf. Kennedy 1997, p. 51).

52:3. This sentence is very similar to the first paragraph of chapter 24 in Ibn al-Bannā''s zij (Vernet, 1951, pp. 141–142). Nallino (1903–1907, 1:308 n. 1) already noticed the relationship between that text and chapter 52 of the Castilian Alfonsine canons.

Chapter 53

53:1. This chapter concerns the solar depression angle corresponding to the beginning of morning twilight (daybreak) and the end of evening twilight (nightfall). This topic was frequently treated by Muslim astronomers because three of the five canonical prayer times are determined by these two moments in time. For religious purposes the Muslim day begins at sunset: evening prayer must be performed in the interval between sunset and nightfall (i.e., during dusk); night prayer in the interval between nightfall and daybreak; and morning prayer in the interval between daybreak and sunrise (i.e., during dawn): see King 1973, p. 346. The solar depression angle allows us to determine the durations of both dawn and dusk, i.e., morning and evening twilights. The title corresponds to that in Ibn al-Raqqām's zij, chapter 141: "On Determining the Hours of the Beginning of the Dawn and the Disappearance of Twilight by Means of the Depression of the Sun Below the Horizon Along a Circle of Altitude, and Vice Versa" (cf. Kennedy 1997, p. 54). Ibn Bāṣo (d. 1316), the chief timekeeper at the Grand Mosque of Granada, wrote a treatise on a universal astrolabe plate, completed in 1274, in which there are 12 chapters concerning different ways to calculate these times (Calvo 2000, p. 284).

53:2. A cross-reference is given to chapter 49 for the convention that solar longitude is counted from the vernal equinox, here called *ygualdad tropica*. The value for the solar depression angle is given as

19°, as in the treatise by Ibn Muᶜādh (Goldstein 1977a; Smith 1992; Smith and Goldstein 1993). Ibn Bāṣo's value for the solar depression angle is 18° (cf. Goldstein 1976).

53:2–6. The procedure described here appeals to the same formula used in chapters 43 and 44 for finding time from solar altitude and vice versa (see also Samsó 1987a, p. 29).

Although chapter 53 does not mention a table, we reproduce a table for the times of daybreak and nightfall, measured in seasonal hours from sunset, found in Madrid, Biblioteca Nacional, MS 10053, f. 16r: see Table 53A. The expression *ascensio aurore* refers to daybreak, and *occasus radiorum* to nightfall. Note that the durations of dawn and dusk are symmetrical, indicating that the same solar depression angle was used for both; moreover, the arguments in this table should be symmetric: Gem 30° should correspond to Cnc 0°, etc. For the twilight tables of Ibn Yūnus (d. 1009) with different solar depression angles for daybreak and nightfall, see King 1973, pp. 365–368, 382.

Chapter 54

54:1. According to the title, this chapter deals with finding the dates of Ash Wednesday and Easter, both by computation and by tables. This discussion belongs with the chapters on calendars, as is the case in other zijes.

54:2–5. A table is mentioned for finding the weekday for the first day of January, the date of Ash Wednesday (sometimes called *ayuno*, that is, "fasting"), and the date of Easter. The table is arranged for a 19-year cycle with years counted according to the era of Alfonso on the one hand, and the era of Caesar (see 1:5) on the other. A similar table is found in Zacut's *Ḥibbur* (Chabás and Goldstein 2000, pp. 58–59).

54:6–12. We are given rules for computing the dates of Ash Wednesday and Easter.

Table 53A: Twilight

				ascensio aurore et occasus radiorum	
Nomina signorum		*Gradus linee secunde*	*Gradus linee prime*	*ascensio aurore* **h m**	*occasus radiorum* **h m**
		6	30	10;51	1; 9
C	S	12	24	10;51	1; 9
A	G	18	18	51	9
P	R	24	12	50	10
		30	6	50	10
		6	30	49	11
A	S	12	24	49	11
Q	C	18	18	49	11
U	O	24	12	48	12
		30	6	10;47	13
		6	30	46	14
P	L	12	24	44	16
S	I	18	18	42	18
C	B	24	12	40	20
		30	6	38	22
		6	30	35	25
A	V	12	24	33	25
R	I	18	18	31	29
I	R	24	12	28	32
		30	6	25	35
		6	30	21	39
T	L	12	24	18	42
A	E	18	18	15	45
U	O	24	12	9	41
		30	6	5	1;55
		6	30	10; 0	2; 0
G	C	12	24	9;58	2
E	N	18	18	55	5
M	C	24	12	53	7
		30	6	9;51	2; 9

CHAPTER 5

CONTEXT

When Alfonso became king of Castile and León in 1252, he began to rule a territory that had recently grown dramatically, for the cities of Córdoba, Murcia, Jaén, and Seville had been conquered by his father, Ferdinand III (*ca.* 1201–1252). This expansion towards the south of the Iberian peninsula gave Christian scholars easier access to Andalusian culture that had been flourishing in the previous centuries (for an account of Muslim science in al-Andalus, see Samsó 1992). All modern authors seem to agree that the reign of Alfonso X was marked by failure in politics and success in cultural affairs. This success depended on surrounding himself with competent intellectuals and acting as a patron of learning, offering them the material conditions they needed to pursue major cultural projects. This attitude towards culture was not something new at the court, for King Ferdinand had already sponsored translations of scientific treatises. Alfonso, however, went far beyond that: he was not a mere patron, but played an active role in many of the cultural enterprises that he sponsored for, according to Procter (1945, p. 22), "he sought for books, initiated projects, allotted work among his collaborators, gave them instructions, and to some extent revised their work; finally he was a scholar who could appreciate the results of their labours."

An important figure in the intellectual life of Toledo at the time of Alfonso deserves to be mentioned here. Gonzalo Pétrez, otherwise known as Gonzalo García Gudiel or Gonzalo Gudiel or simply Don Gonzalo, was dean of Toledo in 1262, and then archbishop of Toledo from 1280 until his death in 1299 (see Gonzálvez 1997, pp. 299–416; Millás 1942, pp. 16–19). He participated in the scientific work of Alfonso X and was probably one of the collaborators in preparing the *Siete Partidas*. Around 1273, Don Gonzalo had his own scriptorium where books were copied and translations were made from Arabic. He had also spent some time in Italy at the papal court where he had been in contact with Campanus de Novara (d. 1296). An inventory of Gonzalo's possessions, dated 1273 and preserved in Madrid, Biblioteca Nacional, MS 13022, includes a list of some fifty volumes among which are quite a few devoted to the exact sciences. Item no. 35 is "Almagest, tablas d'astronomía d'Abençayt" (Gonzálvez 1997, p. 442). Could the tables of this author, whose name was transcribed by Millás (1942, p. 18) as "Auenzait", be those compiled by Ibn Sid? Similar information is also mentioned in the zij of Judah ben Asher II: "The scholar, R. Isaac ben Sid, peace be upon him, investigated [this subject] with his colleagues at the command of the king, and composed the zij of Alfonso ..." (Vatican, MS Heb. 384, f. 285b:2–3). In both cases, the reference seems to be to the Castilian Alfonsine Tables (compiled by Isaac ben Sid and Judah ben Moses ha-Cohen) and Gonzalo's list would then be the earliest citation of these tables. Moreover, on the basis on Gonzalo's list, it seems most likely that the original zij was in Castilian, rather than in Hebrew.

The activity that took place under the patronage of Alfonso resulted in a great variety of treatises, mainly written in Castilian, which made of this romance language a vehicle of cultural expression, and contributed to the transmission of Muslim civilisation to Christian Europe. These works can be grouped in five categories:

- Historical: the *General Estoria*, and the *Cronica General*.
- Legal: the *Siete Partidas,* the *Setenario*, the *Fuero Real,* the *Especulo*, and the *Ordenamiento de las tafurerias*.
- Literary: the *Cantigas de Santa Maria,* a collection of over 400 poems in praise of the Virgin, and *Calila e Dimna*.
- Religious: the translations of the *Bible* and the *Kitāb al-mi^crāj* (*The Book of Muḥammad's Ladder*).
- Scientific: mainly "astronomy and its medieval bedfellow astrology" (Procter 1945, p. 15).

To this list, one should add various works that have been attributed to the Alfonsine collaborators, such as the *Libro de los caballos*, the *Libro de los juegos,* and the Castilian version of a treatise on agriculture.

When considering the scientific works listed below, it is readily seen that almost all of them are concerned with astronomy/astrology and related disciplines with little attention to other basic fields of scientific knowledge and practice such as medicine.

Among the numerous books produced by Alfonso's collaborators, many were translations from Arabic treatises composed in al-Andalus, others were adaptations of previous Arabic material, and still others were original works. The increasing percentage of original treatises represents a departure from previous scientific activity in the Christian part of the peninsula where translations had predominated. Significantly, at the time of Alfonso original astronomical texts were produced within a pre-established and long-range program under the general supervision of the king.

Another characteristic of the scientific endeavors under Alfonso's patronage is that most of the works were collaborative rather than individual compositions. In almost all the translations more than one collaborator is named and this is also true for original works, including the *Libro de las tablas alfonsies*. The king acted as a general editor for the scientific corpus, but none of the prologues to these treatises indicates that he participated directly in the work itself. To be sure, Alfonso had many scientific collaborators and 15 have been identified, of whom five were Jews (see, e.g., Procter 1951, Romano 1971, Gil 1985, Roth 1990, Romano 1992). Among them, the authors of the *Libro de las tablas alfonsies*, Judah ben Moses ha-Cohen and Isaac ben Sid, played a crucial role because of the nature and the number of the works ascribed to them (see our comments in chapter 4 on the canons to the Castilian Alfonsine Tables, chapter 0:1). Three other Jewish collaborators are named:

- "The Jew Abraham", also called Abraham of Toledo or Don Abraham, i.e., Abraham Ben Waqār, later served as the court physician of Alfonso's son, King Sancho IV. He was responsible for the translation from Arabic into Castilian of two works, the *Kitāb al-mi*[c]*rāj* (*The Book of Muḥammad's Ladder*), *ca.* 1263, and Ibn al-Haytham's *Kitāb fī hay'at al-*[c]*ālam* (*On the configuration of the world*), probably after 1270. Around 1263, Alfonso ordered the translation of the *Kitāb al-mi*[c]*rāj* into three languages: Castilian, French, and Latin (the Castilian version is now lost; for an edition of the French and Latin texts, see Muñoz Sendino 1949, in which this work is called *Escala de Mahoma*; for a modern translation into English, see Hyatte 1997). In 1277, Abraham also participated in the revision of the translation made by Fernando de Toledo of the *Libro de la açafeha*.
- Samuel ha-Levi was the author of the *Libro del relogio de la candela* and, in 1276, he collaborated in the revision of the Castilian version of al-Ṣūfī's star catalogue, the *Libro de las estrellas de la ochaua espera*.

- "Don Xosse [or Mosse] nuestro alfaquim" is mentioned in the *Libro de la alcora* as the author of an added chapter (Roth 1990, p. 64). Some scholars read "Xosse", others "Mosse", and still others have identified him to Judah ben Moses ha-Cohen, a suggestion rejected by Romano (1992, p. 138). On the other hand, Roth (1990, p. 64) argued that he is to be identified as the physician Ziza Ibn Susan, also in the service of the king.

The Christian collaborators were Spanish and Italian. Among the Spaniards we find:

- Fernando de Toledo translated into Castilian the *Libro de la açafeha* in 1255–1256. The translation did not satisfied the king who ordered a new translation in 1277.
- Garci Perez helped Judah ben Moses ha-Cohen in the translation into Castilian of the *Lapidario* (between 1243 and 1250) .
- Guillén Arremón Daspa helped Judah ben Moses ha-Cohen in the translation into Castilian of the *Libro de las estrellas de la ochaua espera* (1256).
- Johan Daspa helped Judah ben Moses ha-Cohen in the translation into Castilian of the *Libro de las cruzes* and the *Libro de la alcora* (both in 1259), and contributed to the revision of *Libro de las estrellas de la ochaua espera* in 1276.
- Bernardo "el arauigo", presumably a Christian who had converted from Islam, participated in 1277, together with Abraham, in revising the translation of the *Libro de la açafeha* made by Fernando de Toledo more than 20 years earlier.
- Álvaro de Oviedo, also known as Álvaro de Toledo (see Gonzálvez 1997, pp. 602–616), wrote a refutation of Averroes's *De substantia orbis* and was responsible for one of the two Latin translations of the *Libro conplido en los iudizios de las estrellas* by ᶜAlī Ibn Abi l-Rijāl (11th century; also known as Abenragel) previously translated into Castilian by Judah ben Moses ha-Cohen. His annotations of several astronomical treatises, in a handwriting that Millás considered extremely difficult to read, are preserved in manuscripts at the Biblioteca Nacional de Madrid, together with some manuscripts copied by him (see Millás 1942, pp. 34–35, 191).

The group of Italian scientific collaborators of the king consisted of at least four scholars:

- John of Messina and John of Cremona participated in the revision of the *Libro de las estrellas de la ochaua espera* in 1276.

• Egidius de Tebaldis of Parma, together with Petrus de Regio, translated into Latin in about 1257 the *Libro conplido en los iudizios de las estrellas* which, in turn, was a translation from Arabic into Castilian made by Judah ben Moses ha-Cohen. Between 1271 and 1275 Egidius translated into Latin a Castilian version of Ptolemy's *Tetrabiblos,* with a commentary by ᶜAlī ibn Riḍwān (11th century).

To this list of Alfonso's Italian collaborators one could add Bonaventura de Siena, who translated into French, *ca.* 1263 in Seville, the Castilian version of the *Kitāb al-miᶜrāj,* previously translated from Arabic by Abraham.

In contrast to all previous scientific works produced in the Christian part of the Iberian peninsula, the bulk of the Alfonsine corpus was written in the vernacular rather than Latin, the language of scholarship. Perhaps Alfonso aimed at reaching a wider audience among his subjects, as the prologue to the *Lapidario* indicates (Rodríguez Montalvo 1981, p. 19): "[Alfonso] commanded him to translate it from Arabic into the Castilian language, so that everyone might better understand it" ([Alfonso] *mando gelo trasladar de arauigo en lenguaie castellano por que los omnes lo entendiessen meior*). Some books were also translated into languages other than Castilian, but these translations were almost always from Castilian. The decision to use Castilian as a language of science had important implications. In the case of astronomy/astrology, the translators of Arabic texts had to face the difficult task of creating *ex nihilo* terminology and phraseology in all these domains. As shown in chapter 3, most of the technical terms appearing in the *Libro de las tablas alfonsies* were adapted from Latin, a language closer to Castilian than Arabic. However, as we have already noted, there are examples where an Arabic term has been introduced into Castilian. In the process of constructing a new vernacular scientific language the astronomers/translators paid attention to some linguistic subtleties whereas, as Kasten (1990, pp. 39–40) has put it, "orthography was a matter that Alfonso could have taken into account, but the practice in the various texts he sponsored has not been consistent. (...) Spelling appears not to have been a matter of deep concern to medieval people."

It has been argued that, as a result of using Castilian as a scientific language, the corpus of works produced by Alfonso's astronomers was not as widely diffused as it would have been in Latin. We cannot know. Despite this generalization, it is undeniable that the *Libro del saber de astrologia* was translated into the Florentine dialect of Italian in 1341 and the *Libro de las tablas alfonsies* reached Paris around 1320, if not before, whence it was diffused in Latin throughout Europe (see chapter 6). Moreover, Alfonso's enduring fame was not derived from the translations or the literary, legal, or historical works he sponsored, but rests on the scientific treatises in the fields of astronomy and astrology compiled by his scientific team, and most importantly on the *Libro de la tablas alfonsies*. In contrast, the translations and original works in literary, legal, and historical genres have largely been forgotten except by specialists.

The scientific corpus produced under Alfonso's sponsorship comes to about 30 treatises, 14 of which are dated (Romano 1992). They fall into three categories:

(i) separate works on astronomy and astrology, mainly translations from Arabic into Castilian;
(ii) the Alfonsine Tables;
(iii) three collections of works on astronomy, astrology, and magic.

Most of the scientific corpus, together with some historical books such as the *General Estoria*, but not the *Libro de las tablas alfonsies*, have been edited in electronic format by Kasten, Nitti, and Jonxis-Henkemans (1997).

Of special interest is the possibility that a translation of the *Almagest* was made for Alfonso. In his well-documented history of Alfonso X, Ballesteros (1984, p. 816) mentions a commentary by the Marquis of Mondéjar (18th century) stating that King Alfonso ordered the translation into Castilian of an Arabic version of the *Almagest*. According to the Marquis (Mondéjar 1777, p. 453),

> *fue su interprete Isaac, hijo de Sid, comunmente conocido con el sobrenombre de Hazan, porque era Cantor de la Synagoga de Toledo: i de cuya versión Castellana formò la suya Latina, aunque sumamente barbara, Egidio Tebaldino (...).*

Unfortunately, no trace of this translation, if it really existed, by Isaac ben Sid has been found. Therefore, we do not list it in Alfonso's scientific corpus.

(i) Separate treatises

- *Libro conplido en los iudizios de las estrellas*

  This is a translation from Arabic into Castilian of a work by Abū l-Ḥasan ʿAlī ibn Abi l-Rijāl. The translator was Judah ben Moses ha-Cohen, who began this task in Toledo on March 12, 1254, according to the horoscope cast for that occasion. The *Libro conplido* was edited by Hilty (1954; see also Hilty 1955), based on a 13th-century manuscript that only preserves the first 5 books: Madrid, Biblioteca Nacional, MS 3605. A partial version in Castilian is also extant in Barcelona, Biblioteca de Catalunya, MS 981, and there is a Judeo-Portuguese version in Oxford, Bodleian Library, MS Laud. Or. 310, based on the Castilian text. Valladolid, Biblioteca de Santa Cruz, MS 253, a manuscript written entirely in Castilian, contains 13 chapters of Book VI hitherto considered lost, as well as most of Book V (the text begins at the middle of the second chapter). On the other hand, Segovia, Catedral, MS 115, contains a copy in Castilian of Book VIII (Beaujouan 1968, p. 17). The Castilian version was translated into Latin by Egidius de Tebaldis and Petrus de Regio, also in Alfonso's service.

- *Libro de las cruzes*

  This astrological treatise was translated from Arabic into Castilian by Judah ben Moses ha-Cohen with the help of Johan Daspa. The colophon indicates that the translation was finished in 1259. According to Samsó (1992, pp. 27–35), the *Libro de las cruzes* is evidence for the survival of Latin Visigothic astrology in 8th-century al-Andalus. It has been edited by Kasten and Kiddle (1961) from Madrid, Biblioteca Nacional, MS 9294. As pointed out by Beaujouan (1968, pp. 17–18), another copy in Castilian is extant in Segovia, Catedral, MS 115. A summary of this work is also found in a 16th-century manuscript: Madrid, Real Academia de la Historia, MS D 181.

- The Castilian translation of Ibn al-Haytham's treatise

  The Castilian version of the *Kitāb fī hay'at al-ʿālam* (*On the configuration of the world*) by Ibn al-Haytham (965–*ca*. 1040) is not extant. There is, however, an adaptation of it into Latin (see Millás 1942, pp. 285–312; Mancha 1990), but it is not known whether this Latin adaptation was due to the Alfonsine astronomers. The text in Latin explains that the translator from Arabic into Castilian was "the Jew Abraham". Samsó has suggested that the Castilian version was probably intended to be a part of the collection known as *Libros del saber de astrologia* (see Samsó 1990, p. 131).

- *Quadripartitum*

  According to Millás (1942, p. 159), Judah ben Moses ha-Cohen was responsible for the translation from Arabic into Castilian of Ptolemy's astrological treatise, the *Tetrabiblos*

(*Quadripartitum* in Latin). The Castilian text is not extant but, according to Procter (1945, p. 26), it was translated into Latin between 1271 and 1275 by Egidius de Tebaldis. The Latin text includes a translation of the commentary by ᶜAlī ibn Riḍwān on the *Tetrabiblos*. (Editions of this text and commentary in Latin were published in Venice in 1493 and 1519.)

- *Canones de Albateni*

This is the translation from Arabic into Castilian of the canons to al-Battānī's zij made by an unnamed collaborator of Alfonso. Bossong (1978) edited the text, adding a glossary, based on a late 13th-century manuscript: Paris, Bibliothèque de l'Arsenal, MS 8322.

- *Almanaque de Azarquiel*

The Almanac composed by Azarquiel consists of canons (Millás 1943–1950, pp. 72–148) and tables. A Castilian version of the tables, intended for Alfonso, is uniquely preserved in Paris, Bibliothèque de l'Arsenal, MS 8322. It has been edited by Millás (1943–1950, pp. 149–151), who believed that Isaac ben Sid was responsible for the translation into Castilian (Millás 1960, p. 225).

- *Libro del quadrante sennero*

As stated in the prologue, this book was written by Isaac ben Sid at the request of the king. The text was divided into two parts, but only a fragment of the first part is preserved (8 chapters). Millás (1956) edited it based on the unique copy: Paris, Bibliothèque de l'Arsenal, MS 8322. The extant text deals mainly with the use of this instrument, not its construction, and has been described as a collection of astronomical problems of increasing difficulty (Ausejo 1984). According to Samsó (1987a, p. 33), this instrument has a certain similarity with the double quadrant described by Mu'ayyad al-Dīn al-ᶜUrḍī (13th century).

(ii) *Libro de las tablas alfonsies*, compiled between 1263 and 1272 (Rico Sinobas 1863–1867, vol. 4, pp. 111–183).
(iii) Collections of works on astronomy (*Libros del saber de astrologia*), astrology (*Libro de las formas et de las ymagenes*), and magic (*Picatrix*).

1. Astronomy: *Libros del saber de astrologia*

This collection consists of 16 treatises. In addition to a catalogue of stars, it contains a series of works of Arabic origin on the construction and use of astronomical instruments and clocks. The presentations of the treatises on instruments are all similar: they are each divided into two books, one for the construction of the instrument and another for its use. The collection was completed in 1276–1277, although many of the treatises belonging to it were already translated between 1255 and 1259. The *Libros del saber de astrologia* are extant in a manuscript copied at the royal scriptorium, now Madrid, Universidad Complutense, MS 156. All the treatises except the star catalogue are also preserved in a later manuscript: Madrid, Biblioteca Nacional, MS 3306. The star catalogue, however, is partially preserved in Madrid, Biblioteca Nacional, MSS 1197 and 18668(54). We note that MS 18668(54) is a very partial copy made in the 18th century that reproduces the first pages of the star catalogue. The title page reads: "Copia de el manuscripto de las obras astronómicas de el Rei Don Alfonso El Sabio X de León i IV de Castilla como se halla aunque maltratado por descuido en la bibliotheca de el Colegio Maior de San Ildefonso de Alcalá con el título de *Tablas de el Rei Don Alonso*". Comparison with Madrid, Universidad Complutense, MS 156, indicates that MS 18668(54) is a partial copy of the Complutense manuscript, at that time kept in Alcalá.

The *Libros del saber de astrologia* were included by Rico Sinobas in the 5 volumes he published under the general title of *Libros del Saber de Astronomía*. Cárdenas (1974) reedited them with the title *Libros del Saber de Astrología* and listed some descendants of the Complutense manuscript, in addition to those mentioned above, that contain part of the *Libros* (Madrid, Real Academia de la Historia, MS 9/5707, only contains the *Libro de las estrellas de la ochaua espera*; Escorial, Biblioteca del Monasterio, MSS h-I-1 and V-II-9; Vatican, Biblioteca Apostolica, MS lat. 8174; Seville, Biblioteca Colombina, MS 4126; Oxford, Bodleian Library, MS Can. Misc. 340). This collection, as found in Madrid, Biblioteca Lázaro Galdiano, MSS 553, 554, and 555, only contains 11 treatises (Yeves 1998, pp. 557–559).

There is also a version in the Florentine dialect of the *Libros del saber de astrologia* extant in Vatican, Biblioteca Apostolica, MS 8174 (see Section 6.4.2).

The list of the treatises below follows their order in Madrid, Universidad Complutense, MS 156, rather than Rico's order, or that in Madrid, Biblioteca Nacional, MS 3306, where the *Libro del ataçir* follows the *Libro de la açafeha*, and the *Libro del quadrante* precedes the *Libro de las laminas de los siete planetas*.

- *Libro de las estrellas de la ochaua espera* (Rico Sinobas 1863–1867, vol. 1, pp. 5–145), also known as *Libro de las XLVIII figuras de la VIII spera* or even as *Libro de las estrellas fixas*

  This is an adaptation of the star catalogue for 964 AD by the Persian astronomer Abu l-Ḥusain ᶜAbd al-Raḥmān ibn ᶜUmar al-Ṣūfī (903–986) which, in turn, depended on the star catalogue in Ptolemy's *Almagest* (see Comes 1990). This work was compiled in 1256 by Judah ben Moses ha-Cohen with the help of Guillén Arremón Daspa. It was later revised (1276) by the king himself, in collaboration with Judah ben Moses ha-Cohen, Samuel ha-Levi, Johan Daspa, John of Messina, and John of Cremona. Curiously, the compilers of this text did not include any new data based on observations, despite the many observational instruments described in the corpus. The total precession in this text is 17;8° (see Madrid, Universidad Complutense, MS 156, *passim*), whereas in al-Ṣūfī's catalogue the total precession was 12;42° (Samsó and Comes 1988, p. 69). While the total precession in this Castilian text agrees with the total precession in the Latin catalogue commonly appended to the Parisian Alfonsine Tables, they have little else in common (Kunitzsch 1986, p. 95 n. 4). The Latin catalogue derives from the translation of the *Almagest* by Gerard of Cremona in about 1175, and is not based on either al-Ṣūfī or our Castilian catalogue. Moreover, the Latin catalogue is specifically for the epoch of the Alfonsine era used in Paris, June 1, 1252 (see our comments in chapter 4 to 1:11-16), whereas the epoch of the Alfonsine era in the Castilian Alfonsine Tables is Jan. 1, 1252 (see sentences 1:11–16 in chapter 2, and our commentary in chapter 4). In our view, it is likely that the Parisians took the total precession of 17;8° from a Castilian source, and applied it to their epoch for Alfonso (cf. Samsó and Comes 1988, p. 76). We think it most unlikely that the 17;8° was derived from the theory of trepidation as presented in the Parisian Alfonsine Tables (for which there is no evidence in Castilian); rather, it was one of the parameters incorporated in that theory, fixing the total precession for the epoch of the Parisian Alfonsine era. On the theory of trepidation in the Parisian Alfonsine Tables, see Mercier 1977, pp. 58–60. For previous derivations of the 17;8°, see Poulle 1988 (who assumed that this parameter is related to a pre-Alfonsine theory of trepidation with a mistake in the computation), and Samsó and Castelló 1988 (who assumed that the Alfonsine astronomers had the wrong date for Ptolemy's star catalogue).

It remains for us to explain the origin of 17;8° for the total precession in this text. We claim it is based on the total precession since the epoch of al-Ṣūfī, 964 AD. According to al-Ṣūfī, precession takes place at the rate of 1° in 66 years, even though Ibn Ezra, William of Saint-Cloud, and John of Murs take Ṣūfī's rate to be 1° in 70 years (Millás 1947, p. 78; Poulle 1980, p. 262; Samsó and Comes 1988, p. 68; see also Section 6.2.8). Al-Battānī also took the rate of precession to be 1° in 66 years (Nallino 1903–1907, 1:124 (line 33), 292) and, again, we note that al-Battānī's zij was translated into Castilian at the court of Alfonso. The difference between 17;8° and 12;42° is 4;26°. At the rate of 1° per 66 years, this total precession corresponds to about 292;30y (≈ 4;26° · 66y) which, when added to 964, yields 1256;30—and 1256 is the year of composition of the first version of the *Libro de las estrellas de la ochaua espera*.

According to al-Ṣūfī and al-Battānī, the epoch of Ptolemy's star catalogue is 138 AD (see Samsó and Castelló 1988, p. 117) whereas, in fact, the epoch is 137 AD (Toomer 1984, p. 340, n. 91). With either date, the rate of precession that results from dividing the number of years between Ptolemy's catalogue and al-Ṣūfī's catalogue (826 or 827 years) by 12;42° is 1° in about 65 years (rather than 66 years as explicitly stated by al-Ṣūfī). On the other hand, if we consider that 1118 years (= 1256 – 138) elapsed from the date of Ptolemy's star catalogue to the date when the Castilian catalogue was compiled, and take a rate of precession of 1° in 66 years, then the total precession in 1118 years would only be about 16;56°. In our view, the discrepancy between 16;56° and 17;8° is due to approximation in al-Ṣūfī's procedure, not to an error in dating Ptolemy's star catalogue greater than 1 year, and not to a miscalculation on the part of the Alfonsine astronomers.

Note that in his *Ḥibbur* Abraham Zacut (1452–1515) dated the *Libro de las estrellas de la ochaua espera* to 1256, and specified that in 1256 the longitude of Regulus was Leo 19;38°, i.e., 17;8° greater than its longitude in Ptolemy's star catalogue (Cantera 1931, pp. 193, 298; Cantera 1935, p. 160). In his *Sefer Yuḥasin* Zacut added the following (Freimann 1924, pp. 221b-222a):

> In the year 5012 [= 1252 AD] don Alfonso began to reign on the last day of May. He was a lover of the sciences, in particular of astronomy. At that time the scholar, Rabbi Isaac ben Sid, ḥazzan of Toledo, prepared (*tiqqen*) tables for the host of the heavens with great precision at the [command] of the king, and no other tables or books of astronomy had their precision. Indeed, Rabbi Isaac Israeli and Rabbi Levi ben Gerson and all who came after them found no fault in [these tables]; on him [Isaac ben Sid] it is said "with your wisdom and discernment", etc. [cf. Deut. 4:6 "Observe them (i.e., these laws) faithfully, for that will be proof of your wisdom and discernment to other peoples who, on hearing of all these laws, will say, 'Surely, that is a great nation of wise and discerning people.'"]. They are called the zij (*zij*) of Alfonso, and from the east to the west, [in] Germany, France, England, all Italy, and Spain, they [the astronomers] broke all their former tables and used these tables [in their stead] until this day.
>
> I, Abraham Zacut, the author, corrected [?] (*tiqqanti*) all the books on the basis of the tables that I made, and they [the tables?] are diffused in all the Christian (*Edom*) and Muslim lands. In the fourth year of his reign, year 5016 [= 1256 AD], he [i.e., Alfonso] commanded the scholar R. Judah ben Moses ha-Cohen of Toledo to translate into the common language [Castilian] the most precious book that Abu l-Ḥasan [al-Ṣūfī], the Muslim, wrote concerning the 1022 great stars—

> that are designated by name—on the eighth orb in the 12 zodiacal constellations (*mazzalot*) and the 36 [other] constellations (*mazzalot*), of which 15 lie to the south of, and 28 [read: 21] to the north of, the 12 zodiacal constellations, for a total of 48 [constellations] that were depicted in the sky [as seen] in the city of Egypt [by Ptolemy]....

- *Libro de la alcora* (Rico Sinobas 1863–1867, vol. 1, pp. 153–209)

  This book, also called *Libro de la fayçon dell espera et de sus figuras et de sus uebras*, is the Castilian translation by Judah ben Moses ha-Cohen of an Arabic treatise on the use of a celestial globe (*alcora*, from the Arabic *al-kura*, meaning "sphere"), written in the 9th century by Qusṭā ibn Lūqā (see Samsó 1982; Savage-Smith 1985, pp. 21-22). The translation, in collaboration with Johan Daspa, is dated 1259, and a revised version was completed in 1277. The first four chapters of this *Libro* were added by the astronomers in the service of the king; Hilty (1955, pp. 38–40) considered Judah ben Moses to be the author of these added chapters, whereas Romano (1971, p. 690) suggested it was Isaac ben Sid.
- The general prologue to the collection mentions a *Libro de las armellas del ataçyr*, but this is really a chapter appended to the *Libro de la alcora*. The beginning of this chapter reads (Madrid, Universidad Complutense, MS 156, f. 38r):

> *Et porque fuesse esta obra de la espera mas complida mandamos nos Rey don ALFONSO el sobre dicho annader hy este capitulo pora fazer armillas en la espera pora saber ell ataçir e egualar las casas segund la oppinion de Hermes. Et mandamos a don Mosse nuestro alfaquim que lo fiziesse. Capitulo de saber como se fazen las armillas dell ataçyr en la espera et egualar las casas segund la oppinion de Hermes et como obren con ellas.*

- *Libro del astrolabio redondo* (Rico Sinobas 1863–1867, vol. 2, pp. 112–222)
  According to the prologue Book I, on the construction of the instrument, was compiled by Isaac ben Sid. Book II is an adaptation of the treatise on the astrolabe by Abu l-Qāsim ibn al-Samḥ (d. 1035), a disciple of Maslama al-Majrīṭī (see Viladrich 1987).
- *Libro del astrolabio llano* (Rico Sinobas 1863–1867, vol. 2, pp. 224–292)
  This book, attributed to Isaac ben Sid (Roth 1985, p. 447), is probably based on materials derived from Maslama al-Majrīṭī and his group (see Martí and Viladrich 1983).
- *Libro de la lamina universal* (Rico Sinobas 1863–1867, vol. 3, pp. 1–132)
  This is a treatise on the construction and use of a universal astronomical instrument, using meridian projections. This book was compiled by Isaac ben Sid and it is based on a work by ᶜAlī ibn Khalaf, a Cordoban astronomer at the time of Azarquiel (see North 1986, p. 34; Calvo 1990).
- *Libro de las armellas* (Rico Sinobas 1863–1867, vol. 2, pp. 1–82)
  This book is an adaptation by Isaac ibn Sid of a treatise by Azarquiel. An incomplete copy of the Castilian version is also found in Oxford, Bodleian Library, MS Can. Misc. 340.
- *Libro de la açafeha* (Rico Sinobas 1863–1867, vol. 3, pp. 135–237)
  This book consists of two treatises by Azarquiel on the construction and use, respectively, of a *saphea* of the *zarqāliyya* type (see Puig 1987a and 1987b). It was first translated into Castilian in 1255–1256 by Fernando de Toledo but, as the king was not satisfied with the translation, a new version was prepared in 1277 by Bernardo "el arauigo" and Abraham of Toledo. A translation from Arabic (Escorial, MS Ar. 962) into modern Castilian has been published by Puig (1987b). There is also a Latin version, apparently finished in 1231, by Judah ben Moses ha-Cohen with the help of Gillelmus Anglicus (see Millás 1943–1950, pp. 453–454; Hilty 1955, pp. 16–18).

- *Libro de las laminas de los siete planetas* (Rico Sinobas 1863–1867, vol. 3, pp. 241–284)
  Book I of this work contains the Castilian version of the treatise on the construction of an equatorium by Ibn al-Samḥ (see Samsó 1983; Mancha 1987) and Book II is a Castilian version of Azarquiel's treatise on the construction of an equatorium (see E. Millás 1983).
- *Libro del quadrante con que rectifican* (Rico Sinobas 1863–1867, vol. 3, pp. 288–316)
  This treatise was revised (or written) in 1277 by Isaac ben Sid, and it is the only dated work by him other than the *Libro de las tablas alfonsies*.
- *Libro del relogio de la piedra de la sombra* (Rico Sinobas 1863–1867, vol. 4, pp. 1–23)
  The author seems to be Isaac ben Sid (Romano 1992, p. 158).
- *Libro del relogio del agua* (Rico Sinobas 1863–1867, vol. 4, pp. 24–64)
  The author of this work was Isaac ben Sid.
- *Libro del relogio del argent vivo* (Rico Sinobas 1863–1867, vol. 4, pp. 61–75)
  The author of this work was Isaac ben Sid.
- *Libro del relogio de la candela* (Rico Sinobas 1863–1867, vol. 4, pp. 77–93)
  The author of this work, Samuel ha-Levi of Toledo, is explicitly mentioned in the prologue. According to Comes (1990, pp. 32–33), he was a member of the eminent Abulafia family but, according to Roth (1990, p. 70), he was not.
- *Libro del relogio del palacio de las horas* (Rico Sinobas 1863–1867, vol. 4, pp. 94–107)
  The author of this work was Isaac ben Sid.
- *Libro del ataçir* (Rico Sinobas 1863–1867, vol. 2, pp. 295–309)
  This treatise has been attributed to Isaac ben Sid (Viladrich and Martí 1983, p. 76).

2. Astrology: *Libro de las formas et de las ymagenes*

This collection of treatises on the virtues and powers of stones and minerals was compiled between 1276 and 1279 and is no longer extant. A list of its contents has survived in

Escorial, MS h-I-16 and, according to it, the collection consisted of 11 books. A treatise associated with this collection bears the title *Lapidario*, and it is preserved in another manuscript written at the royal scriptorium: Escorial, Biblioteca del Monasterio, MS h-I-15. Moreover, there are two known 16th-century copies: Madrid, Biblioteca Nacional, MS 1197; and Escorial, Biblioteca del Monasterio, MS &-II-16. This is a treatise on the magical properties of stones and metals in 360 chapters, each describing a stone corresponding to a degree of the zodiac, with 30 chapters for each sign. The original Arabic version is attributed to an unidentified author, "Abolays" (see our comments in chapter 4 on the canons to the Castilian Alfonsine Tables, chapter 0:1). According to the prologue of the Castilian version, it was translated between 1243 and 1250 by Judah ben Moses ha-Cohen (*mucho entendudo en la arte de astronomia et sabie et entendie bien el arauigo et el latin*) with the help of Garci Perez (*otrossi mucho entendudo en este saber de astronomia*). The revised version was completed between 1276 and 1279. The *Lapidario* has been edited by Diman and Winget (1980) and Rodríguez Montalvo (1981). Escorial, Biblioteca del Monasterio, MS h-I-15, also contains three other lapidaries, one of which is called *El libro de las fazes*. An incomplete version of one treatise of the *Libro de las formas et de las ymagenes* is extant in Vatican, Biblioteca Apostolica, MS Reg. lat. 1283.

3. Magic: *Liber Picatrix*

The *Ghāyat al-Ḥakim*, a book on talismanic magic attributed to a disciple of Maslama al-Majrīṭī, was translated at the royal scriptorium into Castilian and later from Castilian into Latin. The Castilian version was completed between 1256 and 1258 (see Pingree 1981) and Judah ben Moses ha-Cohen was probably responsible for it (Forcada 1990): it is preserved in Vatican, Biblioteca Apostolica, MS Reg. lat. 1283. A facsimile edition of this manuscript in now available (Biblioteca Apostolica Vaticana 2000).

The astronomical and astrological works produced under the patronage of King Alfonso can be grouped in three periods (Romano 1971):

1. Before 1250 to 1259: translations,
2. Between 1262 and 1272: astronomical tables,
3. Between 1276 and 1279: compilations.

Based on the previous list of original or translated treatises, it is clear that the two authors of the *Libro de las tablas alfonsies*, Judah ben Moses ha-Cohen and Isaac ben Sid, played a crucial role in many of these scientific endeavors. Judah's participation is attested over a long period from before 1231 to 1276, i.e., his activity began prior to the reign of Alfonso and continued in all three periods mentioned above. Although we are told that he was a physician, Judah is never cited for his medical skills. Rather, he is frequently mentioned as a translator and a reviser of astronomical treatises. Other translators were associated with the royal court, but none of them came close to producing the number of works attributed to Judah. He is certainly to be considered as the main figure in scientific translation at the Alfonsine court—a difficult task, indeed, if we keep in mind that it involved adapting Castilian to serve as a scientific language. On the other hand, Isaac ben Sid's participation is only attested between 1263 and 1277, that is, during two of the periods above, and he was probably much younger than his senior collaborator, Judah. Isaac ben Sid was the most productive scientist in Alfonso's team of scholars, but his role was not that of a translator, as is clear from the list of his works. Rather, he was mainly responsible for original treatises on astronomical instruments, and he was involved in those astronomical matters that required a high degree of technical competence. Moreover, he made observations of solar and lunar eclipses, as attested in contemporary sources. His skills in this domain probably made him the astronomer in whom Alfonso had the most confidence.

The *Libro de las tablas alfonsies* is thus the outcome of the collaboration of the most outstanding scientists in Alfonso's entourage, and the result of a long astronomical tradition.

CHAPTER 6

THE LEGACY OF ALFONSINE ASTRONOMY

6.1 Introduction

In chapter 5 we discussed the astronomical works produced under the patronage of Alfonso X of Castile. In this chapter we focus on the diffusion of the Alfonsine Tables after their compilation in Toledo in about 1270. Surprisingly, very few references to them are preserved in the astronomical literature prior to their adaptation in Paris in the 1320s. This state of affairs has led some scholars to claim that the Parisian astronomers in the 1320s had no intellectual debt to the astronomers who worked for Alfonso. We will argue, however, that this claim is unfounded and based on several unwarranted assumptions. The fundamental text for the Parisians was the Alfonsine Tables in Castilian of which only the canons survive (see chapter 2). Although the transmission of these tables from Castile to Paris has not been documented, the fact remains that they indeed reached Paris. Given this gap in the historical record, it was perhaps inevitable that the character of the Castilian Alfonsine Tables would be disputed and, indeed, their relationship with the Parisian Alfonsine Tables has been the subject of controversy. Poulle (1984) restricted the Parisian Alfonsine Tables to those that correspond to the canons of John of Saxony (1327), leaving only chronological tables, radices, as well as mean motions and equations for the Sun, Moon, and planets. He excluded tables concerning the daily rotation, equation of time, syzygies, eclipses, planetary latitudes, and planetary velocities, among others, that we believe belong to the Alfonsine corpus in Latin for reasons that will become clear in this chapter. Poulle has also argued for the independence of the Parisian Alfonsine Tables, i.e., that they did not originate in Castile, remarking that "all we know for sure about the new Alfonsine astronomy is that it had Paris as its birthplace and 1321 as its birthdate", and calling the attribution to Alfonso a "literary fiction" (Poulle 1988, pp. 105, 110; Poulle and Savoie 1998, p. 201). But, in our view, the Parisian version of the Alfonsine Tables has strong roots in Castile, and the astronomical tradition of Castile is evident throughout it.

A set of astronomical tables is best characterized by its underlying parameters as well by its completeness and internal coherence; changing the presentation of a table may sometimes be very helpful for the computer, but it is not a very significant contribution, as North (1996, p. 455) has emphasized. In the case of the Parisian version of the Alfonsine Tables, there are some changes in presentation, possibly a new model for trepidation (but see Section 6.3.1), and a few new parameters. A special problem with these tables is that there is no adequate survey of the hundreds of manuscripts preserved in a great many libraries, and it is unlikely that such a survey will become available in the foreseeable future. Further, there is no systematic survey or description of the tabular material generated in Paris in the early 14th century: the tables that later appeared in the *editio princeps* of the Alfonsine Tables (1483) have been the focus of attention, while other tabular material has largely been ignored. Related to this problem is one of definition: which tables belong to the Parisian version of the Alfonsine Tables? For the moment we postpone responding to this question: see Section 6.3, below, where the most important texts in the Alfonsine corpus from the early 14th century are described; and Section 6.4.3 where tables in the Alfonsine corpus are listed.

In examining the Parisian Alfonsine Tables we will pay close attention to the chronological tables; those for the solar, lunar, and planetary equations; those for lunar and planetary velocities; and those for the motion in trepidation. In many cases we will find strong links with the antecedent tradition coming from Castile. The use of the era of Alfonso (1252) and the meridian of Toledo in the Parisian Alfonsine Tables surely reflects dependence on a Castilian source, but this is only a small part of the evidence that allows us to draw this conclusion.

Before discussing the characteristic features of the Parisian Alfonsine Tables, we note that at the turn of the 14th century Parisian astronomers generally depended on the Toledan Tables or the Tables of Toulouse, an adaptation of the Toledan Tables for the Christian calendar (F. S. Pedersen 1998; Poulle 1994). Nevertheless, some of the astronomers working in Paris found occasions to criticize these tables while others defended them.

William of Saint-Cloud was a leading astronomer in Paris in the 1290s (see McCluskey 1998, pp. 200–202). His precise dates are not known and, according to Poulle (1981, p. 389), the first activity he reports is an observation he made in 1285. Only three works of his are known: the *Directorium*, a treatise on the compass; the *Kalendarium Regine*, a calendar with base year 1292; and the *Almanach planetarum* beginning in 1292. Saint-Cloud was certainly well connected to the royal court. His calendar was dedicated to Queen Marie of Brabant, widow of King Philip III of France (d. 1285); William himself translated it into French at the request of Queen Jeanne of Navarre (d. 1305), wife of King Philip IV of France (reigned 1285–1314), and a copy of this translation in Rennes, Bibliothèque municipale, MS. 593, is dated 1303 (Shore 1989, p. 305). The text that introduces the calendar begins with "Testante Vegecio libro suo de re militari..." (see, e.g., Toledo, Catedral, MS 99-5, f. 26r; cf. Millás 1942, p. 258). His almanac displays the positions of the luminaries and the five planets for 20 years and, according to Poulle (1981, p. 390), these positions are based on the Tables of Toulouse with slightly modified mean motions. The almanac is preceded by a prologue in which William criticizes the Tables of Toulouse concerning planetary mean motions and the theory of trepidation attributed to Thābit ibn Qurra (Millás 1943–1950, pp. 496–509; Neugebauer 1962b). In this prologue William also discusses the camera obscura and reports a few astronomical observations: a solar eclipse, solar altitudes, conjunctions of the superior planets with each other, and some with fixed stars (Mancha 1991). Both the calendar and the prologue to the almanac are preserved in Paris, Bibliothèque nationale de France, MS lat. 7281, ff. 141r–148v, a 15th century copy including annotations by an otherwise unknown "Jo. B." (Boudet 1994, pp. 179–180). In a note to Saint-Cloud's *Kalendarium*, Jo. B. says: *Iste Guillelmus probavit et bene vidit ad oculum quod tabule antique de motibus Ptholomei, Alexandri et Toletane et Tolosane et alie etc. non continent veritatem in motibus planetarum et 8^e^ spere. Ideo ex se correxit aliquos medios motus ut melius vidit... Nec habuit notitiam tabularum Alfonsi que corrigunt defectus huiusmodi tabularum, quia nondum erant in manus hominum*. This dissatisfaction with Pseudo-Thābit's model for trepidation may have provided the motivation for constructing a new model for the same purpose, prior to the Parisian Alfonsine Tables. As we shall see, in the *Expositio*, John of Murs cites Saint-Cloud for an observation of the Sun and paraphrases a passage in his prologue to the *Almanach planetarum* (see Section 6.2.8).

One of the astronomers who relied on the Toledan Tables was John of Sicily (active in Paris *ca.* 1291), and he wrote a commentary on the canons to the Toledan Tables ascribed to Azarquiel. He predicted a solar eclipse and a lunar eclipse for Paris, both of which were to take place in 694 AH, i.e. 1295; and he was concerned about precession, computing a value of 9;22,10° for "the present" (F. S. Pedersen 1986, pp. 8–9).

An unnamed astronomer, active in Paris in 1290, was the author of a different commentary on the canons to the Toledan Tables. According to F. S. Pedersen (1984), this text was probably intended for university teaching, and was likely used by John of Sicily for his own commentary. The value for precession in this text is 9;22,20° for year 1290.

Peter of Saint Omer (Petrus de Sancto Ademaro) wrote a treatise on eclipses and revised a treatise on the *New Quadrant* by Profatius Judaeus (Jacob ben Makhir) in 1309 in Paris (Thorndike 1948, p. 203; Thorndike 1959, p. 37; Thorndike and Kibre 1963, col. 18).

Peter Nightingale, also known as Peter of Dacia and Peter Philomena, arrived in Paris in 1292 from Bologna and composed a calendar that became very popular (North 1994, p. 239; cf. F. S. Pedersen 1983–1984). It has been suggested that Peter collaborated with William of Saint-Cloud, based on common features of their calendars (O. Pedersen 1974, p. 541). A French translation of Peter's work, entitled *La lettre a savoir le vrai cours de la lune par le qualendrier mestre Pierre de Dace, dit Rosignol*, is preserved in Rennes, Bibliothèque municipale, MS 593; the same manuscript contains Saint-Cloud's French translation of his own work and it is likely that Saint-Cloud was also responsible for this translation (Shore 1989, pp. 305–306). Moreover, this manuscript preserves a French version of part of the *Almanac* by Profatius Judaeus (Jacob ben Makhir) which had been composed in southern France only a few years earlier (Boffito and Melzi d'Eril 1908, p. 2). Another manuscript that has been dated to *ca.* 1300 contains a translation into French of the *Centiloquium* ascribed to Ptolemy (Paris, Bibliothèque nationale de France, MS fr. 1349: Shore 1989, p. 309). These texts provide evidence for intense interest in astronomy and astrology in the vernacular at the royal court in Paris during the reign of Philip IV.

One of those who opposed the use of the Alfonsine Tables was Geoffrey of Meaux, the author of another calendar, also extant in Paris, Bibliothèque nationale de France, MS lat. 7281, and in Toledo, Catedral, MS 99-5. This calendar starts with January 1321 ("1320 completo") and is preceded by an introduction beginning *Cunctis Solis et Lune scire desiderantibus....* A feature of this calendar is the use of Roman style dating with respect to the calends and the ides, in addition to the usual listing of saints' days. In the introduction we are told that the author has specifically rejected the radices of Alfonso and has depended on those "by Azarquiel found in the Toledan Tables" (Toledo, Catedral, MS 99-5, f. 14r). This is surely evidence that the Alfonsine radices were known in Paris prior to 1321.

Solid evidence for the Alfonsine Tables outside Spain is lacking before about 1320, and Geoffrey of Meaux is the earliest datable reference to them that we have found. Another fixed point is John of Lignères's canons, dated 1322 in the colophon, in which Alfonso is specifically mentioned. Moreover, his *Tabule magne* that depend on the Parisian Alfonsine Tables are dated approximately 1325, based on the dedication to Robert of Florence, Dean of Glasgow (see Section 6.3.3). In his *Canones tabule tabularum,* dated 1321, John of Murs refers to the *Expositio tabularum Alfoncii,* and this title certainly indicates that John of Murs knew something about Alfonso in 1321. The author of an undated text entitled, *Expositio intentionis Regis Alfonsii circa tabulas eius,* is clearly by John of Murs because it includes the phrase "ego Johannes de Muris" (e.g., Poulle 1980, p. 265). John of Murs's *Expositio* cannot be earlier than 1319 because an observation made in that year is cited in it. But it is difficult to determine at what precise date it was written because, in our view, the identification of this text with the reference to an *Expositio* in a treatise by John of Murs in 1321 is uncertain; these problems will be addressed in Section 6.3.2. The tables of John Vimond (see Section 6.3.1) provide new evidence for the arrival of Alfonsine material in Paris although Alfonso's name does not appear in the unique copy available to us. In any event, although Vimond says that his tables are for 1320, we are unable to confirm this precise date on the basis on the radices he gives. Despite these difficulties in dating the various texts, it is clear that Alfonsine material had reached Paris by 1320 and it is entirely possible that it had already been there for a few years.

North (1996, p. 470) also called attention to a passage in a text from about 1347 by a writer who "evidently was a Parisian scholar (...). In the course of comparing star and solar aux movements he tells us that he saw a *Spanish* version of the Alfonsine book of the fixed stars, taken from the king's bookshelves (or cupboard, *de armario regis Alfontii*), according to the person who acquired it." It is unclear whether this book was seen in Castile or in Paris. Although there is no evidence in Parisian sources for the influence of this book (except for the total precession of 17;8°), it indicates a possible route for the Castilian Alfonsine Tables to have reached Paris. As noted in chapter 5, a text, entitled *The Book of Muḥammad's Ladder*, was translated from Castilian into French *ca.* 1263 at the court of Alfonso, and this suggests that there may have been some cultural exchanges between the royal courts of Castile and France before we find the first traces in Latin of the Castilian Alfonsine Tables in Paris (see chapter 1 for information on such exchanges). Hence we do not exclude, among other possibilities, that there was even a (lost) French version of the Castilian Alfonsine Tables.

Another reference, independent of the Parisians, comes from Andaló di Negro of Genoa who, in 1323, gives the date of the Alfonsine Tables as 1272, information that was not available in any Parisian text and that must have come from the prologue in the canons to the Castilian Alfonsine Tables (North 1977, p. 270).

The fact that many astronomers in Paris were unaware of the Castilian Alfonsine Tables before 1320 does not mean that no copy was there, for the links between Alfonso X and Philip III, and between Sancho IV and Philip IV, provide possible routes of transmission in the late 13th century to scholars associated with the king of France. So, given all the evidence, we offer as a conjecture that the court of Philip III or Philip IV served as the initial reception point in Paris for these tables with William of Saint-Cloud playing a prominent role in the process that led to their adoption by the astronomical community.

6.2 The characteristics of Alfonsine astronomy in Paris

In modern times the expression "Alfonsine Tables" usually refers to the *editio princeps,* which includes the 1327 canons by John of Saxony. But it also contains many tables which are not described in those canons such as the star catalogue and the list of cities with their geographical coordinates. The edition of 1483 is the consolidation of a manuscript tradition that had gathered Alfonsine material labeled "Alfonsine Tables", "tables of John of Lignères", and a star catalogue (according to the translation of Gerard of Cremona made in Toledo in about 1175 with longitudes increased by 17;8° over those in the *Almagest*: Kunitzsch 1986, p. 89), together with some other tables. It is often assumed that subsequent editions are identical with it, but they are not. For example, the planetary velocity table, described in Castilian Alfonsine Tables (see chapter 27) and in the canons of John of Lignères (1322), is not found in the 1483 edition, although it is in the edition of 1492.

For this reason it is helpful to define the Alfonsine corpus more comprehensively than simply the "Alfonsine Tables". By Alfonsine corpus we mean the broadest category of astronomical texts beginning in 13th-century Castile under the patronage of Alfonso X and all texts that derive from them in some significant way.

For the first half of the 14th century, we include in this corpus the astronomical works produced by the Parisians, notably all the Johns (John Vimond, John of Murs, John of Lignères, John of Saxony, John of Montfort, and John of Genoa) and works produced in Oxford, notably the tables of 1348 by William Batecombe. All these works show the ongoing legacy of Alfonsine astronomy as it was first elaborated in Castile. In later sections of this chapter we will briefly mention some subsequent developments of the Alfonsine corpus that largely depend on the tradition that spread from Paris: notably, their impact in England and Italy, and copies of the Alfonsine tables in Spain. This corpus is vast and much of it remains to be explored by modern scholars, for it is almost co-extensive with European astronomy until the middle of the 16th century. For this reason we will only discuss a few selected items in this corpus.

6.2.1 Chronology

In the category of chronology there are three relevant characteristics that have been ascribed to the Parisian Alfonsine Tables: multiple calendars, differences between eras, and counting days sexagesimally. But we note that multiple calendars are found everywhere in medieval astronomy, including the Toledan Tables and the Castilian Alfonsine Tables. The discussion of a variety of calendars makes sense in Spain where many calendars were used. In Paris, however, dates were only given in the Christian calendar with the era of Incarnation; other calendars had no cultural context at the time. The differences between the eras expressed in sexagesimal days that appear in the Parisian Alfonsine Tables (for convenience, references will be to Ratdolt 1483, unless otherwise specified) are very similar to, and overlap with, those in some manuscripts of the Toledan Tables: see, e.g., Naples, MS VIII-C-49, f. 5r; Toomer 1968, p. 18; and F. S. Pedersen 2002, pp. 899–901. Indeed, the canons to the Castilian Alfonsine Tables, chapter 10:15, give the difference in days between the era of the Arabs and the Persian era sexagesimally. There is also a list of differences between eras in the Khwārizmī zij (in the Latin translation of Maslama al-Majrīṭī's version), expressed in years, months, and days (Suter 1914, p. 109). Such lists are found elsewhere in the Muslim world as well: for the list by Ibn Yūnus (*ca.* 1000), see Delambre 1819, p. 96; in a text of the 11th century, al-Bīrūnī's *Chronology of Ancient Nations* (trans. Sachau 1879, pp. 133, 404), the differences between eras are expressed sexagesimally in a table. So, in this respect, the Parisian Alfonsine Tables belong to the antecedent tradition. For example, consider the following data in a 13th-century century copy of the Toledan Tables, Naples, MS VIII-C-49, f. 5r:

inter adam et dilluvium*	4, 3,47,11d
(mg.: "2226 years, 1 month, 23 days, [?] hours")	
inter diluvium et nabo [i.e., Nabonassar]**	3,58,56,13d
inter nabo [i.e., Nabonassar] et alexandrem	0,44,11,41d
inter alexandrem et christum	0,31,34,45d
inter adam et arabes	10,21,33,35d
inter christum et machometum	1, 3, 3,35d
inter arabes et gezdargirch [i.e., Yazdegird], persarum rex	0, 1, 0,24d
inter alexandrem et arabes	1,34,38,20d

*The manuscript has "dilluvium" here; "diluvium", spelled correctly, occurs in the next entry.

**Nabonassar's name appears in various forms in the Alfonsine corpus, e.g., Nabuchodonosor (Ratdolt 1483, f. c1r; see also our comments in chapter 4 on 1:11–16).

In another copy of the Toledan Tables (Paris, Bibliothèque nationale de France, MS lat. 7198, f. 26r), the same interval between Adam and the Diluvium, 4,3,47,11d (≈ 2403 Julian years), is reported (F. S. Pedersen 2002, p. 900) and, according to Abū Maʿshar, this interval is 2226 years 1 month 23 days 4 hours (Pingree 1968a, p. 38; cf. al-Bīrūnī, trans. Sachau 1879, p. 25). With the exception of "era adam" all these data are found in the Parisian Alfonsine Tables. Moreover, in the second edition of the Parisian Alfonsine Tables (Venice 1492, f. a1v) the difference in time from Adam to the Flood is given as 3882 years and 267d or, equivalently, as 6,33,56,7d. It is also of interest that a number of different eras, including the era of Adam, are used in the general prologue to the *Siete Partidas,* produced at the court of Alfonso, to give the date when the king began his reign (Craddock 1974, pp. 367–368).

In the Parisian Alfonsine Tables sexagesimal notation for days and fractions of days is used consistently, rather than years, months, days, and hours, as found elsewhere. But tables for converting dates in the Christian calendar to a sexagesimal number of days since some epoch are already found in the Toledan Tables (see Toomer 1968, p. 14; and F. S. Pedersen 2002, pp. 879, 881, and 883; cf. Poulle 1984, p. 109), and sexagesimal fractions of a day had long been used for the length of the mean synodic month and tables based on it (see, e.g., *Almagest*, VI.3; Toomer 1984, pp. 278–280; and Nallino 1903–1907, 2:84–87; cf. Goldstein 2003b). Hence the only innovation in the

Parisian Alfonsine Tables related to sexagesimal notation for days is the arrangement of the mean motion tables; this is quite helpful for the computer but it does not introduce any new concepts.

6.2.2 Signs of 60°

The Parisian Alfonsine Tables also display a new presentation of radices and mean motions using so called "physical signs" of 60°, rather than "zodiacal signs" of 30°. This presentation became a characteristic feature of Parisian Alfonsine astronomy as it spread all over Europe, and it parallels the sexagesimalization of days. However, zodiacal signs of 30° were never entirely displaced and even appear in works by the principal collaborators who produced the Parisian Alfonsine Tables. Moreover, the idea of using signs of 60° is not uniquely associated with these Parisian astronomers, for it appears in a Spanish text, namely the Hebrew zij of Judah ben Asher II (d. 1391), that is unrelated to the Parisian Alfonsine Tables and that also mentions Isaac ben Sid as the author of Alfonso's zij (Vatican, MS Heb. 384, f. 285b), information unavailable in Latin.

6.2.3 The maximum solar equation

Since we have already considered John of Murs's discussion of 2;10° for the maximum solar equation as well as his discussion of the solar mean motion (Goldstein and Chabás 2001), we will only summarize our argument here. In the *Expositio*, John of Murs asserts that there were two observations of autumnal equinox, one by Ptolemy on September 25, 132 at 2 p.m. (*Almagest* III.7; Toomer 1984, p. 168) and the other by Alfonso on September 16, 1252, without specifying the hour. From these two observations he claims to derive a daily solar motion of 0;59,8,19,37,19,13,56°, for between these two dates there were 409,071 days, equivalent to 1120 revolutions of the Sun or 403,200° (accurately 403,200°/409,071d = 0;59,8,19,57,40,...°/d). Corresponding to this daily solar motion is the length of the tropical year that John of Murs ascribes to Alfonso (Poulle 1980, p. 251), namely, 365;14,33,9,59,20,7,30d (accurately 360°/0;59,8,19,37,19,13,56 = 365;14,33,9,57,...d). Later in the *Expositio*, the solar year is given as 365d 6h – 0;10,44h, and this is equivalent to 365;14,33,10d (Poulle 1980, p. 265).

The first observation took place on JDN 1769539 and the second on JDN 2178610. As for the number of revolutions cited, John of Murs assumes that the Sun returned to the same mean position. He correctly reports that in the *Almagest* the mean position of the Sun at the time of the first observation was 182;10° (= 116;40° + 65;30°, where the solar apogee is 65;30°), from which it follows that at the time of the second observation the mean Sun was again at 182;10°. Now, at autumnal equinox, the true Sun has to be at 180°, hence the solar equation is –2;10°. We have no independent evidence that there was an observation of equinox on September 16, 1252, although the canons to the Castilian Alfonsine Tables, chapter 0:12, indicate that some observations were made at equinoxes, solstices, and the mid-points between them. On the other hand, it happens that computing with the Toledan Tables one finds that there was an equinox on that date. In any event, John of Murs asserts that he has seen this observational report in what he calls the "Tables of Alfonso" and that the daily solar motion appeared there at the head of the table for the solar motion (*in capite tabule motus solis*: Poulle 1980, p. 253). If one further assumes, as does John of Murs, that the solar apogee at his time is at about 90°, it follows that 2;10° is the maximum solar equation. As we noted (Goldstein and Chabás 2001, pp. 346–347), the "derivation" of the maximum solar equation in the *Expositio* is clearly based on computed data, not on observations (and no observations are cited in the relevant passage). Therefore, we infer that the maximum solar equation of 2;10° was in John of Murs's source, as did North (1996, p. 461). In addition to the table by John Vimond (see Section 6.3.1), there is evidence from 13th-century Castile that the maximum solar equation was already taken to be 2;10° (see the comments in chapter 4 on 16:6).

6.2.4 The maximum lunar equation

There is no discussion of lunar parameters in the *Expositio*, but in the Parisian Alfonsine Tables the maximum lunar equation of center is 4;56°, whereas in the Toledan Tables it is 5;0,59° and in al-Battānī's zij it is 5;1,0° (as in the *Almagest*). Poulle argued that this lunar parameter, 4;56°, is a characteristic feature of the Alfonsine Tables of 1327, that is, those defined by the canons of John of Saxony (Poulle 1987, pp. 52, 69; Poulle 1988, pp. 99–101). However, it is not a new parameter, for it already appeared in Hindu astronomical texts that were transmitted to the Islamic world in the late 8th century. It also appears in the Latin version of al-Khwārizmī's zij that was in the Hindu tradition (Neugebauer 1962a, p. 96; Suter 1914, p. 134). Moreover, in the 11th century this parameter was adapted in al-Andalus by Ibn Muᶜādh to an eccentric model in the Ptolemaic tradition (Samsó 1992, p. 157). Hence, it is clear that no new observations were required in the 13th (or 14th) century to replace Ptolemy's 5;1° with 4;56°. Poulle (1987, p. 63) also claimed that the tables in Paris, Bibliothèque nationale de France, MS lat. 7286C (ff. 9r–58v) are those described in John of Lignères's *Priores astrologi* of 1322, and noted that for the Moon there are two columns in the same table in this manuscript, one with a maximum of 5;0,59°, and the other with a maximum of 4;56° explicitly attributed to "Alfonso" in the heading. (Due to an isolated scribal error, the manuscript gives 4;55,0° as the lunar equation for argument 3s 5° instead of 4;56,0°.) Poulle interprets this as a juxtaposition of old and new values, or as an intermediate stage ("un état intermédiaire"). But, as we have seen, the value 4;56° is certainly not new: his suggestion of an intermediate stage is not supported in the corresponding canons of the *Priores astrologi* and there is also no hint in them of two values for the maximum lunar equation. We note that Vimond's lunar tables, in the very same manuscript, have a maximum equation of 4;56° (see Section 6.3.1).

6.2.5 The maximum equation of center for Venus and Jupiter

In the tables associated with John of Lignères in Paris, Bibliothèque nationale de France, MS lat. 7286C, there is also a table for the equation of center of Venus with a maximum of 1;59°, whereas the solar equation has a maximum of 2;10°. Poulle interprets this as support for an intermediate stage before the introduction of 2;10° as the maximum equation of center for Venus which is the value that appears in the Parisian Alfonsine Tables. However, in the tables associated with John Vimond in the very same manuscript, there is a table giving the equation of center for Venus (f. 5v), with the same format as that for the equation of the Sun and with the same entries, although presented at 6°-intervals instead of 3°-intervals, where the maximum equation of center is 2;10° (see Section 6.3.1). We are thus persuaded that there was no intermediate stage in Paris, for the maximum equation of center for Venus was probably 2;10° already in Toledo. As we noted in the comments on the canons to the Castilian Alfonsine Tables, chapter 18, it was common practice to set the apogee of Venus equal to that of the Sun, and to set the maximum equation of center for Venus equal to the maximum solar equation (Goldstein 1977). For example, in the Toledan Tables the apogees of the Sun and Venus are both 2s 17;50°, whereas the maximum solar equation is 1;59,10° and the maximum equation of center for Venus is 1;59° (Toomer 1968, pp. 45, 56, 65). In fact, John of Lignères holds the same view, for in canon 11 of *Priores astrologi* he says that the argument of the Sun and the center of Venus are the same (*et habebis argumentum solis in principio illius anni et centrum Veneris que semper sunt idem*), i.e., they have the same apogee; and in canon 16 he indicates that the equations are also the same (*et scias quod centrum medium Veneris est idem cum argumento solis; cum centro igitur medio cuiusvis eorum intra tabulas equationis eiusdem querendo simile in lineis numeri...*). No parameters are given in canons 11 and 16 (Saby 1987, pp. 197–201). According to Poulle, Paris, Bibliothèque nationale de France, MS lat. 7286C, is the only manuscript to preserve John of Lignères's original planetary equation tables of 1322, for they were generally displaced by the corresponding tables associated with John of Saxony's canons of 1327. But the fact that this manuscript, or a part of it, is exceptional is not in itself a sufficient reason to believe it is more faithful to the archetype of the tables of 1322, and Saby's description of this manuscript (1987, pp. 516–520) offers no support for Poulle's conjecture.

Moreover, it is significant that the maximum equation of center for Jupiter is 5;57° both in Vimond's tables and the Parisian Alfonsine Tables, for this value was not known in any text prior to the Parisian Alfonsine Tables, indicating a strong connection between the tables of John Vimond and the work of his Parisian contemporaries. The origin or derivation of this parameter for Jupiter is not described in any extant text, and it is likely that this value was simply taken from an earlier work: the most reasonable candidate is the Alfonsine Tables as they existed in Castile. Changing parameters for maximum equation is well attested in al-Andalus and the Maghrib and, in fact, Ibn al-Bannā' (d. 1321) has a table where the maximum equation for Jupiter is 5;41° instead of 5;15°, the value given by Ptolemy and al-Battānī (Samsó and Vernet, 1998, p. 273). But in the Alfonsine tradition in the 14th and 15th centuries that was diffused from Paris we have no examples of replacing a parameter for maximum equation. In sum, we think it most unlikely that John of Lignères was responsible for introducing any new values for maximum equation.

6.2.6 Planetary velocities

As we already indicated in the comments on the canons to the Castilian Alfonsine Tables, chapter 27, "On the daily unequal motion of the planets", the table for this purpose in manuscripts of John of Lignères follows the rules given here. Moreover, it has been demonstrated that canon 44 in John of Lignères's *Priores astrologi* (1322) agrees very closely with the Castilian instructions in chapter 27 (Goldstein *et al.* 1994, especially pp. 85–87). As F. S. Pedersen informed us, this table appears in a number of 13th-century copies of the Toledan Tables, but no detailed canon for them is known in the corpus of the Toledan Tables that he has examined exhaustively (F. S. Pedersen 2002). Hence, we conclude that John of Lignères had access to the canons of the Castilian Alfonsine Tables in some form.

6.2.7 Tropical and sidereal years

The canons to the Castilian Alfonsine Tables fix the sidereal year and vary the tropical year (chapter 49), as in the Toledan Tables and elsewhere. However, we will argue that the model for trepidation described in these canons is not the model ascribed to Thābit that underlies the Toledan Tables; rather, it is preserved in the periodic term in the trepidation model of the Parisian Alfonsine Tables, reinterpreted to conform with a variable sidereal year (see Section 6.2.8, below). Indeed, John of Murs indicates that the radices of the apogees and the fixed stars in the Castilian Alfonsine Tables were given on the 8th sphere, i.e., in sidereal coordinates: *Alfonsius considerans motus augium et fixarum non posuit ad 9am speram sed eorum radices posuit ad 8am* (Poulle 1980, p. 263).

The length of the tropical year given in the Parisian Alfonsine Tables first appears in the *Expositio*, and John of Murs explicitly says that he took it from his copy of Castilian Alfonsine Tables (Poulle 1980, p. 251 *et passim*). This value for the tropical year, 365;14,33,9,59,...d, is very close to that appearing in Pseudo-Thābit's *On the solar year* (Neugebauer 1962b, p. 285; and Morelon 1987, p. 60): 365;14,33,12d. The motion of access and recess goes back to Pseudo-Thābit's *On the motion of the eighth sphere* (Neugebauer 1962b) with subsequent variants by Azarquiel and others, in which the sidereal year was fixed and the tropical year varied. Historically, the reason for introducing access and recess, i.e., trepidation, was that the value for the tropical year in the *Almagest* (365;14,48d) differed substantially from the value for the tropical year found by Arab astronomers in the 9th century. While these astronomers recognized that Ptolemy's value for the tropical year was not very good for their epoch, they were reluctant to say Ptolemy was wrong about his own epoch. In *On the solar year* by the Banū Mūsā (9th century), but later ascribed to Thābit ibn Qurra (d. 901), the length of the sidereal year is given as 365;15,23,34,33d (Morelon 1987, p. 56; Neugebauer 1962b, p. 280: in Latin the last place is 43 instead of 33), and this is close to the implied length of the sidereal year, 365;15,24,31,...d, that follows from Ptolemy's tropical year and his value for the rate of precession (see Section 6.2.8, below). In order to harmonize the data, they introduced a variable term for the difference between the fixed sidereal year and the tropical year. In other words, there was evidence in the historical record for varying the tropical year, but nothing in that record implied that the sidereal year varies. Note that no evidence is presented by John of Murs (or anyone else) to support the claim that the sidereal year varies, and it is this claim that underlies the trepidation model in the Parisian Alfonsine Tables. Therefore, maintaining a fixed tropical year and varying the sidereal year is to misunderstand the problem, and using an improved value for the tropical year is not a good reason to vary the sidereal year (Goldstein 1994). In fact, with good values for both year lengths there is no need to vary either of them.

6.2.8 Precession and trepidation

Precession is a term used to describe a constant rate of increase in the tropical positions of the fixed stars, and trepidation refers to models according to which precession varies (and can even be negative for certain periods of time). The accumulated precession (or trepidation) between two epochs is called the total precession, and some models, such as the one ascribed to Thābit, yield a value for the total precession at a given time directly. In all cases it is assumed that the sidereal positions of the fixed stars are constant, i.e., that the positions of the fixed stars relative to one another do not change over time. Properly speaking, the radix in a theory of trepidation refers to the time when the vernal equinox is assumed to have coincided with the zero point of sidereal longitude, called in some texts "the head of Aries" but, for convenience in computation, radices (i.e., values for the arguments of the terms in the theory) may be given for other epochs. However, for those who assumed that there is only a fixed rate of precession (e.g., Ptolemy), there is no radix and the epoch for precession can be any date for which a star list is available (e.g., Ptolemy's star catalogue for 137 AD); in this case the total precession is simply the product of the rate of precession and the time difference between two dates. Ptolemy assumed that the solar apogee was fixed tropically but, in the 9th century, Arab astronomers realized that fixing the solar apogee sidereally was preferable and, in the 11th century, Azarquiel noted that the solar apogee has a proper motion as well (see, e.g., Samsó and E. Millás 1989, pp. 7–9). As we shall see, Vimond's tables also include a proper motion for the solar apogee (Section 6.3.1).

Ptolemy fixed the rate of precession at 1° per 100 years. With this rate of precession (equivalent to 0;0,36°/y, corresponding to the solar motion in 0;0,36,31d) and his tropical year of 365;14,48d, the implied sidereal year is 365;15,24,31,...d (cf. Neugebauer 1975, p. 902), and this value for the sidereal year is considerably better from a modern point of view than is his value for the tropical year. The modern values for the tropical and sidereal years are 365.24219d ≈ 365;14,31,53,...d and 365.25636d ≈ 365;15,22,53,...d, respectively (Ridpath 1989, p. 58). There were many values for precession in the Middle Ages: in the *Expositio*, we find 1° in 100 years ascribed to

Ptolemy, 1° in 66 years ascribed to al-Battānī, and 1° in 72;30 years ascribed to Alfonso (Poulle 1980, p. 258; cf. Nallino 1903–1907, 1:124, 292). Elsewhere in the *Expositio*, the value 1° in 70 years is ascribed to al-Ṣūfī (Poulle 1980, p. 262) although, in fact, al-Ṣūfī's value was 1° in 66 years (Samsó and Comes 1988, p. 68). In this case, John of Murs's source, on which he depended heavily, was the prologue to William of Saint-Cloud's *Almanach planetarum* (see, e.g., Paris, Bibliothèque nationale de France, MS lat. 7281, f. 141r, as noted in Section 6.3.2). The value ascribed to Alfonso can be derived from the data given in *On the solar year*, for there we are told that an increment in stellar longitudes of 13;12° took place in 957 years, and this is exactly the same as 1° in 72;30 years (13;12 · 72;30 = 957: cf. Goldstein 1994, p. 190). The *Expositio* also has another value for precession: 1;22,40,53° in 100 years which, according to the text, is equivalent to 0;0,49,28°/year (accurately, 0;0,49,36,32°/year or 1° in about 72;34 years: Goldstein 1994, p. 196 n. 33), whereas 1° in 72;30 years corresponds to 0;0,49,39°/year. This computation may have been inspired by comparing the length of the sidereal year implicit in the Toledan Tables with the length of the tropical year in the *Expositio*: 365;15,23,29d – 365;14,33,10d = 0;0,50,19d, and 0;0,50,19d · 0;59,8,19°/d ≈ 0;0,49,36° or 1° in about 72;35 years (cf. Goldstein 1994, p. 193).

Despite the report in the *Expositio* that Alfonso had a value for precession (and a very good one at that), a model for trepidation is also ascribed to him. It is difficult to reconcile these mutually incompatible claims and to determine what exactly was in John of Murs's source. The standard model for trepidation in the Parisian Alfonsine Tables has two components, one linear and the other periodic, with a fixed tropical year and a variable sidereal year, and it is also described by John of Lignères in canon 10 of his *Priores astrologi* of 1322. We have discovered that the same model for trepidation with two such terms underlies the tables of John Vimond (see Section 6.3.1; Chabás and Goldstein 2003). The relationship between John Vimond and his two better known contemporaries is not yet clear, and it is possible that they all depended on a common source. There is no evidence for a model with two components in the Castilian Alfonsine Tables, but there are discussions of two component models in the earlier literature. In particular, al-Battānī describes a combination of Theon of Alexandria's trepidation model that has a maximum increment ("access") or decrement ("recess") of 8° together with a linear term (Ragep 1996, pp. 271–272). Moreover, al-Biṭrūjī (*ca.* 1200) describes a model with two components: one of precession at Ptolemy's rate of 1° in 100 years, and the other of trepidation with Theon's amplitude of 8° (Goldstein 1971, 1:90); and Azarquiel says much the same thing (Millás 1943–1950, pp. 275–278). But the model described in the *Expositio* is different, for the linear term is based on the difference between a fixed tropical year and the calendrical year of 365;15d with a period of 49,000 years. The period of 49,000 years may simply be a rounding, for

$$360° / [0;59,8 \cdot (365;15d - 365;14,33,10d)] \approx 49,006 \text{ years},$$

or it may have been derived from rounding a computation that invoked the tropical year given in *On the solar year:*

$$360° / [0;59,8 \cdot (365;15d - 365;14,33,12d)] \approx 49,067 \text{ years}$$

i.e., no new observations were needed to arrive at this period of 49,000 years. This term depends on a fixed tropical year and so it is most unlikely to have been represented in the Castilian Alfonsine Tables which have a variable tropical year. Regardless of the historical origin of this periodic term, it is devoid of astronomical meaning, for the calendrical year is merely conventional and the difference between it and the tropical year corresponds to nothing in nature (cf. North 1996, p. 463).

In the Parisian Alfonsine Tables one finds a list of radices for the motion of the apogee of the Sun and Venus "without the motion of the 8th sphere"; a table for the mean motion in "access and recess"; and a table for the "equation of motion of access and recess of the 8th sphere". The list of radices takes into account the linear term in the Parisian model for trepidation, whereas the tables for the mean motion and equation of access and recess represent the periodic term in this model. The total precession that results from combining these two terms is then to be applied to the longitudes of the planetary apogees and to the longitudes of the fixed stars to yield their tropical positions.

The linear term, p_1, can be expressed as

$$p_1 = (y - r) \cdot 360 / 49,000,$$

where y is a given year and r is the year of the radix. If y – r is given in days, then

$$p_1 = (y - r) \cdot 360 / (49,000 \cdot 365;15).$$

In the *Expositio* (Poulle 1980, p. 257), the daily increment for the linear term is given as 0;0,0,4,20,41,17,...°, which is equal to 360/(49,000 · 365;15), and it is ascribed to Alfonso (this parameter is also in the *editio princeps*: Ratdolt 1483, f. d4v). This implies that the source used by John of Murs for Alfonso differed in some respects from the canons to the Castilian Alfonsine Tables. North (1996, p. 463) has suggested that the linear term in a trepidation model can be concealed in the list of radices for the solar apogee (as it is in the Parisian Alfonsine Tables where the radices are different, depending on the epoch), in which case it need not be discussed in the canons. But if the apogees are fixed sidereally (i.e., with respect to the 8th sphere) and the sidereal year is fixed, there is no motion of the apogees in sidereal coordinates and no need for a list of radices for the solar apogee. Note that the solar apogee in the Toledan Tables is fixed at 77;50°, and there is no motion associated with it because these tables use sidereal coordinates and a fixed sidereal year. As we indicated above, the Castilian Alfonsine Tables also had a fixed sidereal year; hence, North's argument does not apply to them. On the other hand, the proper motion of the apogees could be concealed in the radices of the apogees (in sidereal coordinates), and this may have been the case with the Castilian Alfonsine Tables. In fact, Azarquiel introduced a proper motion for the solar apogee and it was applied to planetary apogees by several astronomers in Muslim Spain (Samsó 1992, p. 212; Chabás and Goldstein 1994, p. 28; Samsó and E. Millás 1998, pp. 268–270; cf. Goldstein 2001, pp. 235–236; and Goldstein 2003a, p. 160); nevertheless, there is no proper motion for the solar or planetary apogees in the Parisian Alfonsine Tables and no discussion of this motion in the canons to the Castilian Alfonsine Tables. (For Vimond's inclusion in his tables of a proper motion for the solar and planetary apogees, see Section 6.3.1.)

What appears to be a derivation of the linear term in the *Expositio* is really a circular argument. John of Murs begins with the daily solar motion, equivalent to the tropical year length mentioned earlier in the text (that also was inserted in a circular argument), and then finds that the motion of the Sun in a calendrical year of 365;15d is equal to 360;0,26,27° (Poulle 1980, p. 260). The excess over 360° is 0;0,26,27° and it accumulates to 1° in about 136 years. This value can be compared with the period for the linear term: 49,000y / 360° = 136;6,40 years or 0;0,26,26,56,...° per year (see Poulle 1980, p. 254). But this argument does not justify taking the tropical year as fixed (the underlying assumption for the linear term), for the conclusion here is identical with the starting point.

In sum, there is no reason to believe that the linear term in the trepidation theory first presented in the *Expositio* appeared in the Castilian Alfonsine Tables. On the other hand, in John Vimond's tables a model for trepidation has a linear term of about 1° in 136 years (Section 6.3.1), that is, its period is very close to the period of the linear term in the Parisian Alfonsine Tables (49,000 years). Since we argue that Vimond's tables are independent of the *Expositio*, it is entirely possible that Vimond influenced John of Murs or, more likely, that they both drew from a common source.

The periodic term presents different problems. In the *Expositio* there is no derivation of this term and there is no attempt to justify it by showing that it accounts for any data, in contrast to the discussion by Azarquiel (see Samsó 1990b). Therefore, it is reasonable to assume that this periodic term was simply taken over from John of Murs's source, and there is no obstacle to assuming that it was taken from the trepidation model in the Castilian Alfonsine Tables. Again, John Vimond's tables have a periodic term such that its cycle of 360° is completed in about 7,000 years, which is close to the value in the Parisian Alfonsine Tables and implied in the *Expositio*; this also points to a common source. Although the periodic term would have been associated with a fixed sidereal year in the Castilian Alfonsine Tables, the appropriate formula allows it to be used with either a fixed or a variable sidereal year. In the corresponding table in the Parisian Alfonsine Tables (Ratdolt, f. d3v), this term, p_2, is approximately sinusoidal (deviating at most by 0;0,55°) with a maximum of 9° and a period of 7,000 years (Poulle 1980, p. 258; North 1996, p. 464), and it can be represented by the formula:

$$p_2 = 9° \sin [(y - r) \cdot (360/7{,}000)],$$

where y is a given year and r is the epoch of the radix. Similar models for trepidation were used previously: in *On the motion of the eighth sphere*, ascribed to Thābit and only extant in Latin, the table for trepidation has a maximum of 10;45° (Neugebauer 1962b, p. 297); a model with a maximum of 10;24° is associated with the tables of Ibn Isḥāq and Ibn al-Bannā' (Samsó and E. Millás 1989, p. 15); and Ibn al-Kammād has a sinusoidal term with a maximum of 9;59° (Chabás and Goldstein 1994, pp. 25–27). In other words, a sinusoidal trepidation term makes sense in a Spanish context. In the *Expositio*, Thabit's model is said to have a maximum of 16;45° (Poulle 1980, p. 258), but this is probably a scribal error for 10;45° (see the discussion of William of Saint-Cloud's remarks, below).

There has been some discussion of the epoch of the radix for this term. Using the data in the Parisian Alfonsine Tables, the epoch, when $p_2 = 0°$, is May 15, 16 AD (Mercier 1977, pp. 58–59). In other words, for the epoch of the Incarnation $p_2 = -0;47,26°$ which appears in the tables as 359;12,34°. Since this date in 16 AD is the time when star positions should agree with those in Ptolemy's star catalogue in the *Almagest*, it has led to speculation that Alfonso's astronomers did not realize that this catalogue in fact is dated 137 AD (cf. Samsó and Castelló 1988). The *Expositio* does not provide any data that would indicate how the epoch was determined, and this suggests that the tables and its epoch were already in the source used by John of Murs.

In the *Expositio*, there is a computation that depends on this model, for we are told that for 1300 *perfecto* (i.e., the beginning of year 1301) the "motion of the 8th sphere" is 8;13,13° (Poulle 1980, p. 261). With the model and the Parisian Alfonsine Tables, the motion in 1300 years in access and recess is:

$1300 \cdot 0;3,5,8,34,17,...°/y \approx$	66;51,26°
Radix Incarnation:	359;12,34°
Total	66; 4, 0°

and

$$p_2(66;4,0°) = 9° \sin 66;4° = 8;13,34°.$$

However, if we interpolate for p_2 in the table for the equation of access and recess that appears in the *edito princeps* (d3v), we get 8;13,13°, as in the text. As already noted by Poulle (1984, p. 131), this table is not exactly a sine function. Indeed, the entries in the table are systematically smaller (in the range from 0° to 90°) than those obtained from a properly computed sine table; the maximum discrepancy is 0;0,55° and occurs at 37°. Hence, there is no reason to doubt that John of Murs had the table for the periodic term that appears in the Parisian Alfonsine Tables.

The complete model for trepidation in the Parisian Alfonsine Tables yields a total precession, $p = p_1 + p_2$, and it produces a value for p of about 17;8° for the epoch of Alfonso used in Paris (i.e., June 1, 1252), in agreement with the total precession in the Alfonsine star catalogue (Samsó and Castelló 1988, pp. 116–117):

$$p_1 = (1252 - 16) \cdot (360°/49{,}000) \approx 9;5°$$

and, from the list of radices in the *editio princeps* (c8r),

$$p_2 = 8;4°.$$

Hence p = 17;9° (or 17;8° with the correction to p_2 suggested by Samsó and Castelló 1988, p. 176). But the total precession of 17;8° in the star catalogue (which is dependent on the catalogue of al-Ṣūfī) probably is based on a very different computation, as we indicated in chapter 5, above. According to al-Ṣūfī, the total precession for his epoch, 964 AD, is 12;42° (to be added to the longitudes of the fixed stars in Ptolemy's star catalogue). The difference between 17;8° and 12;42° is 4;26°. At the rate of 1° per 66 years, this total precession corresponds to about 292;30y (≈ 4;26° · 66y) which, when added to 964, yields 1256;30—and 1256 is the year of composition of the first version of the *Libro de las estrellas de la ochaua espera*. Thus, the total precession, 17;8°, results from adding the amount of precession given by al-Ṣūfī for his time and the amount of precession computed from then to the time of Alfonso at a rate of 1° per 66 years.

In any event, if the total precession of 17;8° was determined independently of the trepidation model, it would serve as a fixed point for that model, leading to an epoch of 16 AD that was probably unintended. The Parisian Alfonsine Tables do not give a sidereally fixed solar apogee, for the list of radices already incorporates the linear term, p_1, and the sidereal year varies. It is not clear which of the radices serves as the epoch for the sidereal longitude of the solar apogee, but the value for the epoch of the Hijra in 622 AD, 75;59,21° (≈ 76°), is a likely candidate. While this parameter is not found in any previous text, it is not far from the value in the Toledan Tables, 77;50° (Toomer 1968, p. 45).

The solar mean motion in the *Expositio* was found by comparing an observation of autumnal equinox made by Ptolemy in 132 AD with and the autumnal equinox of 1252 (see Section 6.2.3). This computation invoked the position of the mean Sun at the autumnal equinox in 132 AD, and the value in the *Almagest* cited by John of Murs depended on taking the solar apogee at that time to be 65;30°. The tropical longitude of the solar apogee at the time of Alfonso, according to the Parisian Alfonsine Tables, is 88;41° (= 80;37° + 8;4°); hence the difference in 1120 years (= 1252 – 132) is 23;11°, and this value for the total precession is incompatible with the Parisian model. So, at the very least, the value for the solar apogee at the time of Ptolemy was ignored in constructing the Parisian trepidation model. But this citation of an observation in the *Almagest* from 132 AD indicates that Ptolemy's date was known (to within narrow limits), and certainly not in error by more than 100 years, as suggested by the epoch of the periodic term.

In the *Expositio*, John of Murs appeals to William of Saint-Cloud's criticism of Pseudo-Thābit's model for trepidation, and cites the following data from him for the vernal equinox of 1290: it was observed on March 12 at 16h after noon; according to the Tables of Toulouse, the position of the Sun in the 8th sphere was 11s 19;47° (= 349;47°) when its position on the 9th sphere was 0° or 360°, i.e., the value for precession at that time was 10;13° (= 360° – 349;47°); but according to Thābit, the value for precession at that time was only 9;23° (Poulle 1980, p. 262). With the Tables of Toulouse (cf. Poulle 1994, tables 1 and 6), in which the year begins on March 1, we compute:

Radix for "1272" (= Feb. 28, 1273):	11s 5;40,13°
17 years:	11s 29;38,39°
12d*	11;49,38°
16h*	0;39,25°
Total:	347;47,55°

*For these two values we use the recent edition of the Toledan Tables (F. S. Pedersen 2002, p. 1148).

If we add to it 1;59°, the value for the solar equation in the Toledan Tables, we get 349;46,55° ≈ 349;47° for the true solar longitude on the 8th sphere, as in the text. Curiously, John of Murs uses the solar equation as it appeared in the Tables of Toulouse without changing it to 2;10° which is the maximum solar equation he "derived" in the *Expositio*. But it is now clear that he was simply copying the computation he found in Saint-Cloud's prologue (Paris, Bibliothèque nationale de France, MS lat. 7281, f. 141r) that included a value for the maximum solar equation taken from the Toledan Tables, 1;59°.

As mentioned before, the value of trepidation in the *Expositio* for 1290 according to Thābit is said to be 9;23° (as it is in Saint-Cloud's prologue), and we can confirm this value. The daily motion of the argument in the Toledan Tables is 0;0,0,52,28,37°/d (see Toomer 1968, pp. 118–120). For 1290, we subtract the epoch year 622 (for the Hijra), and the remainder is 668 years. Then the argument for trepidation results from adding the radix for the Hijra (1;34,2°) to 668 · 365;15 · 0;0,0,52,28,37°/d, that is, 1;34,2° + 59;16,35° = 60;50,37°. Hence

$$p = 10;45° \sin (60;50,37°) = 9;23,16,...° \approx 9;23°,$$

using the sine function to approximate the table ascribed to Thābit. John of Murs remarks that the value for the amount of trepidation according to Thābit is in error by about 1° in his time, i.e., *ca.* 1320, and the same error is said to apply to 1290 (Poulle 1980, pp. 258, 262). Now, the periodic term in the Parisian Alfonsine Tables for 1290 is

$$p_2 = 9° \sin[(1290 - 16) \cdot 360/7{,}000] = 8;11°,$$

whereas for 1320 it is

$$p_2 = 9° \sin[(1320 - 16) \cdot 360/7{,}000] = 8;17°$$

which is about 1° less than the value attributed to Thābit.

We have been able to confirm that the data in this section of the *Expositio* related to 1290 were taken from Saint-Cloud. But it is by no means clear how these computations could help John of Murs determine the parameters of the Parisian model for the periodic term. William of Saint-Cloud is known to have criticized Pseudo-Thābit's model for trepidation (Mancha 1992, p. 277 n. 5), but John of Murs does not indicate how these criticisms might justify the model he proposes. Note that Pseudo-Thābit's model produces a total precession of 9;23°, a value intermediate between William of Saint-Cloud's 10;13° and the periodic term according to the Parisians of 8;11° (for 1290).

John of Murs cites an observation he made of the vernal equinox on March 13, 1319 (see Fig. 6.1). The passage is confusing, for he gives the wrong year ("1318 imperfecto": corrected by Poulle [1980, pp. 266–267] to "1318 perfecto") and he gives the co-latitude (that is, the complement of the geographical latitude) of Paris as 41;26,40°, whereas he labels it the distance of Paris from the west which should mean its geographical longitude, not its co-latitude:

> *inveni altitudinem solis meridianam in dicta civitate Ebroicensi* [Evreux] *que ad modicum distat a Parisius ab occidente 41;26,40°, a qua substraxi altitudinem equinoctialem in meridie que est 41;10° sicut et Parisius et probavi, et remanent 0;16,40° in differentia qua fuit illa altitudo maior altitudine equinoctialis.*

If the co-latitude of Paris is 41;26,40°, its latitude is 48;33,20° (= 90° – 41;26,40°), whereas in the *editio princeps* it is 48;0° (Ratdolt 1483, f. m5r), in a star list of 1223 it is 48;30° (Kunitzsch 1966, p. 69), and John of Lignères gives 48;50° (Saby 1987, p. 189). The meridian altitude of the Sun on that date was observed to be 41;10° (i.e., its declination was –0;16,40°) from which John of Murs concluded correctly that the entry of the Sun into Aries was 16;40h after noon of March 13, 1319. He then indicates that, according to Alfonso, the vernal equinox in that year took place in Castile at 15;41h after noon of March 13. To get this result (accurately 15;43h), John of Murs (Poulle 1980, p. 264) depended on a correction of 0;20° to the radix of the era of Alfonso (a correction that is then applied to all solar radices), based on his belief that the copy he had of the Alfonsine Tables was corrupt (for the historical significance of this remark see Section 6.3.2). He adds that according to William of Saint-Cloud it took place in Paris at 16;47h

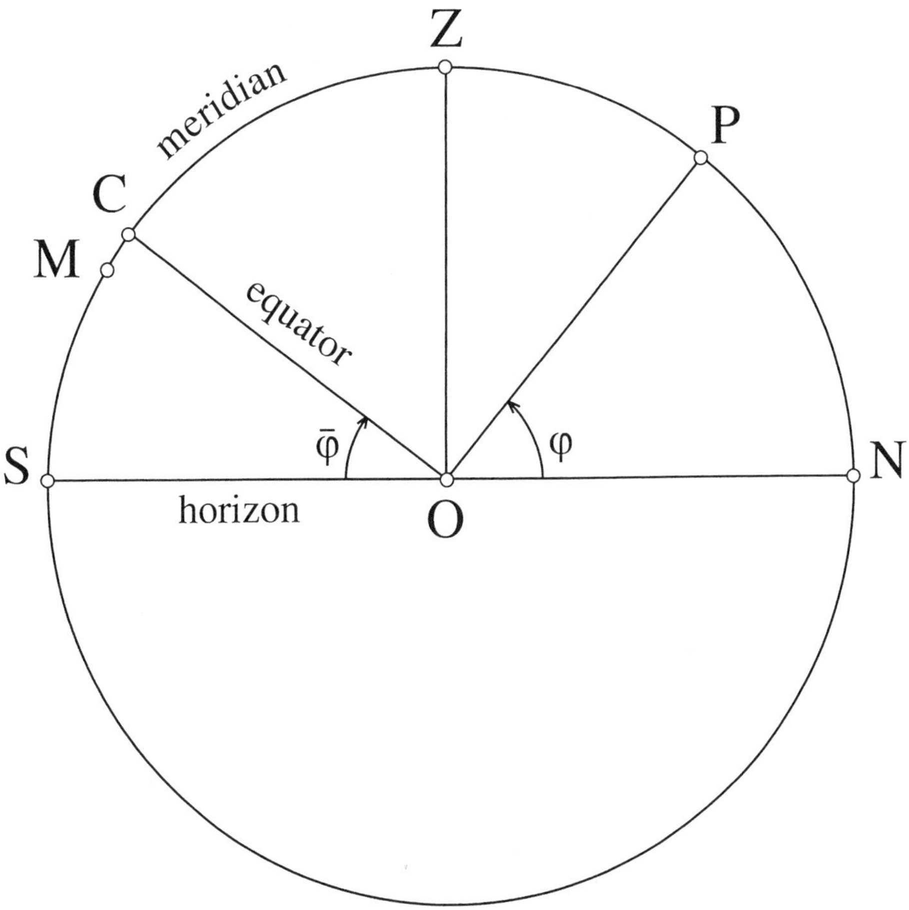

Fig. 6.1. The celestial sphere in cross-section: S is the south point on the horizon, N is the north point on the horizon, P is the north celestial pole, and Z is the zenith for an observer at O. Arc NP is the terrestrial latitude φ, and arc SC is the co-latitude $\bar{\varphi}$. According to the *Expositio*, at noon on March 13, 1319 the Sun culminated at M where arc SM = 41;10° and arc SC = 41;26,40°; hence, arc MC is 0;16,40°. Point M is south of the equator, i.e., the Sun has not yet reached the vernal equinox which is the place where the Sun crosses the equator from south to north.

after noon of March 13, and asserts that the difference of 1;6h (= 16;30°) is due to the difference between the meridians of Toledo and Paris. Note that in the *editio princeps* the difference between these two meridians is 11;20° or 0;45,20h (Ratdolt 1483, f. m5r), and in most manuscripts of the Alfonsine corpus it is 12° (= 0;48h) with values ranging from 11;55° to 18;6° but not including 16;30° (Kremer and Dobrzycki 1998, p. 194). Although this observation serves to confirm the maximum solar equation in the Alfonsine Tables, John of Murs does not use it to derive, or even to confirm, the parameters of his trepidation theory.

Despite all these difficulties with the Parisian model for trepidation, it was rarely challenged in the Latin West before Copernicus who restored the sidereal year and introduced a new model for trepidation (Goldstein 1994). The Parisian model succeeded in producing reasonable values for the solar apogee (*ca.* 90° for the 14th century), and acceptable tropical longitudes for the fixed stars. However, all trepidation models were ultimately rejected, largely due to the efforts of Tycho Brahe at the end of the 16th century, and precession was fixed at close to its modern value of about 0;0,50° per year or 1° in 72 years (Dreyer 1890, pp. 354–356), a value that had already appeared in the Tables of Barcelona (Chabás 1996a, pp. 486–487).

The modern formula for the annual rate of precession is (Smart 1965, p. 238):

$$p = 50''.2564 + 0''.000222\, t,$$

where t is the number of years after 1900. For epoch 1300, t = –600 years. Hence,

$$p = 50''.1232 = 0;0,50,7...^\circ.$$

Ironically, the rate of precession ascribed to Alfonso in the *Expositio* is very close to the value for his time that follows from a modern formula, and the Parisian theory of trepidation was not an improvement on it.

6.3 The astronomers in the Alfonsine tradition in Paris

The evidence in the early 14th century for the impact of the Castilian Alfonsine Tables in Paris is very extensive in contrast to the paucity of evidence in contemporary Spain. The astronomers in Paris usually associated with the introduction of the Alfonsine Tables are John of Murs, John of Lignères, and John of Saxony. It has been claimed that the *Expositio intentionis regis Alfonsii circa tabulas eius* by John of Murs was written in 1321 and that it is the first evidence of the arrival of the Alfonsine Tables in Paris (Poulle 1988). For problems with this dating of the *Expositio*, see Section 6.3.2. However, as mentioned above, Geoffrey of Meaux was already aware of "Alfonsine radices" no later than 1320. Moreover, John Vimond composed a set of tables based, in part, on parameters that probably came from the Castilian Alfonsine Tables or a tradition closely associated with them. The exact date of composition of this set of tables is uncertain, and the author is barely known, but it was produced shortly before or after 1320. This set of tables has not previously been described, and its significance has not been appreciated. In fact, it convinced us that the transmission of astronomical material from Castile to Paris is complex and probably involved more than one text, for Vimond's tables reflect a variety of traditions from Muslim Spain, some of which are otherwise unknown in Latin.

In Paris, due to the efforts of several astronomers, the Castilian Alfonsine Tables were recast into a set of tables that we call the Parisian Alfonsine Tables which proved to be most attractive to scholars all over Europe, including Spain, and they largely displaced the Toledan Tables that preceded them. In this section we offer an overview of this transformation and of those principally responsible for it. The evidence suggests that the Parisian version of these tables was a product of a significant level of collaboration, although the details are lacking. Shortly after this recasting was completed, in 1327 John of Saxony composed a set of canons that played a central role in their diffusion.

6.3.1 John Vimond

Vimond's tables are uniquely extant in Paris, Bibliothèque nationale de France, MS lat. 7286C (ff. 1r–8v), but no canons for them have been identified (see Chabás and Goldstein 2003). In a brief text at the end of this set of tables they are attributed to Iohannes Vimundus who made them "to be used by students at the University of Paris" (f. 8v):

> *Et in hoc terminatur opus Iohannis Vimundi baioceum dyocesis de disposicionibus planetarum et stellarum fixarum (...) et de disposicionibus eclipsalibus solis et lune et sequentibus, et de aliis disposicionibus ipsorum et aliorum corporum celestium, ad utilitatem scolarium universitatis parisiensis et omnium aliorum.*

> [Here ends the work by John Vimond of the diocese of Bayeux on the dispositions of the planets and the fixed stars, (...) and on the dispositions of solar and lunar eclipses and those following them (i.e., other syzygies), and on the other dispositions of these and other celestial bodies, for use by students at the University of Paris and all others.]

In the paragraph preceding his tables, Vimond tells us that they were compiled for Paris with 1320 as epoch (f. 1r):

> *(...) anno domini nostri Ihesu Christi 1320 secundum numeracionem annorum romanorum qui incipiunt ex inicio diei circoncisionis domini nostri Ihesu Christi existentis ad longitudinem civitatis Parisius que distat a medio mundi per 49 g et 30 min ita quod illa civitas est in parte occidentali et distat ab illo medio per 8 min 15 sec diei equalis.*
>
> [(...) in the year of our Lord Jesus Christ 1320 according to the count of Roman years (which start from the beginning of the Day of the Circumcision of our Living Lord Jesus), for the longitude of the city of Paris which is thus distant from the middle of the world by 49 degrees and 30 minutes because that city is in the western direction and distant from that middle by 8 minutes and 15 seconds of an equinoctial day.]

We note that both values given for the longitude of Paris from Arin, called "the middle of the world", are equivalent. Arin, a corruption of Ujjain (a city in India), was thought to be halfway between the eastern and western limits of the world (Neugebauer 1962a, p. 11, n. 2). The distance from Arin to Toledo was taken to be 61;30° and, since Paris was generally said to be 0;48h or 12° to the east of Toledo, its longitude from Arin is 49;30°, as in the passage above (Millás 1943–1950, p. 49; Kremer and Dobrzycki 1998, p. 194; and F. S. Pedersen 2002, p. 431). Moreover, when a day is taken to be 360°, it follows that 49;30° corresponds to 0;8,15d, for 49;30°/360° = 0;8,15. We also note that, according to this text, Vimond's tables were computed for Paris whereas in the early 1320s other Parisian astronomers who recast the Alfonsine Tables computed them for Toledo, as is the case for the tables with epoch 1321 by John of Murs (see, e.g., Lisbon, Biblioteca de Ajuda, MS 52-XII-35).

This John Vimond is mentioned by Simon de Phares among the most outstanding astronomers of his time, in his *Recueil des plus celebres astrologues* (1494–1498) recently edited by Boudet (1997–1999, 1:467):

> *Maistre Jehan Vymond fut a Paris, homme moult singulier et grant astrologien, lequel eut en ce temps grant cours pour la science des estoilles. Entre ses euvres, fist une verifficacion de la conjunction des lu[mi]naires, aussi des eclipses et estoilles fixes pour plusieurs ans. Cestui predist les grans vens qui furent en son temps et fist plusieurs beaulx jugemens, dont il acquist grant loz et renommee en France et fut moult devost en Nostre Seigneur.*
>
> [Master John Vimond lived in Paris, a most singular man and a great astrologer, who had at that time much prestige because of (his knowledge of) the science of the stars. Among his works is a verification of the conjunction of the luminaries, as well as eclipses and the fixed stars, for many years. He predicted the great winds which took place in his time and made many fine judgments for which he acquired great praise and renown in France and he was most devoted to our Lord.]

As we shall see, the *verifficacion de la conjunction des lu[mi]naires* refers to his tables.

There is no evidence that John Vimond had any relationship with John of Murs or John of Lignères and, indeed, Vimond's tables are arranged very differently from those of his Parisian contemporaries. Vimond tells us that he came from the diocese of Bayeux, Normandy, and his only other known work is a short treatise on the construction of an astronomical instrument, extant in Erfurt, MS F.377, and beginning *Planicelium vero componitur ex eis que sunt diversorum operum*… (Thorndike and Kibre 1963, col. 1050; Saby 1987, pp. 471, 474). He is also mentioned in Vatican, Biblioteca Apostolica, MS Ottob. lat. 1826, f. 153r, in a text that begins on f. 148ra, *Canon tabulae sequentis quae intitulatur tabula motus diversi solis et lunae in una hora et semidiametrorum secundum tabulas Alfonsi* (cited in F. S. Pedersen 2002, p. 177).

Vimond's tables are peculiar in several ways and retain signs of 30° rather than the physical signs of 60° that are characteristic of the Parisian version of the Alfonsine Tables. The organizing principle for the mean motion tables rests on syzygies, and the basic unit of time is the mean synodic month. As a result, most of his tables have formats which differ substantially from those found in the Toledan

Tables, the Parisian Alfonsine Tables, etc. In this set we find some tables based on parameters used by John of Murs and John of Lignères in their respective sets of tables, whereas others depend on parameters that differ from those in the Parisian Alfonsine Tables. John Vimond does not mention the names of any predecessors, and we can only surmise that he depended on Castilian sources in addition to the Toledan Tables that were already available in Paris.

A résumé of Vimond's 28 tables follows. The first table contains an annual calendar beginning on the day of Saint Matthew (February 24) and lists the dates associated with several saints, as well as the dates and times of 25 consecutive mean syzygies. In chapter 3 of the canons of 1327 by John of Saxony, it is said that the leap day is Saint Matthew's day, i.e., every four years there are two days associated with this saint (Poulle 1984, p. 36). In some calendars the first Saint Matthew's day is the last day of the year and the second Saint Matthew's day is the first day of the following year such that all years in the 4-year cycle begin on this saint's day. Hence it was not unreasonable of Vimond to use an annual calendar that begins on this day. The calendars by Geoffrey of Meaux and William of Saint-Cloud who were active in Paris shortly before Vimond also include astronomical data together with saints' days, but they depended on the Toledan Tables for their computations. According to our computations, the epoch for Vimond's tables seems to be Feb. 24, 1324 (rather than 1320, the date that appears in the text), and the radices elsewhere in his tables present difficulties which will be discussed below. In particular the solar apogee at epoch is given as 2s 29;56,15° (f. 1v), and it differs slightly from the values computed according to the Parisian Alfonsine Tables for either 1320 or 1324. In general, radices and mean motions are given to seconds, but equations are given only to minutes. In the Parisian Alfonsine Tables the equations for the Sun, the Moon, and the motion in access and recess are given to seconds, but planetary equations are given only to minutes.

The mean motion tables are arranged at intervals of half a mean synodic month where the value for the mean synodic month is 29;31,50,7,44,35d ± 0;0,0,0,0,4d. In contrast, in the Parisian Alfonsine Tables, the mean synodic month is 29;31,50,7,37,27,8,25d. These tables have entries for 1, 2, 3, and 4 years; for multiples of 4 years until 76 years; and then for 152, 304, 608, 1216, and 2432 years.

The mean motion of the Sun, 0;59,8,8,23,17°/d, that we derive from his tables implies a year length of 365;15,42,33,...d which is sidereal. In fact, all the mean motions are sidereal. We note that the value for the motion of the solar apogee, 0;0,0,11,13,34°/d, includes precession as well as its proper motion for, if we add the value for the (sidereal) mean motion of the Sun to the motion of the solar apogee, we find 0;59,8,19,36,51°/d, in close agreement with the corresponding value of the (tropical) mean motion of the Sun in the Parisian Alfonsine Tables (0;59,8,19,37,19,13,56°/d). In the *Almagest*, the planetary apogees are sidereally fixed whereas the solar apogee is tropically fixed. In the 9th century, astronomers in Baghdad fixed the solar apogee sidereally so that it too was subject to precession (or trepidation). But in the 11th century Azarquiel realized that the solar apogee had a proper motion in addition to precession, and fixed its amount as 1° in 279 Julian years or about 0;0,0,2°/d. In one Andalusian tradition, this proper motion of the solar apogee was applied to the planetary apogees as well (see Samsó and E. Millás 1998, p. 269). If we take al-Battānī's value for precession of 1° in 66 years, or about 0;0,0,9°/d, and add it to the proper motion of the solar apogee, the result is about 0;0,0,11°/d. There is no hint of this proper motion for either the solar apogee or the planetary apogees in the Parisian version of the Alfonsine Tables where these apogees are all sidereally fixed.

For the three superior planets the mean daily motion in longitude resulting from the entries for year 8 are shown below under the heading "Vimond". If we add the daily motion of the apogees (0;0,0,11,13,34°/d), as we did in the case of the Sun, we obtain the entries displayed in the second column, in very good agreement with the values for the mean motions in longitude in the Parisian Alfonsine Tables.

	Vimond	**Including the motion of the apogee**	**Ratdolt 1483**
Saturn	0; 2, 0,24, 3,56°/d	0; 2, 0,35,17,30°/d	0; 2, 0,35,17,40,...°/d
Jupiter	0; 4,59, 4, 1,19	0; 4,59,15,14,53	0; 4,59,15,27,...
Mars	0;31,26,27,26,34	0;31,26,38,40, 8	0;31,26,38,40, 5

The tables for the solar, lunar, and planetary equations differ in presentation from those in the Parisian Alfonsine Tables, even where the underlying parameters are the same. For the Sun the first column gives the distance from the solar apogee at intervals of 3° and the second column gives the algebraic sum of the solar equation and the argument in column 1. The maximum equation is 2;10° as in the Parisian version of the Alfonsine Tables. For the Moon the first column gives the argument of anomaly at intervals of 1° and the second column gives the corresponding equation of center. The maximum equation of center is 4;56° as in the Parisian version of the Alfonsine Tables. As noted above, both values appear in Spanish texts prior to 1300. The maximum values for the equation of center in Vimond's planetary tables are the same as in the *editio princeps* of the Alfonsine Tables:

	al-Battānī	Toledan Tables	Vimond	Ratdolt 1483
Sun	1;59,10°	1;59,10°	2;10°	2;10, 0°
Moon	5; 1, 0°	5; 0,59°	4;56°	4;56, 0°
Mercury	3; 2°	3; 2°	3; 2°	3; 2°
Venus	1;59°	1;59°	2;10°	2;10°
Mars	11;24°	11;24°	11;24°	11;24°
Jupiter	5;15°	5;15°	5;57°	5;57°
Saturn	6;31°	6;31°	6;31°	6;31°

For each of the five planets there are two tables of equations: one for the columns that depend on the argument of center, and one for the columns that depend on the argument of anomaly. In most zijes in the Ptolemaic tradition, there is only one such table for each planet, although this twofold presentation is already found in Andalusian and Maghribi zijes (e.g., the zij of Ibn Isḥāq of the early 13th century, described in Mestres 1996). So, in this case, the arrangement of Vimond's tables follows a tradition otherwise unattested in Latin prior to 1320, but we do find this twofold presentation for the planetary equations in a 14th-century manuscript containing a set of Alfonsine Tables (Erfurt, MS Q.362) and in a printed book: Alfonso de Córdoba's *Tabule astronomice Elisabeth Regine* (1503). The entries in Vimond's tables for the equation of center for Mercury, Mars, Jupiter, and Saturn are shifted by about 119°, 45°, 82°, and 162°, respectively, in relation to those in the zij of al-Battānī and the Toledan Tables. No such shift appears in the table for Venus. If we add the solar apogee (about 90°) to the values for the shifts listed above, we find that the planetary apogees are 209° for Mercury, 90° for Venus, 135° for Mars, 172° for Jupiter, and 252° for Saturn. In the Toledan Tables, the apogees of the Sun and of Venus are both 77;50° (Toomer 1968, p. 45), that is, about 12° less than 90°. Adding this difference of 12° to the planetary apogees in the Toledan Tables rounded to degrees, we find 210° for Mercury, 90° for Venus, 134° for Mars, 176° for Jupiter, and 252° for Saturn. The agreement of Vimond's data with the apogees in the Toledan Tables shows that Vimond has included the motion of the solar apogee (both precession and its proper motion) in the motions of the planetary apogees, thus following a theory for which there was no previous evidence outside al-

Andalus and the Maghrib. We know of no other set of planetary equation tables arranged in this way.

For each planet the table with argument of anomaly has 7 columns. The entries in column 2 give the correction due to the argument of anomaly at maximum distance in degrees and minutes and represent the difference between the equation of anomaly and the correction for maximum distance (cf. *Almagest*, XI.11, c_6 and c_5; and Neugebauer 1975, pp. 183–184). The entries in column 5 are given in minutes and seconds and result from adding the correction for maximum distance to the correction for minimum distance (cf. *Almagest*, XI.11, c_5 and c_7). The only other text of which we are aware that treats the equation of anomaly in this way is the zij of Ibn al-Bannā', and he applies this presentation in his tables for Saturn and Jupiter but not for the other planets (see Samsó and E. Millás 1998, pp. 278–285); this reinforces the connection with Andalusian and Maghribi sources. Ptolemy's eccentricities underlie the entries in these columns even though, in the case of Venus and Jupiter, the eccentricities were modified for computing the equation of center, as is also true for the Parisian Alfonsine Tables (cf. North 1976, 3:196).

For each planet there is a column in both tables for a component of velocity in minutes of arc per day. In the table for the equation of center the entries in this column are the same, but for rounding to minutes, as those in the column labeled *motus centri* or *motus puncti* in the table for planetary velocities. Analogously, in the table for the equation of anomaly the entries in the corresponding column are the same, but for rounding to minutes, as those in the column labeled *motus argumenti* in the table for planetary velocities. These columns for planetary velocities are described at length in the Castilian canons, chapter 27 (see our comments in chapter 4, above), but we have not found this presentation anywhere else.

The tables for the planetary latitudes, both for the superior and the inferior planets, are in the style of *Almagest* XIII.5, the zij of al-Battānī (Nallino 1903–1907, 2:140–141), and the Parisian Alfonsine Tables (Ratdolt 1483, f. h1v), as opposed to those in the *Handy Tables* and those in the zij of al-Khwārizmī. In all cases column 1 displays the argument at 12°-intervals from 0s 12° to 12s 0°. Column 2 is only found in the tables for the inferior planets: the entries, in minutes, give the latitudinal component due to the inclination of the plane of the deferent with respect to the ecliptic (called β_o in Neugebauer 1975, p. 224):

for Mercury, $\beta_o = -0;45 \cdot c_5$
and for Venus, $\beta_o = +0;10 \cdot c_5$

where c_5 is the column for sixtieths in the table for planetary latitude in *Almagest* XIII.5. We know of no other table before 1400 that has a column for this third component of latitude for the inferior planets, but we note that chapter 22 of the Castilian Alfonsine Tables mentions a table with such a column for Venus (22:12) and for Mercury (22:28). The inclination and the slant for the inferior planets are given in degrees and minutes in separate columns, whereas for each of the superior planets there are two columns that display the latitude, in degrees and minutes, under the assumption that the center of the epicycle is located at the extremal northern and southern latitudes of the deferent, respectively (Neugebauer 1975, p. 217). The extremal values in Vimond's tables agree with those in the Toledan Tables with two exceptions, one of which is a trivial variant for Mercury (a difference of 0;1°). But, as far as we know, the maximum value for the inclination of Venus in John Vimond's table (7;12°) is not attested in any previous text. It is probably significant that this value later appeared in the *editio princeps* of the Alfonsine Tables (1483).

Vimond's model for trepidation with two terms is identical with that underlying the Parisian Alfonsine Tables although the presentation is different. The mean motions of each component are displayed in the same way that Vimond used for the other mean motions, but the underlying parameters are slightly different from the standard tables as they appear in the *editio princeps* (Ratdolt 1483, f. d4r–v). The period we derive for the mean motion of the linear term is about 49,000 years to complete one revolution as in the standard Alfonsine model for trepidation, and it corresponds to the difference between the calendrical year of 365;15 days and a fixed tropical year. The value we derive for the periodic term is about 7,000 years to complete one revolution as in the standard Alfonsine model for trepidation (cf. Section 6.2.8). This periodic term corresponds to the difference between a variable sidereal year and the calendrical year of 365;15 days. However, in his other tables Vimond has used a fixed sidereal year: we are unable to account for this inconsistency. To be sure, Vimond's canons may have explained what he intended.

The equation of access and recess is rounded to minutes: it has a maximum of 17;17° for argument 204° and a minimum of –0;43° for argument 24°. These extremal values are exactly 18° apart (= 17;17° + 0;43°); hence the amplitude of the sinusoidal curve corresponding to Vimond's table is 9°. This is indeed the characteristic parameter of the table for the equation of access and recess in the Parisian Alfonsine Tables, whose maximum is 9° for argument 90°. Comparison of the entries in both tables shows that the curve representing John Vimond's table is the same as that representing the table used by other Parisian astronomers of his time but shifted in two ways: 247°

for the argument and –8;17° for the equation (see Table 6.1 and Fig. 6.2). In fact, the entries in Vimond's table can be derived from those in the Parisian Alfonsine Tables by taking an argument and its corresponding equation in the latter (where they are given to seconds) and then adding 113° to the argument and 8;17° to the equation. Moreover, in Vimond's table we find the same divergence from a sinusoidal function (here rounded to minutes) that is in the corresponding table in the Parisian Alfonsine Tables. Apparently, the value for the equation of access and recess that Vimond thought correct for his time was 8;17°, and he shifted the curve (i.e., the entries in the table) accordingly; indeed, calculation of the periodic term in trepidation with the parameters for 1320 in the Parisian Alfonsine Tables yields 8;17°:

1320 · 0;3,5,8,34,17°/y ≈	67;53°
radix Incarnation	359;13°
Total	67; 6°

and

$$9° \sin 67;6° = 8;17°.$$

Note that 67;6° + 180° = 247;6° or about 247°, and 360° – 247° = 113° which is the phase angle of the shift introduced by Vimond.

This table establishes a strong connection between Vimond and the Parisian Alfonsine Tables, for this theory of trepidation is not found in any previous text. But again, since the mean motions are different, we see no reason to assume that Vimond based his theory on the Parisian Alfonsine Tables.

In Vimond's star list the amount for the total precession is 17;52° which argues against dependence on the Parisian Alfonsine Tables. At a rate of 1° in 66 years, the total precession of 17;52° corresponds to 1179 years (= 66 · 17;52), and 1320 – 1179 = 141 AD, a plausible date for Ptolemy, indicating that no trepidation is required. Moreover, this list demonstrates that the Alfonsine corpus is not homogeneous with respect to star lists, and that it is excessively simple to assume that the Alfonsine Tables are characterized by the frequently discussed list with a total precession of 17;8°.

Table 6.1: Vimond's periodic term*

argumentum s	(°)	*motus* (°)
0	6	–0;19
0	12	–0;33
0	18	–0;41
0	24	–0;43
1	0	–0;39
1	6	–0;29
1	12	–0;13
1	18	0; 8
1	24	0;35
2	0	1; 6
...		
3	0	4;47
...		
6	0	16;34
6	12	17; 7
6	18	17;15
6	24	17;17
7	0	17;13
7	6	17; 3
...		
9	0	11;47
...		
11	18	0;55
11	24	0;19
12	0	0; 0

*There is no indication in the MS that some of the entries in this table are negative.

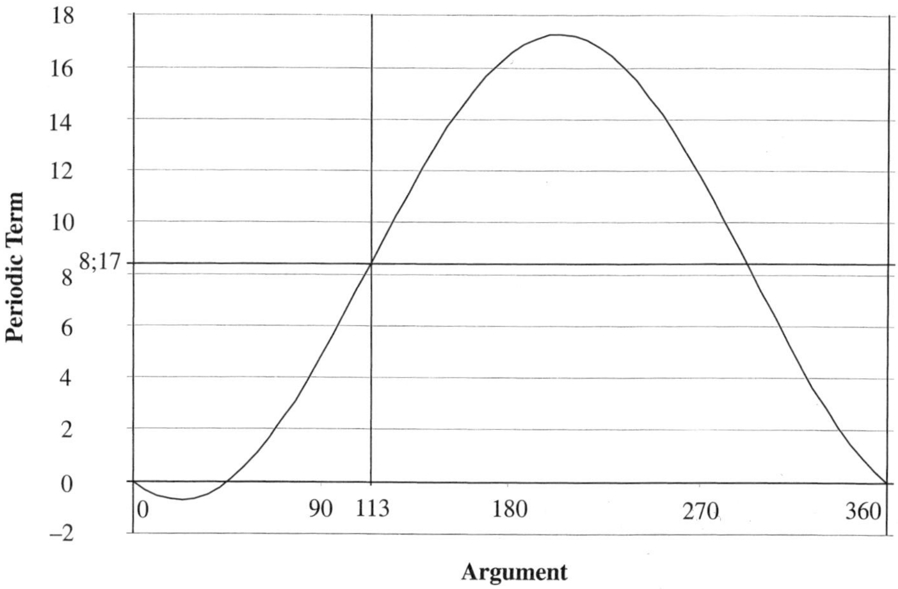

Fig. 6.2. A graph of John Vimond's periodic term for trepidation (i.e., the equation of access and recess): the entries in Vimond's table are shifted with respect to those in the corresponding standard Alfonsine table by –113° on the x-axis and by –8;17° on the y-axis.

6.3.2 John of Murs

John of Murs was born in the diocese of Lisieux in Normandy in about 1295, but the date is uncertain. A man with this name together with his father, Henri, attacked a cleric and killed him at midnight on Sept. 7, 1310. If this is the same John of Murs as the astronomer (and his association in the aftermath of this incident with a John of Lignères makes this identification almost certain), it seems most unlikely for him to have been born in 1300 or later, as some scholars had believed before the relevant documents were discovered. John of Murs, among others, was then banished to Cyprus for a period of 3 to 7 years, but it is not known if he actually went there (Huillier 1980).

He was an astronomer of note, for in 1344 Pope Clement VI called him to Avignon (then the papal residence) to discuss the reform of the calendar (Schabel 1996). John of Murs is not known to have engaged in any astronomical activity after 1348 but he may have survived another 10 years or so (Gushee 1969). In his earliest known astronomical work of 1317, beginning *Autores calendarii...*, John of Murs praises the tables of Toulouse and seems to be unaware of Alfonsine material (Poulle 1973a, p. 129).

The contributions of John of Murs to the Alfonsine corpus have been described by Poulle in several publications. Poulle gives a pivotal role to the *Expositio intentionis regis Alfonsii circa tabulas eius* in establishing the Parisian Alfonsine tradition, as noted above, and we have discussed many of its features in the previous section. The date of this treatise has been deduced by Poulle from a remark by John of Murs in another text, dated 1321, the *Canones tabule tabularum* (Poulle 1980, p. 244; cf. Gushee 1969, p. 7):

> *Eodemque anno artis musice noticia profunde figurande tam mensurabilis quam plane ad omnem modum discantandi non solum per integra sed usque ad minimas fractiones, cognitioque circuli quadrature perfectissime demonstrata, expositio quoque tabularum Alfoncii regis Castelle, ac calogia (sic) astronomie, nobis claruerunt, altissimo collaudato, que tanquam inaudita et ignota ceteris anni precedentibus quasi sopita in sapientie thezauro iacuerunt.*

In this passage John of Murs says that the *Expositio tabularum Alfoncii,* among other things, became clear to us (or became known to us: *nobis claruerunt*), where "us" means "me". At least two of the items mentioned here ("on squaring the circle" and "a genealogy [?] of astronomy") do not correspond to any extant work by John of Murs (Poulle 1973a, p. 128), and so it is by no means evident that John of Murs's intention was to present a list of his writings. Poulle interprets this passage to mean that John of Murs composed the *Expositio* in 1321, and that he "invented" the Alfonsine Tables at the same time, for there is no evidence of his knowledge of them in his previous works. We think the passage is ambiguous and can be interpreted to mean that John of Murs first became aware of a previous exposition of the Alfonsine Tables (not extant) in 1321, elaborating what he found there in his own version that may have been written somewhat later. In this treatise there are several dates with which computations or observations are associated, notably: 534, 1290, 1300, and 1319. We are told that in 1290 William of Saint-Cloud observed a vernal equinox (Poulle 1980, p. 262), and in 1319 John of Murs speaks in the first person of having observed a vernal equinox (pp. 265–266). The year 1300 appears several times, and in one place computations for the years 534 *perfecto* and 1300 *perfecto* (i.e., 1301) are presented in a discussion of precession in the 766 years between them (p. 259). Poulle explains the year 1300 *perfecto* as a "round" number (rather than a date shortly before the time of composition of an earlier version), but the year 534 *perfecto* remains unexplained. Perhaps a reason for it was given in John of Murs's source that has not survived. Duhem (1916, 4:19–21; cited in Poulle 1980, p. 214) ascribed all but the last few sections of the *Expositio* to William of Saint-Cloud, but the extant version is surely not by him, for in it he is called *vir magnus nomine* (p. 261), which is unlikely to be a self-designation. Moreover, as we have noted, a whole section of the *Expositio* is copied from Saint-Cloud's prologue to his *Almanach planetarum* (see Section 6.2.8, above); nevertheless, it is clear that John of Murs's text is different from his source.

The purpose of John of Murs's *Expositio* was to explain parameters and models that are already found in the tables he had in hand. We agree with North that in the *Expositio* he intended to be "faithful" to his Alfonsine source. North adds: "There are those who are uneasy with this new figure [2;10° as the maximum solar equation] and cannot believe that it is Castilian since they cannot find it in any *earlier* document — a very strange criterion" (North 1996, p. 461). The same can be said for John of Murs's other references to Alfonso.

In this work John of Murs uses the expressions "tables of Alfonso" (*Alfonsius ... in capite tabule motus solis*: Poulle 1980, p. 253) and "introduction to the tables of Alfonso" (*ut invenit Alfonsius causa introductionis tabularum eius...*: Poulle 1980, p. 256) to describe his source, but the manuscript(s) on which he depended are not known. According to him, this Alfonsine material contained a value for the length of the tropical year, a daily solar motion, a rate of precession, an equinox observation from the time of Alfonso, a value for the solar eccentricity, etc. The easiest way to account for John of Murs's attribution of data in the *Expositio* to Alfonso is to say that the Alfonsine material available to him included a Latin version of the Castilian canons, the tables which are described in them, and some additional material. We see no reason to assume that there was a major revision of the Castilian Alfonsine Tables, whether in Latin or Castilian, made in Toledo.

In previous accounts of John of Murs's *Expositio* it has not been remarked that a whole section depends (sometimes verbatim) on William of Saint-Cloud's prologue to his *Almanach planetarum* of 1292 (the relevant part is unpublished). As we indicated in Section 6.2.8, the *Expositio* includes a discussion of an observation by Saint-Cloud of a vernal equinox in 1290 and the details all agree with those in Saint-Cloud's text (note that the unique copy of the *Expositio* is in the very same manuscript as a copy of Saint-Cloud's text!). Moreover, John of Murs also copied other parts of Saint-Cloud's discussion including Saint-Cloud's computation of a solar position with a maximum solar equation of 1;59° as in the Toledan Tables (despite John of Murs's advocacy of 2;10° elsewhere in the *Expositio*) and a computation according to Pseudo-Thābit's model for trepidation. And a few sentences in the *Expositio* (p. 262), particularly that concerning the rates of precession according to Ptolemy, al-Battānī, and al-Ṣūfī, are taken almost word for word from Saint-Cloud. Since John of Murs was faithful to his source in the case of William of Saint-Cloud, we see no reason to doubt his faithfulness to his source in the case of "Alfonso".

Another passage in the *Expositio* (Poulle 1980, p. 264) reinforces our claim that John of Murs depended on "Alfonso". The radix of the solar motion for the epoch of Alfonso in the Parisian Alfonsine Tables (June 1, 1252) is 1,16;37,12,38,43° (≈ 76;37,13°), but John of Murs rejects this number, suggesting it is a scribal error for 76;17,13° (= 2s 16;17,13°) that he computed backwards from Saint-Cloud's observation of 1290 (here with a maximum solar equation of 2;10°), i.e., he believed that the difference of 0;20° was due to a corrupt reading in the text he had before his eyes (*Quod ideo dico quia in libris nostris et eis quos vidimus scriptum est 2[s] 16;37,13° quam de 20 minutis unius gradus estimamus corruptam vel ex ymaginatione ut pluries accidit aut ex vicio scriptoris mutatam ut, credens scribere 17, scripsit 37...*). It would be most peculiar for John of Murs to pretend he had a book before him that was corrupt and such a pretense is made even more unlikely from the fact that the Parisian Alfonsine Tables do not agree with John of Murs's preferred number but preserve the number he rejected (see, e.g., Poulle 1984, p. 124).

To summarize: the data in the *Expositio* which come from the Parisian Alfonsine Tables are: (1) the daily mean motion of the Sun; (2) the maximum solar equation; (3) the daily motion of the linear term for trepidation; (4) the value for the periodic term in trepidation for 1300 *perfecto*; and (5) the radix of the solar motion for the epoch of Alfonso. These data provide clear evidence for John of Murs's dependence on the Parisian Alfonsine Tables that were already in place.

For these and other reasons we consider John of Murs's *Expositio* to be a work of textual analysis: it is not a report of scientific discovery and the parameters in it had already appeared in the sources used by John of Murs. In particular, these parameters were not determined from observations cited in the *Expositio*.

In 1339, John of Murs composed canons for the Parisian Alfonsine Tables and he compiled various tables for epoch 1321 that were probably written at the end of the 1320s (Poulle 1973a, pp. 130, 133; North 1977, pp. 284–285; Poulle 1988, pp. 103–104). As was the case with John Vimond's tables, those by John of Murs are mainly devoted to conjunctions and oppositions of the Sun and the Moon, as well as to conjunctions of the Sun with the other planets. John of Murs is also the author of two other sets of tables, both focusing on syzygies. The first one is associated with the text beginning *Patefit ex Ptolomei disciplinis in libro suo...* and gives a list of mean conjunctions for 76 years beginning in 1321, as well as a list of 251 true conjunctions and 235 true oppositions (Lisbon, Biblioteca de Ajuda, MS 52-XII-35, ff. 9r–12v). The other set of tables is called *Tabulae permanentes* and it is associated with the text beginning *Omnis utriusque sexus armoniam celestem....* These tables address the difficult problem of finding the time between mean and true syzygies (Porres and Chabás 2001). According to the *explicit*, the *Tabulae permanentes* were the result of John de Murs's collaboration with Firmin of Bellavalle, who collaborated with John of Murs in another work, a treatise on calendar reform (Schabel 1996). Moreover, John of Murs and Firmin of Bellavalle each wrote a treatise on the conjunction of Jupiter and Saturn in 1345 (Goldstein and Pingree 1990, p. 7).

John of Murs made a number of astronomical observations between 1321 and 1344, preserved in Escorial, Biblioteca del Monasterio, MS O-II-10 that contains, among other items, the Toledan Tables and their canons (F. S. Pedersen 2002, pp. 114–115). The notes in the hand of John of Murs include observations of meridian altitudes of the Sun and of solar eclipses in 1333 and 1337. In these notes John of Murs refers to "Tabule Alfonsi" for the calculation of the times of eclipses (Gushee 1969; Poulle 1973a, p. 130; Beaujouan 1975). In other words, John of Murs, one of the principal figures in the compilation of the Parisian Alfonsine Tables, considered eclipse tables to be part of the Alfonsine Tables. But this view differs from the later manuscript tradition in which such tables were placed in the section ascribed to John of Lignères rather than in the section labeled Alfonsine Tables (see Section 6.4.3).

6.3.3 John of Lignères

John of Lignères was from the diocese of Amiens, and lived in Paris from about 1320 to 1335 (Poulle 1973b, p. 122). He compiled two sets of tables based on the Alfonsine Tables. The *Tabule magne* (*ca.* 1325: Poulle 1988, p. 103) are notable for the use of double argument equation tables for the planets (North 1977, pp. 273–274, 278); these tables are associated with the *incipit*, *Multiplicis philosophie variis radiis...*, and were addressed to the dean of Glasgow, Robert the Lombard of Florence, indicating a possible route of transmission of the Parisian Alfonsine Tables to England. In 1322 (a date that appears in the colophon of the manuscripts) John of Lignères compiled another set of tables with canons in 44 chapters for the prime mover (*Cuiuslibet arcus propositi sinum rectum...*), i.e., problems of trigonometry, the daily rotation, etc., and canons in 46 chapters for the motions of the planets and the determination of eclipses (*Priores astrologi motus corporum celestium...*). These canons have been edited, but not published, by Saby (1987). Both canons describe one set of tables computed for the meridian of Paris which have December 31, 1320 as epoch; these canons and tables are extant in many manuscripts. (For a list of these tables, see Section 6.4.3.) In his canons John of Lignères refers to Alfonso in the context of a star catalogue: *Scias quod stelle quarum loca sunt verificata sunt 1022, et earum loca ultimo per Alfoncium fuerunt verificata* (Saby 1987, pp. 273–274). We are also told that a certain Abraham Benthegar (variants: Benthegor, Benbthegar, Benhihegen, Ben Bahegen, etc.) composed canons "that have not yet been translated from Hebrew, although it seems to me from what I have heard that they are the best of all, with the exception of those of al-Battānī" (*Concordat similiter in isto secundo Habraham Benbthegar in canonibus suis qui de hebreo nondum sunt translati licet ut michi videtur per ea que audivi sint meliores omnibus aliis excepto Albategni*: Saby 1987, p. 237). The reference is unclear, for no one with this name has been identified, but surely John of Lignères has the Alfonsine Tables in mind. There are many difficulties with taking this Abraham to be Abraham Ben Waqār, a translator who worked for Alfonso, for he is not known to have composed any works in Hebrew (and the transformation of his name is difficult to explain): see Goldstein *et al.* 1994, pp. 65–66. As noted in chapter 5, Ben Waqār translated a work, *The Book of Muḥammad's Ladder (Miʿrāj Muḥammad)*, from Arabic into Castilian which survives in a French version completed *ca.* 1263 in which he is called "Habraym iuif et fisicien" (Procter 1945, p. 23), and he was probably the only collaborator of Alfonso that John of Lignères knew by name. In any event, it seems that John of Lignères intended to indicate that he admired tables from some place other than Paris. John of Lignères also wrote canons, *Quia ad inveniendum loca planetarum...* (undated, but later than 1322), that apply to tables of the same character as those described in the canons of John of Saxony and those of John of Murs (Poulle 1973b, pp. 124, 127; Poulle 1987, p. 56).

John of Lignères is also the author of an almanac that has never been studied or described by historians of astronomy. This work is uniquely preserved (as far as we know), in an incomplete copy, in Philadelphia, Free Library, MS Lewis E.3 (ff. 3r–10r). Folio 1 gives the name of the owner of the manuscript, and folio 2 is blank. The set of tables begins abruptly in the middle of a table for Saturn, and there are no tables for the positions of the Sun, the Moon, and the lunar node, as in other almanacs. The entries are rounded to the nearest degree, in zodiacal signs of 30°, and the years begin in January. Here is a brief description of the extant copy.

- f. 3r. Table for the true positions of Saturn with entries only for years labeled 40 to 59. The entries are given at 10-day intervals (days 1, 11, and 21). The first entry is Ari 13° (Jan. 1, Year 40) and the last is Sgr 19° (Dec. 21, Year 59).
- ff. 3v–5r. Table for the true positions of Jupiter. The entries are given at 10-day intervals (days 1, 11, and 21 of each month) for 83 years. Surprisingly, for years 62 to 83 the entries are given for days 10, 20, and the last day of each month. The first entry is Lib 14° (Jan. 1, Year 1) and the last is Lib 13° (Dec. 31, Year 83).
- ff. 5v–7r. Table for the true positions of Mars. The entries are given at 10-day intervals (days 10, 20, and the last day of each month) for 79 years. The first entry is Sgr 5° (Jan 1, Year 1) and the last is Sco 27° (Dec 31, Year 79).
- f. 7v. Table for the true positions of Venus. The entries are given at 5-day intervals (days 1, 6, 11, 16, 21, and 26 of each month) for 8 years. The first entry is Sgr 23° (Jan. 1, Year 1) and the last is Sco 23° (Dec. 26, Year 8).
- ff. 8r–10r. Table for the true positions of Mercury. The entries are given at 5-day intervals (days 5, 10, 15, 20, 25, and the last day of each month) for 46 years. The first entry is Cap 4° (Jan. 5, Year 1) and the last is Sgr 28° (Dec. 31, Year 46).

The number of years taken as a cycle for each of the planets is the same as in the Almanac of Azarquiel and the Almanac of Profatius Judaeus (Jacob ben Makhir). The frequency of the entries for each of the planets follows the Almanac of Azarquiel for Saturn and Venus, and the Almanac of Profatius Judaeus for Jupiter, Mars, and Mercury (Millás 1943–1950; and Boffito and Melzi d'Eril 1908).

- f. 10r. A short canon entitled: *Canon magistri Iohannis de Lineriis super almanach,* and beginning: *Subtrahe ab annis Christi 1340 annos Christi in tuo tempore currentibus et cum residuo intra in capite tabule.* Thus the starting date of the almanac of John of Lignères is 1341 (and not 1321, as stated by other historians) and this is confirmed by the recomputation of the entries with the Parisian Alfonsine Tables for the beginning of 1341.

The text also indicates that after a complete cycle the correction for Saturn is +1;30°, for Jupiter –0;30°, and for Mars +1;40°; none is given for Venus and Mercury. A comparison of these corrections with those in other almanacs follows:

	Almanac of J. of Lignères	Almanac of Profatius Judaeus	Almanac of Azarquiel
Saturn	59y + 1;30°	59y + 1;30°	59y + 1;55°
Jupiter	83y – 0;30°	83y – 0;30°	83y – 2°
Mars	79y + 1;40°	79y + 1;40°	79y + 1°

6.3.4 John of Saxony

John of Saxony (Johannes de Saxonia, also known as Johannes de Danecowe) was a student of John of Lignères (Poulle 1973c). His copy of John of Lignères's canons for the 1322 tables is extant (Erfurt, MS F.377) and dated 1323 (Saby 1987, p. 476). The canons of 1327 by John of Saxony, beginning *Tempus est mensura motus...*, were widely copied in the late Middle Ages, and included in the *editio princeps* of the Alfonsine Tables (reprinted, with minor modifications, in Poulle 1984). These canons explain the use of the tables but say nothing about their derivation.

Poulle has argued that the Alfonsine Tables, properly speaking, should only refer to tables described in the John of Saxony's canons. Nevertheless, we note that Poulle excluded from his "Alfonsine Tables", among others, the table for the equation of time, despite its specific mention by John of Saxony in canon 22 (*intra ... in tabulam equationis dierum cum suis noctibus*: Poulle 1984, pp. 5–6, 86). Poulle also excluded eclipse tables from his "Alfonsine Tables" but, as we noted in Section 6.3.2, in reports on the solar eclipses of 1333 and 1337 John of Murs refers to the Alfonsine Tables. Moreover, in John of Saxony's canons (Ratdolt 1483, f. a5v) the eclipse limits for both solar and lunar eclipses are given as ±12° from either node, citing the canons of Alfonso and those of John of Lignères with an explicit reference to chapter (*propositio*) 33. Surprisingly, the relevant sentence was omitted in Poulle's transcription with an unconvincing explanation (Poulle 1984, pp. 22, 60). For example, Poulle finds the reference to chapter 33 "erroneous", but in John of Lignères's *Priores astrologi*, the relevant passage is precisely in chapter 33 (Saby 1987). And it happens that the corresponding passage for solar eclipses is in chapter 33 of the Castilian canons. What seems to be new in this passage in John of Saxony's canons is the (unwarranted) claim for symmetry in the case of solar eclipses. Hence we consider Poulle's selection of tables unjustified and unsupported by the texts of the Parisian astronomers.

John of Saxony contributed to the diffusion of John of Lignères tables in a text known as *Exempla super tabulas primi mobilis et canones Johannes de Lineriis*, beginning *Quia plures astrologorum diversos libros fecerunt*... (*ca.* 1335). It is a collection of numerical applications of John of Lignères's text in 44 chapters for the prime mover (*Cuiuslibet arcus propositi*...) and it too is divided into 44 chapters.

John of Saxony also compiled an almanac as did John of Lignères. The text that explains its use has the incipit: *Cum animadverterem quamplurimos magistros et scholares in studio Parisiensi*.... The tables and/or the text are preserved in various manuscripts: Erfurt, MS F.386; Cambridge, Gonville and Caius College, MSS 110 and 174 (Thorndike and Kibre 1963, col. 282); Erfurt, MSS F.387 and 4° 360 (canons only; Poulle 1973c, p. 141); and Nuremberg, Stadtbibliothek, MS Cent. VI.16 (Kokott 1998, p. 133). To this list six other manuscripts can now be added: Oxford, Bodleian Library, MS Rawlinson D.1227, ff. 5r–31v; Cracow, Jagiellonian Library, MSS 604 (pp. 1–72), 715 (ff. 65r–67v), 1391 (pp. 6–83), and 1849 (1r–60r); and Vienna, Nationalbibliothek, MS Vin. 5192 (ff. 2r–51r).

The almanac lists the true positions of the luminaries and the planets, distributed in nine tables, covering the period 1336–1380. In some manuscripts the initial date is later: 1340 or even 1368. There follows a description of the almanac, based on MSS in Cracow, Oxford, and Vienna.

- The table for the solar correction is headed *Tabula continuationis motus solis Alfoncii* in Vienna, Nationalbibliothek, MS Vin. 5192 (f. 2r), and *Tabula continuationis veri motus solis* in Oxford, MS Rawlinson D.1227 (f. 5r). In fact, the entries in this table (see Table 6.2) are the increments to be added to the daily true longitudes of the Sun in the cycle 1336–1339 (see Table 6.3, below) in order to obtain the true solar longitudes for the period 1340–1387. It is assumed that the increment corresponding to the first day of the month is valid throughout that month.

The increments for the first year are obtained by taking the product of the solar velocity and the time difference between 4 tropical years and 4 Julian years. Since the solar velocity is not constant throughout the year, the increments vary accordingly. The mean increment for each 4-year interval, 0;1,45,32,40°, corresponds to a correction of 1° after about 136 years, as in Table AP 4 of the *Almanach Perpetuum* associated with Abraham Zacut (Chabás and Goldstein 2000, pp. 106–108). This correction is the same as the linear term in the Parisian model for trepidation that first appears in the works of John Vimond and John of Murs, and indicates that John of Saxony was using the length of the tropical year embedded in that model.

To check the entries in the table, we have calculated the solar longitudes for the first day of each month in 1336 and 1340. Table 6.3 shows these values, labeled C(1336) and C(1340), their difference in minutes, and the comparison with the entry in the text (T – C), also in minutes.

- The table for the daily positions of the Sun lists its true position for each day in a period of 4 years beginning in a leap year. All entries are given in zodiacal signs and degrees. No indication of the years is given in the headings, but years 1336–1339 yield full agreement. Table 6.4 displays the entries given in the text for the first day of each month, and the corresponding recomputed values for 1336 in Toledo. Since Paris is 12° = 0;48h to the east of Toledo (corresponding to about 0;2° in solar longitude), the computed values have to be lowered accordingly. With this adjustment, the differences between the text and the computed values vanish, except in three cases, where the difference is 0;1°.
- The table for the daily positions of the Moon lists its true positions for each day in a period whose length differs in the extant copies; for instance, Vienna, Nationalbibliothek, MS Vin. 5192 (ff. 6v-22r), has entries for 1364–1379, whereas Oxford, Bodleian Library, MS Rawlinson D.1227, covers the period 1361–1380. The entries are given in zodiacal signs and degrees and, as was the case with the solar positions, they were calculated according to the Parisian Alfonsine Tables.
- The table for the positions of the lunar node, in zodiacal signs and degrees, lists them at 10-day intervals (days 10, 20, and the last day of each month).
- The table for the true positions of Saturn, in zodiacal signs and degrees, lists them at 10-day intervals (days 10, 20, and the last day of each month).
- The table for the true positions of Jupiter, in zodiacal signs and degrees, lists them at 8-day intervals (days 8, 16, 24, and the last day of each month).
- The table for the true positions of Mars, in zodiacal signs and degrees, lists them at 6-day intervals (days 6, 12, 18, 24, and the last day of each month).

Table 6.2 Increments to the daily true longitudes of the Sun. The entries, in minutes, were copied from Vienna, Nationalbibliothek, MS Vin. 5192.

Year	Jan.	Feb.	Mar.	Apr.	May	Jun.
1340	1;46	1;45	1;45	1;46	1;47	1;48
1344	3;33	3;30	3;31	3;33	3;35	3;36
1348	5;17	5;14	5;16	5;19	5;22	5;23
1352	7; 3	6;59	7; 1	7; 6	7;10	7;11
1356	8;49	8;44	8;47	8;52	8;58	8;59
1360	10;35	10;29	10;32	10;19 *	10;45	10;40 *
1364	12;21	12;14	12;17	12;15	12;32	12;35
1368	14; 6	13;59	14; 3	14;12	14;20	14;23
1372	15;52	15;43	15;48	15;58	16; 7	16;10
1376	17;38	17;28	17;33	17;45	17;54	17;58
1380	19;24	19;13	19;19	19;31	19;42	19;46
1384	21;10	20;58	21; 4	21;18	21;29	21;34

Year	Jul.	Aug.	Sep.	Oct.	Nov.	Dec.
1340	1;47	1;47	1;46	1;45	1;45	1;46
1344	3;33	3;33	3;33	3;31	3;30	3;32
1348	5;20	5;20	5;19	5;16	5;15	5;17
1352	7; 7	7; 7	7; 5	7; 1	7;59 *	7; 3
1356	8;53	8;54	8;15 *	8;46	8;44	8;49
1360	10;40	10;40	10;38	10;32	10;29	10;35
1364	12;26	12;27	12;24	12;17	12;14	12;20
1368	14;13	14;14	14;10	14; 2	14;59 *	14; 6
1372	16; 0	16; 1	16;57 *	15;48	15;44	15;52
1376	17;46	17;43	17;43	17;33	17;28	17;38
1380	19;33	19;34	19;29	19;18	19;14	19;24
1384	21;20	21;21	21;15	21; 4	21;58 *	21; 9

* Based on the surrounding entries, this entry is faulty.

Table 6.3: Recomputation of the solar correction for 1340

Day	C(1336) (°)	C(1340) (°)	Diff. (min.)	T – C (min.)
Jan. 1	289; 7,49	289; 9,33	1;44	+0;2
Feb. 1	320;38,43	320;40,28	1;45	0
Mar. 1	349;40,41	349;42,26	1;45	0
Apr. 1	20; 7,37	20; 9,24	1;47	–0;1
May 1	49; 4,29	49; 6,16	1;47	0
Jun. 1	78;38, 2	78;39,49	1;47	+0;1
Jul. 1	107; 8,18	107;10, 6	1;48	–0;1
Aug. 1	136;45,29	136;47,16	1;47	0
Sep. 1	166;47,22	166;49, 8	1;46	0
Oct. 1	196;23,53	196;25,38	1;45	0
Nov. 1	227;34,39	227;36,23	1;44	+0;1
Dec. 1	258;10, 5	258;11,49	1;44	+0;2

Table 6.4: A comparison of the daily solar positions in the text with recomputation

Date	Text (°)	Comp. (°)
Jan. 1	289; 6	289; 7,49
Feb. 1	320;37	320;38,43
Mar. 1	349;39	349;40,41
Apr. 1	20; 6	20; 7,37
May 1	49; 2	49; 4,29
Jun. 1	78;36	78;38, 2
Jul. 1	107; 7	107; 8,18
Aug. 1	136;44	136;45,29
Sep. 1	166;46	166;47,22
Oct. 1	196;21	196;23,53
Nov. 1	227;33	227;34,39
Dec. 1	258; 8	258;10, 5

- The table for the true positions of Venus, in zodiacal signs and degrees, lists them at 4-day intervals (days 4, 8, 12, 16, 20, 24, 28, and the last day of each month).
- The table for the true positions of Mercury, in zodiacal signs and degrees, lists them at 4-day intervals (days 4, 8, 12, 16, 20, 24, 28, and the last day of each month).

In all cases the entries were computed from the Parisian Alfonsine Tables and, as was true for the Sun, they are well calculated to an accuracy of a minute of arc.

Vienna, Nationalbibliothek, MS Vin. 5192, f. 53r, reads: *Expliciunt tabule almanach cum canonibus quas magister Johannes Danck compilavit.*

John of Saxony's almanac has the same presentation as that of John of Lignères but there are some important differences: the entries here are computed with a greater precision (to minutes) and there are more of them. The almanac of John of Saxony is not organized according to the goal-year periods used by his predecessors, and no increments are given for each planet after these cycles, as is the case with all perpetual almanacs. This work by John of Saxony is closer to what was later called an "ephemeris", giving the positions of the luminaries and the planets for only a limited number of years.

6.3.5 John of Genoa and John of Montfort

According to Boudet (1997–1999, 1:480 n. 36), John of Genoa (Johannes de Janua) was a disciple of John of Lignères. He is the author of several works concerning the motion of the Sun and the Moon and eclipses.

His *Investigatio eclipsis solis anno Christi 1337*, extant in two manuscripts (Paris, Bibliothèque nationale de France, MS lat. 7281, ff. 208v–210; and Cambridge, University Library, MS Ee.III.61, ff. 75r–81r), contains a detailed computation, based on the Parisian Alfonsine Tables for the meridian of Paris, of the solar eclipse of March 3, 1337. This eclipse was observed by John of Murs (see Section 6.3.2, above), as well as by Levi ben Gerson (Goldstein 1979, see pp. 128–130). John of Genoa's *Canones eclipsium* are dated 1332, and, as the explicit says, they are based on the *Almagesti minoris libri VI* and the work by John of Sicily on the Toledan Tables (see Boudet 1994, p. 184; for John of Sicily, see F. S. Pedersen 1986). A short canon by John of Genoa entitled *Canon tabule ad sciendum motum solis et lune in una hora et semidiametros luminarium secundum tabulas Alfontii* is extant in Paris, Bibliothèque nationale de France, MSS lat. 7281, 7282, and 7322. Associated with this short canon is a table for lunar velocity at syzygy in which John of Genoa took account of Ptolemy's final lunar model (Goldstein 1992). On the other hand, the lunar velocity table in al-Battānī's zij that was widely copied in the Middle Ages is based on Ptolemy's simple lunar model (Nallino 1903–1907, 2:88).

A contemporary of John of Genoa, John of Montfort, wrote a *Tabula ad sciendum motum solis et lune* extant in Erfurt, MS Q.364, and Budapest, MS 62, the latter dated 1332 (Thorndike and Kibre 1963, col. 1551). The lunar velocities in this table are based on Ptolemy's final model, and it is very similar to the corresponding table by John of Genoa. A description of this table based on manuscripts preserved in Paris and Oxford was published by Goldstein (1992, pp. 11–13).

6.4 Beyond Paris

The Alfonsine Tables edited by the Parisian astronomers in the 1320s were used as the basis for composing astronomical treatises and tables in a variety of genres, and they were adapted to many other meridians. Indeed, in one form or another, the Alfonsine Tables dominated Latin and vernacular astronomy in Europe from the 14th to the 16th century, and they were unchallenged until Copernicus presented an alternative to the Ptolemaic system. Since many hundreds of manuscripts bear witness to this tradition (most of which have not been examined by historians of astronomy), we will restrict our attention to a few highlights in the period prior to 1400.

Most important for their diffusion is the fact that the almanacs by John of Lignères and John of Saxony gave rise to a vast number of ephemerides and calendars that were to form a significant part of the Alfonsine corpus. In this way the Alfonsine Tables were widely diffused in practical, "user-friendly" forms that served as effective working tools, in contrast to their "canonical" form that appealed to a more limited audience of specialists.

It is not difficult to find examples in the years following shortly after the 1320s. For instance, Vienna, Nationalbibliothek, MS Vin. 5145, ff. 1r–5v, beginning *Ad vera loca cuiuslibet planetarum per tabulas operari...*, contains "canones tabularum almanach" as stated in the explicit. The text gives worked examples for 1343, Saint John's day (June 24), and refers to the meridians of Paris and Erfurt (f. 2r) in what seems to be an explanation of John of Saxony's almanac. An example of an almanac compiled in England is preserved in a manuscript mentioned previously: Oxford, Bodleian Library, MS Rawlinson D.1227, ff. 47–49, which contains an "almanach pro anno 1357 super Oxon".

6.4.1 The Parisian Alfonsine Tables in England

From the detailed account by John North (1977), it is clear that the Alfonsine tables reached England via Paris.

The London Tables, associated with the *incipit, Ad noticiam tabularum sequencium secundum quod...* and the year 1336, are among the earliest witnesses of Alfonsine astronomy. These tables were adapted to the meridian of that city and have parameters already used by the Parisian astronomers, such as 2;10° for the maximum equation of the Sun and 5;57° for the maximum equation of center for Jupiter (North 1977, p. 276). The Parisian Alfonsine Tables reached many other centers of learning in England, but the first place where they were recast was Oxford. In particular, William Batecombe (about whom little is known) composed the Oxford Tables of 1348 which are associated with the canons beginning *Vera locum omnium planetarum in longitudine....* Inspired by John of Lignères's *Tabule magne*, the Oxford Tables of 1348 have double argument tables to facilitate computations of planetary positions in longitude and latitude (North 1977, pp. 278–284). William Rede of Oxford (*ca.* 1340) was another English astronomer who adapted the Parisian Alfonsine Tables.

As in Paris, calendars for sygygies were composed in Oxford. These are calendars which give the positions of the Sun and the length of daylight for each day in a year as well as other information, including the instants of true syzygies for various cycles of 19 years. In the second half of the 14th century several calendars based on the Alfonsine Tables were produced (North 1990, pp. 87–133). John Somer, a Franciscan friar, compiled in 1380 a *Kalendarium* for the latitude of Oxford (51;50°) with tables for the times of conjunctions (in days, hours, and minutes). This calendar covers the four 19-year cycles beginning in 1387 and ending in 1462 (Mooney 1998). Also in Oxford, Nicholas of Lynn, a Carmelite friar, wrote in 1386 a *Kalendarium* for Oxford based on the Alfonsine Tables and for the same period as Somer's, 1387–1462 (Eisner 1980). The *Kalendarium* by Nicholas of Lynn contains more astronomical information than Somer's and gives the times of true conjunctions and true oppositions (in days, hours, and minutes). This material was adapted for York by the Augustinian friar Richard de Thorpe for the same period (Mooney 1998, p. 14). All these calendars were in turn the continuation of the almanac, with a calendar beginning in 1301, compiled by Walter of Odington for Evesham Abbey, near Oxford, and covering the period 1292–1367 (North 1976, 3:238–252).

6.4.2 The Parisian Alfonsine Tables in Italy

In contrast to England, Alfonsine material followed a Mediterranean route to Italy. As already mentioned in chapter 5, the *Libro del saber de astrologia* was translated into the Florentine dialect of Italian (1341), and the manuscript containing this translation is preserved in Vatican, Biblioteca Apostolica, MS 8174 (Narducci 1864). The translation was made in Seville by Guerruccio, a member of the Florentine family Federighi (Beaujouan 1969, pp. 8–9). For an edition of this translation, see Levi (1993).

According to Dreyer (1920, p. 252), Jacopo Dondi of Padua (1298–1359) was the author of tables derived from those of Alfonso and in them the mean motions were given at intervals of 28 years. Jacopo Dondi was a clockmaker, as was his son Giovanni (1318–1389) who built one of the earliest mechanical clocks, possibly with assistance from his father (on both father and son, see Bedini and Maddison 1966), the instructions for which are extant in Padua, Biblioteca Capitolare, MS D.39. In this manuscript, recently published in facsimile (Dondi 1998), the tables by Alfonso are mentioned several times; e.g., on f. 5vb we are told that the solar apogee, computed according to Alfonso's tables for "anno Christo perfecto 1364", was at Gem 29;47°.

6.4.3 The Parisian Alfonsine Tables in Spain

Although the "tables by Alfonso" are often mentioned in the writings of various astronomers working in the Iberian peninsula in the 14th century, we know of no dated copy—or an example of their use—until the very end of that century. Indeed, the earliest such text we have found is Madrid, Biblioteca Nacional, MS 4238, a manuscript containing the canons by John of Saxony dated 1327 and the Parisian Alfonsine Tables adapted to Morella "in the kingdom of Valencia" (*in regno Valencie*), and it mentions the dates 1396 and 1400 (Chabás 2000). In contrast to the vast number of copies in libraries elsewhere in Europe, only a few manuscripts containing the Parisian version of the Alfonsine Tables have been preserved in Spanish collections; in addition to Madrid, Biblioteca Nacional, MS 4238, the only ones we have identified are the following:

1. Madrid, Biblioteca Nacional, MS 7856, is a 14th manuscript containing the tables (ff. 1r–22r) and the canons written by John of Saxony (1327) and beginning *Tempus est mensura...* (ff. 23r–32v). The *explicit* reads: *Et sic est finis canonum tabularum*

illustrissimi regis Alfoncii quos magister Johannes de Mekoys de Saxonia compilaverat deo. Jacobus de Perugia scripsit.

2. According to Millás (1942, pp. 227–230), Madrid, Biblioteca Nacional, MS 10002, is a manuscript written in a 15th-century hand, probably of Italian origin, and it gives radices for Naples and other Italian cities such as Bologna and Perusa. On f. 2v there is a small table for the entry into each of the zodiacal signs for 1364, without indicating the place, and on f. 1v, there is a worked example for 1407 to compute the number of days since the Incarnation.

3. Oxford, Bodleian Library, MS Can Misc. 27, is a 15th-century manuscript of Spanish origin, highly relevant for the history of astronomy in Spain. Among other items, it contains the canons and tables of Jacob ben David Bonjorn (1361); some tables associated with Ibn al-Kammād (12th century); the canons by John of Lignères beginning *Cujuslibet arcus propositi sinum rectum...* and *Priores astrologi motus corporum celestium...*; the *Tabulae Resolutae* for Salamanca (*ca.* 1460); and the canons to these tables by Nicholaus Polonius, the first incumbent of the chair of astronomy/astrology at the University of Salamanca (see Chabás 1998). Beginning on f. 88v there is also the set of tables dated 1322 usually attributed to John of Lignères.

We list below the tables in the manuscripts extant in Spain (Madrid, Biblioteca Nacional, MSS 4238, 7856, and 10002), or of Spanish origin (Oxford, Bodleian Library, MS Can. Misc. 27), indicating the corresponding table in the *editio princeps* by Ratdolt (1483) and in the edition of the Parisian Alfonsine Tables by Poulle (1984): see Table 6.5. The numbering until 32 follows that given by Poulle. In Madrid, Biblioteca Nacional, MS 4238, just after the table we have labeled 32, we read: *Hic finiunt tabule illustris Alfoncii regis Castelle deo gracias. Amen.* After this set of 32 tables, both Madrid, Biblioteca Nacional, MS 4238 and MS 10002, have other tables, and we have assigned them numbers beginning with 33. In Madrid, Biblioteca Nacional, MS 10002, these additional tables have a general title: *Incipiunt tabule magistri Iohannis de lineriis,* and correspond to those associated with the canons by John of Lignères (1322). As is evident from Table 6.6, most of the tables attributed to John of Lignères (beginning with No. 33) are of Toledan origin.

But the singular feature of Madrid, Biblioteca Nacional, MS 4238, is that it reproduces some of the tables of John Vimond (see Section 6.3.1), demonstrating that his tables had a certain diffusion.

On ff. 65r–66v we find a copy of his star list with a total precession of 17;52°. Very few star-names are given, and the signs used for the longitudes are physical signs of 60°. On f. 66v we are told that the radix for conjunctions is 13;54,54d and that for oppositions 14;45,55d, thus corroborating Vimond's data given above. On the same folio it is said that the radices were compiled for Paris and that they apply to the day of Saint Matthew of 1320 "secundum Romanos", in agreement with the interpretation given above. Then follows Vimond's table for mean conjunctions, although the entry for 2432 years is missing. Moreover, Madrid, Biblioteca Nacional, MS 4238, has both sub-tables for finding true syzygies, but the second sub-table has only 9 rows instead of 60.

Even though this list of manuscripts in Spain (or of Spanish origin) is very short and allowing for the discovery of other manuscripts still extant in Spain that may contain different versions of the Parisian Alfonsine Tables, it shows that the term "Alfonsine Tables" did not have a unique meaning for astronomers in the 14th century and later. Some considered the first 32 tables to be an independent set, different from the set of tables that follows them. Indeed, there are manuscripts in which the tables beginning with No. 33 are preceded by the general heading, "Tables of John of Lignères", although this is at best misleading, for the great majority of them are already found in earlier tables of Toledan origin (see Table 6.6). To be sure, we do not exclude the possiblity that John of Lignères participated in the transmission of some of these tables. Other astronomers considered the whole set of tables listed below as the "Alfonsine Tables", and this was the option chosen by their first editor, Ratdolt. We note that, in order to perform most astronomical computations, both sets are needed and that, taken together, they constitute a coherent ensemble. In addition to very different understandings of the "Alfonsine Tables", it is clear from inspecting even a few manuscripts that there were different copying traditions, for not all sets contain exactly the same tables and some entries in the same table differ from one manuscript to another or from one group of manuscripts to another. A comparative analysis of the hundreds of preserved manuscripts containing the Parisian Alfonsine Tables and those attributed to John of Lignères would be necessary to clarify the picture globally. The initial entries for most of the tables listed below are found in Rosińska (1984, pp. 488–498).

Table 6.5: The Parisian Alfonsine Tables in the manuscripts extant in Spain, or of Spanish origin, compared with two editions

No.	Title	Madrid, MS 4238	Madrid, MS 7856	Madrid, MS 10002	Oxford, MS Can. Misc. 27	Ratdolt 1483	Poulle 1984
1	Intervals between eras	24r	1r–v	1r–v		c1r–v	107–108
2	Calendaric I	24r–25r, 31r	1v, 2v	1v–2r		c2r	109
3	Calendaric II	24v–25v	1v, 2r–v	1v, 3v		c2v	110
4	Calendaric III	26r–v, 27v–28r	2r	2r–v		c3r	111
5	Calendaric IV	26v–27v	2r	2r–v		c3v	112
6	Calendaric V	28r–31r	3v–4r	3r–v		c4r–c6v	113–121
7	Weekdays	31v, 34v	3r	4r, 5r		c7r	122
8	Time conversions	32r–v	3r–v	4r–v		c7r–v	123
9	Radices	32v–34v	4r–5r	5r–6r		c8r–d1r	124–126
10	Radices for places other than Toledo	—	—	—		d1v–d2v	127–129
11	Revolutions of the Years and Months	see note	5r–v	see note		d3r	130
12	Equation of access and recess	35v–36r	5v	6r		d3v	131
13	Mean motion in access and recess	36r–37r	6r	6v		d4r	132
14	Mean motion of the apogees and fixed stars	37r–38r	6v	7r		d4v	133
15	Mean motion of the Sun	38r–v	7r	7v		d5r	134
16	Mean motion of the Moon	39r–v	7v	8r		d5v	135

No.	Title	Madrid, MS 4238	Madrid, MS 7856	Madrid, MS 10002	Oxford, MS Can. Misc. 27	Ratdolt 1483	Poulle 1984
17	Mean motion in lunar anomaly	39v–40r	8r	8v		d6r	136
18	Mean motion in lunar argument of latitude	40v–41r	8v	9r		d6v	137
19	Mean motion in elongation	42r–v	12r	10v		d7r	138
20	Mean motion of the lunar node	41r–v	11v	9v		d7v	139
21	Mean motion of Saturn	42v–43r	10r	10r		d8r	140
22	Mean motion of Jupiter	43v–44r	10v	11v		d8v	141
23	Mean motion of Mars	44r–v	11r	11r		e1r	142
24	Mean motion of Venus	45r–v	9v	12r		e1v	143
25	Mean motion of Mercury	45v–46r	9r	12v		e2r	144
26	Solar equation	46v–47v	12v–13r	13r–v		e2v–e3v	145–147
27	Lunar equation	46v–47v	13v–14v	14r–15r		e4r–e6v	148–153
28	Equation of Saturn	49v–51v	15r–16r	15v–16r		e7r–f1v	154–159
29	Equation of Jupiter	51v–53v	16v–17v	17r–18r		f2r–f4v	160–165
30	Equation of Mars	53v–55v	18r–19r	18v–19v		f5r–f7v	166–171
31	Equation of Venus	55v–57v	19v–20v	20r–21r		f8r–g2v	172–177
32	Equation of Mercury	57v–59v	21r–22r	21v–22v		g3r–g5v	178–183
33	Sines	60v–61r		23r–v	88v–89r	—	
34	Solar declination	—		24r	82v	—	
35	Shadows	—		24r	82v	—	
36	Ascensional differences	—		24v	88r	—	
37	List of 24 stars	—		24v		—	

38	Equation of time	61v		25r	79r	k1r–k2r	
39	Normed right ascension	—		25v	79v	k1r–k2r	
40	Oblique ascensions for the 1st *clima*	—		—	95v–96r	k2v–k3v	
41	Oblique ascensions for the 2nd *clima*	—		—	96v–97r	k4r–k5r	
42	Oblique ascensions for the 3rd *clima*	—		—	97v–98r	k5v–k6v	
43	Oblique ascensions for the 4th *clima*	—		—	98v–99r	k7r–k8r	
44	Oblique ascensions for the 5th *clima*	—		see note	100v–101r	k8v–l1v	
45	Oblique ascensions for the 6th *clima*	—		see note	101v–102r	l2r–l3r	
46	Oblique ascensions for the 7th *clima*	—		28r–v	102v–103r	l3v–l4v	
47	Solar meridian altitudes	—		31r–35r	—	—	
48	Planetary latitudes	—		35v	90v	h1v	
49	*Tabula bipertialis*	—		36r–v	108r–109r	—	
50	*Tabula quadripertialis*	—		37r	109v–110r	—	
51	Lunar latitude	—		37v	91r	h1r	
52	Planetary velocities	62r		38r	90r	—	
53	Planetary retrogradations	—		38v	78v	—	
54	Phases of the three outer planets	—		38v	91v	—	

No.	Title	Madrid, MS 4238	Madrid, MS 7856	Madrid, MS 10002	Oxford, MS Can. Misc. 27	Ratdolt 1483	Poulle 1984
55	Phases of the two inner planets	—		39r	91v	—	
56	Distances between the planets	—		39r		—	
57	Mean conjunctions at intervals of 24 years (1321–1609)	—		39v	92r	—	
58	Mean oppositions at intervals of 24 years (1321–1609)	—		39v	92r	—	
59	Mean syzygies in years	—		40r	92v	—	
60	Mean syzygies in months	—		40r	92v	—	
61	Mean motion in elongation	—		41r		—	
62	Solar and lunar equations and hourly velocities	62v–63r		41v–42v	80r–81r	—	
63	Solar and lunar velocities in a minute of a day	63v		43r–v	81v–82r	—	
64	Solar and lunar hourly velocities	63v		44r	88r	—	

65	Parallax for the 1st *clima*	—		—	83r	15r	
66	Parallax for the 2nd *clima*	—		—	83v	15v	
67	Parallax for the 3rd *clima*	—		—	84r	16r	
68	Parallax for the 4th *clima*	—		—	84v	16v	
69	Parallax for the 5th *clima*	—		44v	85v	17r	
70	Parallax for the 6th *clima*	—		45r	86r	17v	
71	Parallax for the 7th *clima*	—		45v	86v	18r	
72	Table for solar eclipses at greatest distance	—		46r	93v–94r	m1v	
73	Table for solar eclipses at least distance	—		46r	93v–94r	m1v	
74	Table for lunar eclipses at greatest distance	—		46v	94r–95r	m2r	
75	Table for lunar eclipses at least distance	—		46v	94r–95r	m2v	
76	*Tabula actatium*	—		47r	93r	18v	
77	Area of the solar or lunar disk obscured during an eclipse (in digits)	—		47r	94r	—	

No.	Title	Madrid, MS 4238	Madrid, MS 7856	Madrid, MS 10002	Oxford, MS Can. Misc. 27	Ratdolt 1483	Poulle 1984
78	*Tabula reflexionis tenebrarum*	—		47r	94r	—	
79	Entry of the Sun into the 12 signs	—		47r		—	
80	Proportions at 2°-intervals	—		47v	93r	m1r	
81	Geographical coordinates	—		48r	—	—	
82	Solar and lunar semidiameters	60r		—	—	—	

Notes

Table 1: These intervals between the eras are given in sexagesimal days. Madrid, Biblioteca Nacional, MS 4238, only gives the last 7 intervals (compared to 45 in Ratdolt 1483), and adds the decimal number of days in each case. MS 7856 only lists 44 intervals, whereas MS 10002 gives 43 intervals, and adds the decimal number of days for each interval.

Table 6: In Madrid, Biblioteca Nacional, MS 10002, sub-table 4 is missing.

Table 9: In Madrid, Biblioteca Nacional, MS 4238, some sub-tables are missing.

Table 10: This table is missing in all the Madrid manuscripts.

Table 11: Madrid, Biblioteca Nacional, MS 4238, f. 35r–v, and MS 10002, f. 4v, have entries that differ from those in the *editio princeps*. See Tables 48A and 48B in chapter 4, above.

Table 12: The heading in Madrid, Biblioteca Nacional, MS 4238, is *Tabula equacionis accessus et recessus capitis arietis seu octaue spere* and in MS 10002, *Equatio accessus*. We note that in both MSS 4238 and 10002, the entry for 60° is 7;47,7° rather than 7;47,10°, as in MS 7856 and the *editio princeps*.

Table 14: Madrid, Biblioteca Nacional, MS 10002, adds 2 extra sexagesimal places to each entry (e.g., the first entry is 0 0 0 0 4 20 41 17 12 26 30), and MS 7856 adds one extra place (0 0 0 0 4 20 41 17 12 27).

Table 27: See Table 17A in chapter 4, above. In Madrid, Biblioteca Nacional, MS 4238, the maximum for the *diversitas diametri* begins at 1s 42°.

Table 28: See Table 18A in chapter 4, above. In Madrid, Biblioteca Nacional, MS 4238, the maximum for the *longitudo longior* ends at 1s 47° as in MSS 7856 and 10002, and the maximum for the *longitudo propior* begins at 1s 43° and ends at 1s 52°.

Table 29: In Madrid, Biblioteca Nacional, MSS 4238 and 10002, the maximum for the equation of anomaly begins at 1s 39° and ends at 1s 43°, and the last entries for the minutes of proportion are 59, rather than 60. In MS 4238 the maximum for the *longitudo propior* begins at 1s

46°. In MS 10002 the equation of anomaly for 2s 59° is 0;19°, rather than 0;15°.

Table 31: In Madrid, Biblioteca Nacional, MS 4238, the value for the equation of the anomaly for 1° is 0;36° as is the case in MS 10002, and the maximum for the equation of center begins at 1s 28° as is also the case in MSS 7856 and 10002.

Table 32: In Madrid, Biblioteca Nacional, MS 4238, the maximum for the minutes of proportion begins at 1s 52°.

Table 33: The heading in Madrid, Biblioteca Nacional, MS 10002, is *Tabula primi mobilis*, and in MS 4238 *Tabula sinus arcus augmentat per dimidium gradum*.

Table 38: See Table 29A in chapter 4, above.

Tables 40–46: In Madrid, Biblioteca Nacional, MS 10002, two consecutive folios are numbered 28. This manuscript has tables for 40°, 42°, 45°, 48°, 50°, and 52°. In Oxford, MS Can. Misc. 27, ff. 99v–100r, there is also a table for the latitude of Toledo, given as 39;54°, and another one for latitude 50°.

Tables 49–50: These tables for planetary latitude appear in the Toledan Tables, and they are in the tradition of al-Khwārizmī (Toomer 1968, pp. 69–70; Suter 1914, pp. 138–167).

Table 52: See the comments on chapter 27 of the canons to the Castilian Alfonsine Tables in chapter 4, above.

Table 54: Almost the same table is found in Madrid, Biblioteca Nacional, MS 10002, f. 40v.

Table 55: Almost the same table is found in Madrid, Biblioteca Nacional, MS 10002, f. 40v.

Table 56: In Madrid, Biblioteca Nacional, MS 10002, only two of the four columns have a heading: distance from Jupiter to Saturn in months (column 1) and distance from Mars to the Sun (column 3).

Table 69: In Oxford, MS Can. Misc. 27, f. 85r, there is also a table for the latitude of Toledo, given as 39;54°.

Table 76: Madrid, Biblioteca Nacional, MS 10002, has only the column headed *circulus brevis* in Toomer (1968, pp. 116–117) in a separate table, and Oxford, MS Can. Misc. 27, has both the table *actatium* and the separate column. This table is used for interpolation in the tables for computing eclipses (Nos. 72–75) and, as noted by Toomer, a number of manuscripts containing the Toledan Tables also give this column separately. On this table, see also Chabás and Goldstein 2000, p. 120.

The Parisian Alfonsine Tables are almost entirely of Toledan origin: in Table 6.6 we list those tables in the Almanac of Azarquiel or in the Toledan Tables which are identical to, or share the same format with, the Parisian Alfonsine Tables associated with John of Lignères. For the Almanac of Azarquiel the reference is given to the page numbers in Millás (1943–1950), and for the Toledan Tables to the page numbers in Toomer (1968) where explanations of the contents of many tables are found. In Table 6.6 a dash indicates that the Parisian Alfonsine Tables do not contain a table (both with respect to the entries and the format) corresponding to any of those in either the Almanac of Azarquiel or the Toledan Tables. In many cases, there are similar tables in both sets, Toledan and Parisian, used for the same purpose, but with different arrangements and different entries (e.g., for weekdays, for planetary retrogradation, for geographical coordinates, for star coordinates, etc.).

Table 6.6: The Parisian Alfonsine Tables and their relationship to tables of Toledan origin

No.	Title	AA: Almanac of Azarquiel TT: Toledan Tables
1	Intervals between eras	Partially in TT: p. 18
2	Calendaric I	TT: p. 17
3	Calendaric II	—
4	Calendaric III	—
5	Calendaric IV	—
6	Calendaric V	—
7	Weekdays	—
8	Time conversions	—
9	Radices	—
10	Radices for places other than Toledo	—
11	Revolutions of the Years and Months	—
12	Equation of access and recess	—
13	Mean motion in access and recess	—
14	Mean motion of the apogees and fixed stars	—
15	Mean motion of the Sun	—
16	Mean motion of the Moon	—
17	Mean motion in lunar anomaly	—
18	Mean motion in lunar argument of latitude	—
19	Mean motion in elongation	—
20	Mean motion of the lunar node	—
21	Mean motion of Saturn	—
22	Mean motion of Jupiter	—
23	Mean motion of Mars	—
24	Mean motion of Venus	—
25	Mean motion of Mercury	—
26	Solar equation	Same format as TT: p. 56
27	Lunar equation	Same format as TT: pp. 58–59
28	Equation of Saturn	Same format as TT: p. 61
29	Equation of Jupiter	Same format as TT: p. 62

30	Equation of Mars	Same format as TT: p. 63–64
31	Equation of Venus	Same format as TT: p. 65–66
32	Equation of Mercury	Same format as TT: p. 67–68
33	Sines	TT: p. 29
34	Solar declination	TT: pp. 30–31
35	Shadows	AA: p. 230; TT: p. 32
36	Ascensional differences	AA: p. 225
37	List of 24 stars	—
38	Equation of time	TT: pp. 34–35
39	Normed right ascension	TT: pp. 34–35
40	Oblique ascensions for the 1st *clima*	TT: pp. 36–43
41	Oblique ascensions for the 2nd *clima*	TT: pp. 36–43
42	Oblique ascensions for the 3rd *clima*	TT: pp. 36–43
43	Oblique ascensions for the 4th *clima*	TT: pp. 36–43
44	Oblique ascensions for the 5th *clima*	TT: pp. 36–43
45	Oblique ascensions for the 6th *clima*	TT: pp. 36–43
46	Oblique ascensions for the 7th *clima*	TT: pp. 36–43
47	Solar meridian altitudes	—
48	Planetary latitudes	TT: pp. 71–72
49	*Tabula bipertialis*	TT: p. 69
50	*Tabula quadripertialis*	TT: p. 70
51	Lunar latitude	TT: pp. 58–59
52	Planetary velocities	Toledan Corpus: see chapter 4, 27:27
53	Planetary retrogradations	—
54	Phases of the three outer planets	AA: p. 219; TT: pp. 73–74
55	Phases of the two inner planets	AA: p. 219; TT: pp. 74–76
56	Distances between the planets	—
57	Mean conjunctions for intervals of 24 years (1321–1609)	Same format as TT: p. 78
58	Mean oppositions for intervals of 24 years (1321–1609)	Same format as TT: p. 78

No.	Title	AA: Almanac of Azarquiel TT: Toledan Tables
59	Mean syzygies in years	Same format as TT: p. 79
60	Mean syzygies in months	Same format as TT: p. 79
61	Mean motion in elongation	—
62	Solar and lunar equations and hourly velocities	TT: pp. 56, 58, 82
63	Solar and lunar velocities in a minute of a day	—
64	Solar and lunar velocities in an hour	TT: p. 82
65	Parallax for the 1st *clima*	TT: pp. 98–112
66	Parallax for the 2nd *clima*	TT: pp. 98–112
67	Parallax for the 3rd *clima*	TT: pp. 98–112
68	Parallax for the 4th *clima*	TT: pp. 98–112
69	Parallax for the 5th *clima*	TT: pp. 98–112
70	Parallax for the 6th *clima*	TT: pp. 98–112
71	Parallax for the 7th *clima*	TT: pp. 98–112
72	Table for solar eclipses at greatest distance	AA: p. 234 TT: pp. 86–90
73	Table for solar eclipses at least distance	AA: p. 234 TT: pp. 86–90
74	Table for lunar eclipses at greatest distance	AA: p. 231 TT: pp. 91–96
75	Table for lunar eclipses at least distance	AA: p. 231 TT: pp. 91–96
76	*Tabula actatium*	AA: p. 233; TT: p. 116
77	Area of the solar or lunar disk obscured during an eclipse (in digits)	AA: p. 233 TT: p. 113
78	*Tabula reflexionis tenebrarum*	TT: p. 115
79	Entry of the Sun into the 12 signs	—
80	Proportions at 2°-intervals	TT: p. 117
81	Geographical coordinates	—
82	Solar and lunar semidiameters	Same format as TT: pp. 157–158

BIBLIOGRAPHY

I. Editions of the Alfonsine Tables until the middle of the 16th century

1. "Canonical" Alfonsine Tables

1483: *Tabule astronomice illustrissimi Alfontij regis castelle*. E. Ratdolt, Venice.
1492: *Tabule Astronomice Alfonsi Regis*. J. L. Santritter, Venice.
1518/1521: *Tabule Astronomice Divi Alfonsi Regis Romanorum et Castelle*. P. Liechtenstein, Venice.
1524: *Alphonsi Hispaniarum Regis Tabule*. L. Gaurico, Venice.
1545, 1553: *Divi Alphonsi Romanorum et Hispaniarum Regis astronomicae tabulae*. P. Du Hamel, Paris.

2. Other presentations of the Alfonsine Tables: the *Tabulae resolutae*

1511: *Tabulae resolutae de motibus planetarum aliorumque super celestium mobilium*. A. Lacher, Frankfurt.
1536, 1551: *Tabulae Astronomicae, quas vulgo (...) resolutas vocant*. J. Schöner, Nuremberg.

II. References

Aaboe, A. 1964. *Episodes From the Early History of Mathematics*. Washington, D.C.
Ausejo, E. 1984. "Trigonometría y astronomía en el *Tratado del Cuadrante Sennero* (c. 1280)", *Dynamis*, 4:7–22.
Ballesteros Beretta, A. 1984. *Alfonso X el Sabio*. Barcelona (1st edition: Barcelona 1963).

Beaujouan, G. 1968. "Manuscrits scientifiques médiévaux de la cathédrale de Ségovie", *Actes du XIe Congrès international d'histoire des sciences*. Reprinted in Beaujouan 1992, Essay IV.

Beaujouan, G. 1969. "L'Astronomie dans la péninsule ibérique à la fin du moyen âge", *Revista da Universidade de Coimbra*, 24:3–22. Reprinted in Beaujouan 1992, Essay X.

Beaujouan, G. 1975. "Observations et calculs astronomiques de Jean de Murs (1321–1344)", in *Proceedings of the XIVth International Congress of the History of Science (Tokyo–Kyoto 1974)*, 2:27–30. Reprinted in Beaujouan 1991, Essay VII.

Beaujouan, G. 1991. *Par raison des nombres: L'art du calcul et les savoirs scientifiques médiévaux*. Aldershot.

Beaujouan, G. 1992. *Science médiévale d'Espagne et d'alentour*. Aldershot.

Bedini, S. A. and F. R. Maddison 1966. *Mechanical Universe: The* Astrarium *of Giovanni de' Dondi*. Transactions of the American Philosophical Society, 56.5. Philadelphia.

Beer, J. (ed.) 1989. *Medieval translators and their craft*. Kalamazoo.

Benjamin, F. S., Jr., and G. J. Toomer 1971. *Campanus de Novara and medieval planetary theory. Theorica Planetarum*. Madison–London.

Biblioteca Apostolica Vaticana 2000. *Tratado de astrología y magia* [A facsimile of Vatican, Biblioteca Apostolica, MS Reg. lat. 1283]. Valencia.

Biblioteca Nacional 1984. *Inventario General de Manuscritos de la Biblioteca Nacional*, vol. 10. Madrid.

Blecua, A. 1987. *Manual de crítica textual*. Madrid.

Boffito, J. and C. Melzi d'Eril 1908. Almanach *Dantis Aligherii sive Profhacii Judaei Montispessulani* Almanach Perpetuum, *ad annum 1300 inchoatum*. Florence.

Bossong, G. 1978. *Los Cánones de Albateni*. Tübingen.

Boudet, J.-P. 1994. *Lire dans le ciel*. Brussels.

Boudet, J.-P. 1997–1999. *Le "Recueil des plus celebres astrologues" de Simon de Phares*, 2 vols. Paris.

Burns, R. I. (ed.) 1990. *Emperor of Culture. Alfonso X the Learned of Castile and His Thirteenth-Century Renaissance*. Philadelphia.

Calvo, E. 1990. "La lámina universal de ᶜAlī b. Jalaf (s. XI) en la versión alfonsí y su evolución en instrumentos posteriores", in Comes, Mielgo, and Samsó 1990, pp. 221–238.

Calvo, E. 1998. "Astronomical Theories Related to the Sun in Ibn al-Hā'im's *al-Zij al-Kāmil fi'l-Taᶜālīm*", *Zeitschrift für Geschichte der Arabisch-Islamischen Wissenschaften*, 12:51–111.

Calvo, E. 2000. "A Study of the Use of Ibn Bāṣo's Universal Astrolabe Plate", *Archives internationales d'histoire des sciences*, 50:264–295.

Cantera Burgos, F. 1931. "El judío salmantino Abraham Zacut", *Revista de la Academia de Ciencias Exactas, Físicas y Naturales de Madrid*, 27:63–398.

Cantera Burgos, F. 1935. *Abraham Zacut*. Madrid.

Cárdenas, A. J. 1974. *A Study and Edition of the Royal Scriptorium Manuscript of* El Libro del Saber de Astrología *by Alfonso X, El Sabio*. University of Wisconsin, Ph. D. dissertation. University Microfilms, No. 74–24,715.

Casanovas, J. 1987. "On the precession problem in the Alfonsine Tables", in Comes, Puig, and Samsó 1987, pp. 79–87.

Casulleras, J. and J. Samsó (eds.) 1996. *From Baghdad to Barcelona: Studies in the Islamic Exact Sciences in Honour of Prof. Juan Vernet*. Barcelona.

Chabás, J. 1996a. "Astronomía andalusí en Cataluña: Las Tablas de Barcelona", in Casulleras and Samsó 1996, pp. 477–525.

Chabás, J. 1996b. "El almanaque perpetuo de Ferrand Martines (1391)", *Archives internationales d'histoire des sciences*, 46:261–308.

Chabás, J. 1997. "Le cahier d'astronomie d'un croisier du XVe siècle", *Nuncius*, 1:3–16.

Chabás, J. 1998. "Astronomy in Salamanca in the Mid-Fifteenth Century: The *Tabulae Resolutae*", *Journal for the History of Astronomy*, 29:167–175.

Chabás, J. 2000. "Astronomía alfonsí en Morella a finales del siglo XIV", *Cronos: Cuadernos Valencianos de Historia de la Medicina y de la Ciencia*, 3:381–391.

Chabás, J. 2003. "Were the Alfonsine Tables of Toledo first used by their authors?" *Centaurus* (in press).

Chabás, J. and B. R. Goldstein 1994. "Andalusian Astronomy: *al-Zīj al-Muqtabis* of Ibn al-Kammād", *Archive for History of Exact Sciences*, 48:1–41.

Chabás, J. and B. R. Goldstein 1997. "Computational Astronomy: Five centuries of Finding True Syzygy", *Journal for the History of Astronomy*, 28:93–105.

Chabás, J. and B. R. Goldstein 2000. *Astronomy in the Iberian Peninsula: Abraham Zacut and the Transition from Manuscript to Print*. Transactions of the American Philosophical Society, 90.2. Philadelphia.

Chabás, J. and B. R. Goldstein 2003. "John Vimond and the Alfonsine trepidation model", *Journal for the History of Astronomy* (in press).

Charette, F. 1998. "A Monumental Medieval Table for Solving the Problems of Spherical Astronomy for all Latitudes", *Archives internationales d'histoire des sciences*, 48:11–64.

Clark, W. E. 1930. *The Āryabhaṭīya of Āryabhaṭa*. Chicago.
Comes, M. 1990. "Al-Ṣūfī como fuente del libro de la 'Ochaua Espera' de Alfonso X", in Comes, Mielgo, and Samsó 1990, pp. 11–113.
Comes, M. 1992. "A propos de l'influence d'al-Zarqālluh en Afrique du nord: l'apogée solaire et l'obliquité de l'écliptique dans le *zīdj* d'Ibn Isḥāq", *Actas del II coloquio hispano-marroquí de ciencias históricas, Granada*. Madrid, pp. 147–159.
Comes, M. 1992–1994. "The 'Meridian of Water' in the Tables of Geographical Coordinates of al-Andalus and North Africa", *Journal for the History of Arabic Science*, 10:41–51.
Comes, M. 1996. "The Accession and Recession Theory in al-Andalus and the North of Africa", in Casulleras and Samsó 1996, pp. 349–364.
Comes, M. 2001. "Ibn al-Hā'im's Trepidation Model", *Suhayl*, 2:291–408.
Comes, M., H. Mielgo, and J. Samsó 1990. *"Ochava espera" y "astrofísica"*. Barcelona.
Comes, M., R. Puig, and J. Samsó 1987. *De Astronomia Alphonsi Regis*. Barcelona.
Córdoba, Alfonso de. 1503. *Tabule astronomice Elisabeth Regine*. Petrus Liechtenstein, Venice.
Craddock, J. R. 1974. "La nota cronológica inserta en el prólogo de las 'Siete Partidas'", *al-Andalus*, 39:363–389.
Daumet, G. 1913. *Mémoire sur les relations de la France et de la Castille de 1255 à 1320*. Paris.
Delambre, J. B. J. 1819. *Histoire de l'astronomie du moyen âge*. Paris.
Diman, R. C. and L. W. Winget 1980. *Lapidario and Libro de las formas & ymagenes*, Madison.
Dondi, G. 1998. *Horologium Amicorum (Emmanuel Poulle)* [A facsimile ed. of] *L'astrarium de Giovanni Dondi: Padoue, Bibliothèque capitulaire, ms D. 39*. Paris
Dorce, C. 1999. "Sobre el cuadrante solar de Alfonso X el Sabio", *Asclepio*, 52:167–184.
Dreyer, J. L.E. 1890. *Tycho Brahe: A Picture of Scientific Life and Work in the Sixteenth Century*. Edinburgh. Reprinted New York 1963.
Dreyer, J. L.E. 1920. "On the Original Form of the Alfonsine Tables", *Monthly Notices of the Royal Astronomical Society*, 80:243–262.
Duhem, P. 1916. *Le système du monde*, vol. 4. Paris.
Eisner, S. 1980. *The* Kalendarium *of Nicholas of Lynn*. London.
Forcada, M. 1990. "El *Picatrix*, fuente del *Lapidario*", in Comes, Mielgo, and Samsó 1990, pp. 209–220.
Freimann, A. (ed.) 1924. *A. Zacut: Sefer Yuḥasin [The Book of Genealogies]*. Frankfurt a/M.

Gandz, S., J. Obermann, and O. Neugebauer 1956. *Maimonides: Sanctification of the New Moon*. New Haven.

Géraud, H. 1843. *Chronique latine de Guillaume de Nangis de 1113 à 1300*, 2 vols. Paris. Reprinted New York 1965.

Gil, J. S. 1985. *La Escuela de Traductores de Toledo y los Colaboradores Judíos*. Toledo.

Goldberg, B. and L. Rosenkranz (eds.) 1846–1848. *Liber Jesod olam seu Fundamentum mundi, Auctore R. Isaac Israeli*, 2 vols. Berlin.

Goldstein, B. R. 1967. *Ibn al-Muthannā's Commentary on the Astronomical Tables of al-Khwārizmī*. New Haven

Goldstein, B. R. 1971. *Al-Biṭrūjī: On the Principles of Astronomy*, 2 vols. New Haven.

Goldstein, B. R. 1974. *The Astronomical Tables of Levi ben Gerson*. Transactions of the Connecticut Academy of Arts and Sciences, 45. New Haven.

Goldstein, B. R. 1976. "Refraction, Twilight, and the Height of the Atmosphere", *Vistas in Astronomy*, 20:105–107. Reprinted in Goldstein 1985, Essay IX.

Goldstein, B. R. 1977a. "Ibn Muᶜādh's Treatise on Twilight and the Height of the Atmosphere", *Archive for History of Exact Sciences*, 17:97–118.

Goldstein, B. R. 1977b. "Remarks on Ptolemy's Equant Model in Islamic Astronomy", in Y. Maeyama and W. G. Saltzer (eds.), *Prismata, Festschrift für Willy Hartner*. Wiesbaden, pp. 165–181. Reprinted in Goldstein 1985, Essay VII.

Goldstein, B. R. 1979. "Medieval Observations of Solar and Lunar Eclipses," *Archives Internationales d'Histoire des Sciences*, 29:101–156. Reprinted in Goldstein 1985, Essay XVII.

Goldstein, B. R. 1985. *Theory and Observation in Ancient and Medieval Astronomy*. London.

Goldstein, B. R. 1992. "Lunar Velocity in the Ptolemaic Tradition", in P. M. Harman and A. E. Shapiro (eds.), *The investigation of difficult things: Essays on Newton and the history of the exact sciences*, Cambridge, pp. 3-17.

Goldstein, B. R. 1994. "Historical Perspectives on Copernicus's Account of Precession", *Journal for the History of Astronomy*, 24:189–197.

Goldstein, B. R. 1998. "Abraham Zacut and the Medieval Hebrew Astronomical Tradition", *Journal for the History of Astronomy*, 29:177–186.

Goldstein, B. R. 2001. "The Astronomical Tables of Judah ben Verga", *Suhayl*, 2:227–289.

Goldstein, B. R. 2003a. "An Anonymous Zij in Hebrew for 1400 A.D.: A Preliminary Report", *Archive for History of the Exact Sciences*, 57:151–171.

Goldstein, B. R. 2003b. "Ancient and Medieval Values for the Mean Synodic Month", *Journal for the History of Astronomy*, 35:65–74.

Goldstein, B. R. and J. Chabás 2001. "The Maximum Solar Equation in the Alfonsine Tables", *Journal for the History of Astronomy*, 32:345–348.

Goldstein, B. R., J. Chabás, and J. L. Mancha 1994. "Planetary and Lunar Velocities in the Castilian Alfonsine Tables", *Proceedings of the American Philosophical Society*, 138:61–95.

Goldstein, B. R. and D. Pingree 1990. *Levi ben Gerson's Prognostication for the Conjunction of 1345*, Transactions of the American Philosophical Society, 80.6. Philadelphia.

Golius, J. (ed.) 1669. *Muhammedis fil. Ketiri Ferganensis, qui vulgo Alfraganus dicitur,* Elementa astronomica, *arabice & latine.* Amsterdam.

Gonzálvez, R. 1997. *Hombres y Libros de Toledo (1086–1300)*. Madrid.

Gushee, L. 1969. "New Sources for the Biography of Johannes de Muris", *Journal of the American Musicological Society*, 22:3–26.

Hilty, G. (ed.). 1954. *Aly Aben Ragel: El libro conplido en los iudizios de las estrellas*. Madrid.

Hilty, G. 1955. "El libro conplido en los iudizios de las estrellas", *al-Andalus*, 20:1–74.

Huiller, G. l' 1980. "Aspects nouveaux de la biographie de Jean de Murs", *Archives d'histoire doctrinale et littéraire du mayen âge*, 47:272–276.

Hyatte, R. 1997. *The Prophet of Islam in Old French*. Leiden.

Kasten, L. A. 1990. "Alfonso el Sabio and the Thirteenth-Century Spanish Language", in Burns 1990, pp. 33–45.

Kasten, L. A. and L. B. Kiddle (eds.) 1961. *Alfonso El Sabio: Libro de las Cruzes*. Madrid-Madison.

Kasten, L. A., J. Nitti, and W. Jonxis-Henkemans 1997. *The Electronic Texts and Concordances of the Prose Works of Alfonso X, El Sabio*. 7 pp. + 1 CD-ROM. Madison.

Kennedy, E. S. 1956a. *A Survey of Islamic Astronomical Tables*. Transactions of the American Philosophical Society, 46.2. Philadelphia.

Kennedy, E. S. 1956b. "Parallax Theory in Islamic Astronomy", *Isis*, 47:33–53. Reprinted in King and Kennedy 1983, pp. 164–184.

Kennedy, E. S. 1958. "The Sasanian Astronomical Handbook *Zīj-i Shāh* and the Astrological Doctrine of 'Transit' (*Mamarr*)", *Journal of the American Oriental Society*, 78:246–262. Reprinted in King and Kennedy 1983, pp. 319–335.

Kennedy, E. S. 1971. "Al-Bīrūnī's Masudic Canon", *al-Abhath*, 24:59–81. Reprinted in King and Kennedy 1983, pp. 573–595.

Kennedy, E. S. 1997. "The Astronomical Tables of Ibn al-Raqqām, a Scientist of Granada", *Zeitschrift für Geschichte der Arabisch-Islamischen Wissenschaften*, 11:35–72.

Kennedy, E. S. and M. Destombes 1966. Introduction to: [al-Ṣūfī's] *Kitāb al ᶜAmal bil-Asṭurlāb*. Hyderabad-Deccan. Reprinted in King and Kennedy 1983, pp. 405–447.

Kennedy, E. S. and M. H. Kennedy 1987. *Geographical Coordinates of Localities from Islamic Sources*. Frankfurt a/M.

Kennedy, E. S. and H. Krikorian-Preisler 1972. "The Astrological Doctrine of Projecting the Rays", *al-Abhath*, 25:3–15. Reprinted in King and Kennedy 1983, pp. 372–384.

King, D. A. 1973. "Ibn Yūnus' *Very Useful Tables* for Reckoning Time from the Sun", *Archive for History of the Exact Sciences*, 10:342–394. Reprinted in King 1986, Essay IX.

King, D. A. 1986. *Islamic Mathematical Astronomy*. London.

King, D. A. 1987. "Some Early Islamic Tables for Determining Lunar Crescent Visibility", in King and Saliba 1987, pp. 185–225.

King, D. A. 1988. "Ibn Yūnus on lunar crescent visibility", *Journal for the History of Astronomy*, 19:155–168.

King, D. A. and M. H. Kennedy (eds.) 1983. *Studies in the Islamic Exact Sciences*. Beirut.

King, D. A. and G. Saliba (eds.) 1987. *From Deferent to Equant: A volume of studies in the history science in the ancient and medieval Near East in honor of E. S. Kennedy. Annals of the New York Academy of Sciences*, 500.

King, D. A. and J. Samsó, with a contribution by B. R. Goldstein, 2001. "Astronomical Handbooks and Tables from the Islamic World (750-1900): An Interim Report," *Suhayl*, 2:9-105.

Kokott, W. 1998. "Syzygies as pivots: an unusual mid-fifteenth-century working ephemeris", *Journal for the History of Astronomy*, 29:129–135.

Kremer, R. L., and J. Dobrzycki. 1998. "Alfonsine meridians: Tradition versus experience in astronomical practice *c*. 1500", *Journal for the History of Astronomy*, 29:187–199.

Kunitzsch, P. 1966. *Typen von Sternverzeichnissen in astronomischen Handschriften des zehnten bis vierzehnten Jahrhunderts*. Wiesbaden.

Kunitzsch, P. 1986. "The Star Catalogue Commonly Appended to the Alfonsine Tables", *Journal for the History of Astronomy*, 17:89–98.

Lértora, C. A. 1989. "El *Ars inveniendi eclipsim solis et lunae* copiado por Roberto Grosseteste", *Mathesis*, 5:371-394.

Levi, J. A. 1993. *Text and Concordance of the Biblioteca Apostolica Vaticana, MS. 8174, Libro di sapere di astronomia*. Madison.

Mancha, J. L. 1987. "Sobre la versión alfonsí del ecuatorio de Ibn al-Samḥ", in Comes, Puig, and Samsó 1987, pp. 117–123.

Mancha, J. L. 1990. "La versión alfonsí del *Fī hay'at al-ʿālam* (*De Configuratione Mundi*) de Ibn al-Hayṯam", in Comes, Mielgo, and Samsó 1990, pp. 133–207.

Mancha, J. L. 1991. "Astronomical Use of Pinhole Images in William of St.-Cloud's *Almanach Planetarum* (1292)", *Archive for History of Exact Sciences*, 43:275–298.

Mancha, J. L. 1998. "On Ibn al-Kammād's Table for Trepidation", *Archive for History of Exact Sciences*, 52:1–11.

Martí, R. and M. Viladrich 1983. "En torno a los tratados de uso del astrolabio hasta el siglo XIII en al-Andalus, la Marca Hispánica y Castilla", in Vernet 1983, pp. 9–74.

Martínez Gázquez, J. 1989. *Las tablas de los movimientos de los cuerpos çelestiales del Iluxtrisimo Rey don Alonso de Castilla seguidas de su Additio*. Murcia.

McCluskey, S. C. 1998. *Astronomies and Cultures in Early Medieval Europe*. Cambridge.

Mercier, R. 1977. "Studies in the Medieval Conception of Precession (Part II)", *Archives internationales d'histoire des sciences*, 26:33–71.

Mercier, R. 1987. "Astronomical Tables in the Twelfth Century", in C. Burnett (ed.), *Adelard of Bath: An English Scientist and Arabist of the Early Twelfth Century*. London, pp. 87–118.

Mestres, A. 1996. "Maghribī Astronomy in the 13th Century: a Description of Manuscript Hyderabad Andra Pradesh State Library 298", in Casulleras and Samsó 1996, pp. 383–443.

Mestres, A. 1999. *Materials Andalusins en el Zīj d'Ibn Isḥāq al-Tūnisī*. Doctoral Thesis, University of Barcelona.

Millás, E. 1983. "Las posiciones de los apogeos planetarios en el ecuatorio de Azarquiel", in Vernet 1983, pp. 119–123.

Millás, J. M. 1942. *Las traducciones orientales en los manuscritos de la Biblioteca de la Catedral de Toledo*. Madrid.

Millás, J. M. 1943–1950. *Estudios sobre Azarquiel*. Madrid–Granada.

Millás, J. M. 1947. *El libro de los fundamentos de las Tablas astronómicas de R. Abraham ibn ʿEzra*. Madrid–Barcelona.

Millás, J. M. 1956. "Una nueva obra alfonsí: el Tratado del cuadrante 'sennero'", *al-Andalus*, 21:59–92. Reprinted in Millás 1960, pp. 221–258.

Millás, J. M. 1960. *Nuevos estudios sobre historia de la ciencia española*. Barcelona.

Mondéjar, Marqués de. 1777. *Memorias historicas del Rei. D. Alonso el Sabio, i observaciones a su crónica: obra postuma de D. Gaspar Ibañez de Segovia, Peralta i Mendoza..., Marques de Mondéjar*. Madrid.

Mooney, L. R. 1998. *The* Kalendarium *of John Somer*. Athens, Georgia.

Morelon, R. 1987. *Thābit Ibn Qurra: oeuvres d'astronomie*. Paris.

Muñoz Sendino, J. 1949. *La Escala de Mahoma: traducción del árabe al castellano, latín y francés, ordenada por Alfonso X el Sabio*. Madrid.

Nadir, N. 1960. "Abū al-Wafā' on the solar altitude", *The Mathematics Teacher*, 53:460–463. Reprinted in King and Kennedy 1983, pp. 280–283.

Nallino, C. A. 1903–1907. *Al-Battānī sive Albatenii Opus Astronomicum*, 2 vols. Milan.

Narducci, E. 1864. "Intorno ad una traduzione italiana fatta nell'anno 1341 di una compilazione astronomica di Alfonsi X. re di Castiglia." *Giornale Arcadico*, 42:81-112.

Neugebauer, O. 1949. "The Astronomy of Maimonides and Its Sources", *Hebrew Union College Annual*, 22:321–363. Reprinted in Neugebauer 1983, pp. 381–423.

Neugebauer, O. 1956. "An Astronomical Commentary", in Gandz *et al.* 1956, pp. 113–149.

Neugebauer, O. 1962a. *The Astronomical Tables of al-Khwārizmī*. Copenhagen.

Neugebauer, O. 1962b. "Thābit ben Qurra 'On the solar year' and 'On the motion of the eighth sphere'", *Proceedings of the American Philosophical Society*, 106:264–299.

Neugebauer, O. 1969. *The Exact Sciences in Antiquity*. 2nd ed. New York.

Neugebauer, O. 1975. *A History of Ancient Mathematical Astronomy*. Berlin.

Neugebauer, O. 1983. *Astronomy and History*. New York and Berlin.

North, J. D. 1976. *Richard of Wallingford*, 3 vols. Oxford.

North, J. D. 1977. "The Alfonsine Tables in England", in Y. Maeyama and W. G. Salzer, (eds.), *Prismata: Festschrift für Willy Hartner*. Wiesbaden, pp. 269–301. Reprinted in North 1989, pp. 327–359.

North, J. D. 1986. *Horoscopes and History*. London.

North, J. D. 1989. *Stars, Minds and Fate: Essays in Ancient and Medieval Cosmology*. London.

North, J. D. 1990. *Chaucer's Universe*. Oxford. (First printing: 1988)

North, J. D. 1994. *The Fontana History of Astronomy and Cosmology*. London.

North, J. D. 1996. "Just whose were the Alfonsine Tables?", in Casulleras and Samsó 1996, pp. 452–475.

O'Callaghan, J. F. 1990. "Image and Reality: The King Creates His Kingdom", in Burns 1990, pp. 14–32.

Oppolzer, T. von 1887. *Canon der Finsternisse*. Vienna.

Pedersen, F. S. 1983–1984. *Petri Philomenae de Daciae et Petri de S. Audomaro, Opera quadrivalia.* Corpus philosophorum Danicorum medii aevi, 10.1–2. Copenhagen.

Pedersen, F. S. 1984. "A Paris astronomer of 1290", *Cahiers de l'Institut du moyen-âge grec et latin*, 48:163–188.

Pedersen, F. S. 1986. "Scriptum Johannis de Sicilia super canones Azarchelis de tabulis Toletanis", *Cahiers de l'Institut du moyen-âge grec et latin*, 51:1–128 and 52:1–268.

Pedersen, F. S. 1998. "The Toulouse Tables: A List of Manuscripts", *Cahiers de l'Institut du moyen-âge grec et latin*, 68:3–12.

Pedersen, F. S. 1999."Editing Numerals: The Toledan Tables, 13th Century", *Classica et Medievalia*, 50:259–271.

Pedersen, F. S. 2002. *The Toledan Tables: A review of the manuscripts and the textual versions with an edition*. Copenhagen.

Pedersen, O. 1974. "Peter Philomena of Dacia", in *The Dictionary of Scientific Biography*, 10:540–542. New York.

Petri, W. 1968. "Colours of Lunar Eclipses According to Indian Tradition", *Indian Journal of History of Science*, 3:91–98.

Pingree, D. 1968a. *The Thousands of Abū Maᶜshar*. London.

Pingree, D. 1968b. "The Fragments of the Works of Yaᶜqūb ibn Ṭāriq", *Journal of Near Eastern Studies*, 27:97–125.

Pingree, D. 1976. "The Indian and Pseudo-Indian Passages in Greek and Latin Astronomical and Astrological Texts", *Viator*, 7:141–195.

Pingree, D. 1981. "Between the *Ghāya* and the *Picatrix*. The Spanish Version", *Journal of the Warburg and Courtauld Institutes*, 44:27–56.

Porres, B. and J. Chabás 2001. "John of Murs's *Tabulae permanentes* for finding true syzygies", *Journal for the History of Astronomy*, 32:63–72.

Poulle, E. 1973a. "John of Murs", in *The Dictionary of Scientific Biography*, 7:128–133. New York.

Poulle, E. 1973b. "John of Lignères", in *The Dictionary of Scientific Biography*, 7:122–128. New York.

Poulle, E. 1973c. "John of Saxony", in *The Dictionary of Scientific Biography*, 7:139–141. New York.

Poulle, E. 1980. "Jean de Murs et les tables alphonsines", *Archives d'histoire doctrinale et littéraire du moyen âge*, 47:241–271.

Poulle, E. 1981. "William of Saint-Cloud", in *The Dictionary of Scientific Biography*, 14:389–391. New York.

Poulle, E. 1984. *Les tables alphonsines avec les canons de Jean de Saxe*. Paris.

Poulle, E. 1987. "Les Tables Alphonsines sont-elles d'Alphonse X?", in Comes, Puig, and Samsó 1987, pp. 51–69.

Poulle, E. 1988. "The Alfonsine Tables and Alfonso X of Castille", *Journal for the History of Astronomy*, 19:97–113.

Poulle, E. 1994. "Un témoin de l'astronomie latine du XIIIe siècle, les tables de Toulouse", in *Comprendre et maîtriser la nature au moyen âge: Mélanges d'histoire des sciences offerts à Guy Beaujouan*, Geneva and Paris, pp. 55–81.

Poulle, E. and D. Savoie 1998. "La survie de l'astronomie alphonsine", *Journal for the History of Astronomy*, 29:201–207.

Procter, E. S. 1945. "The Scientific Works at the Court of Alfonso X of Castille: The King and his Collaborators", *The Modern Language Review*, 40:12–29.

Procter, E. S. 1951. *Alfonso X of Castile*. Oxford.

Puig, R. 1987a. "La proyección ortográfica en el *Libro de la açafeha* alfonsí", in Comes, Puig, and Samsó 1987, pp. 125–138.

Puig, R. 1987b. *Los Tratados de Construcción y Uso de la Azafea de Azarquiel*. Madrid.

Puig, R. 2000. "The theory of the Moon in the *Al-Zīj al-Kāmil fī-l-Taᶜālīm* of Ibn al-Hā'im (*ca.* 1205)", *Suhayl*, 1:71–99.

Ragep, F. J. 1996. "Al-Battānī, Cosmology, and the Early History of Trepidation in Islam", in Casulleras and Samsó 1996, pp. 267–298.

Ratdolt, E. (ed.) 1483. *Tabulae astronomice illustrissimi Alfontij regis castelle*. Venice.

Real Academia Española 1960–1981. *Diccionario Histórico de la Lengua Española*. 2 vols. (15 fasc.) Madrid.

Rico Sinobas, M. 1863–1867. *Libros del Saber de Astronomía del Rey D. Alfonso X de Castilla*, 5 vols. Madrid.

Ridpath, I. (ed.) 1989. *Norton's 2000.0: Star Atlas and Reference Handbook*. Eighteenth edition. Harlow–New York.

Robbins, F. E. (ed. and trans.) 1940. *Ptolemy: Tetrabiblos*. London.

Rochberg-Halton, F. 1988. *Aspects of Babylonian Celestial Divination: The Lunar Eclipse Tablets of Enūma Anu Enlil*. Vienna.

Rodríguez Montalvo, S. 1981. *Alfonso X: "Lapidario"*. Madrid.

Romano, D. 1971. "Le opere scientifiche di Alfonso X e l'intervento degli ebrei", *Oriente e Occidente nel Medioevo: Filosofia e Scienze*. Rome, pp. 677–711.

Romano, D. 1992. *La ciencia hispanojudía*. Madrid.

Rosen, E. 1976. "The Alfonsine Tables and Copernicus", *Manuscripta*, 20:163–174.

Rosenthal, F. 1966. "'Life is short, the art is long': Arabic Commentaries on the First Hippocratic Aphorism", *Bulletin of the History of Medicine*, 40:226-245.

Rosińska, G. 1984. *Scientific Writings and Astronomical Tables in Cracow: A Census of Manuscript Sources (XIVth-XVIth Centuries)*. Wrocław.

Roth, N. 1985. "Jewish Translators at the Court of Alfonso X", *Thought*, 60:439–455.

Roth, N. 1990. "Jewish Collaborators in Alfonso's Scientific Work", in Burns 1990, pp. 59–71.

Sabra, A. I. 1987. "The Appropriation and Subsequent Naturalization of Greek Science in Medieval Islam: a Preliminary Statement", *History of Science*, 25:223–243.

Saby, M.–M. 1987. *Les canons de Jean de Lignères sur les tables astronomiques de 1321*. Unpublished thesis: Ecole Nationale des Chartes, Paris. A summary appeared as: "Les canons de Jean de Lignères sur les tables astronomiques de 1321", *École Nationale des Chartes: Positions des thèses*, pp. 183–190.

Sachau, C. E. (trans.) 1879. *Albīrūnī: Chronology of Ancient Nations*. London.

Sachau, C. E. (trans.) 1910. *Alberuni's India*. London.

Salem, S. I. and A. Kumar (trans.) 1991. *Science in the Medieval World: "Book of the Categories of Nations" by Ṣāᶜid al-Andalusī*. Austin.

Samsó, J. 1982. "El tratado alfonsí sobre la esfera", *Dynamis*, 2:57–73.

Samsó, J. 1983. "Notas sobre el ecuatorio de Ibn al-Samḥ", in Vernet 1983, pp. 105–118.

Samsó, J. 1984. "Algunas notas sobre el modelo solar y la teoría de la precesión de los equinoccios en la obra astronómica de Alfonso X", *Dynamis: Acta Hispanica ad Medicinae Scientiarumque Historiam Illustrandam*, 4:81–114.

Samsó, J. 1987a. "Alfonso X and Arabic Astronomy", in Comes, Puig, and Samsó 1987, pp. 23–38.

Samsó, J. 1987b. "On the Solar model and the Precession of the Equinoxes in the Alfonsine Zīj and its Arabic Sources", in G. Swarup *et al.* (eds.), *Proceedings of an International Astronomical Union Colloquium no. 91, held in New Dehli, India*. Cambridge, pp. 175–183. Reprinted in Samsó 1994, Essay XIX.

Samsó, J. 1990a. "El original árabe y la versión alfonsí del *Kitāb fī hay'at al-ᶜālam* de Ibn al-Hayṯam", in Comes, Mielgo, and Samsó 1990, pp. 115–131.

Samsó, J. 1990b. "Trepidation in al-Andalus in the 11th Century", in Samsó 1994, Essay VIII.

Samsó, J. 1992. *Las Ciencias de los Antiguos en al-Andalus*. Madrid.

Samsó, J. 1994. *Islamic Astronomy and Medieval Spain*. Aldershot.

Samsó, J. 1997. "Andalusian Astronomy in 14th Century Fez: *al-zīj al-muwāfiq* of Ibn ᶜAzzūz al-Qusanṭīnī", *Zeitschrift für Geschichte der Arabisch-Islamischen Wissenschaften*, 11:73–110.

Samsó J. and F. Castelló 1988. "An Hypothesis on the Epoch of Ptolemy's Star Catalogue According to the Authors of the Alfonsine Tables", *Journal for the History of Astronomy*, 19:115–120. Reprinted in Samsó 1994, Essay XX.

Samsó, J. and M. Comes 1988. "Al-Ṣūfī and Alfonso X", *Archives internationales d'histoire des sciences*, 38:67–76. Reprinted in Samsó 1994, Essay XVII.

Samsó, J. and E. Millás 1989. "Ibn al-Bannā', Ibn Isḥāq and Ibn al-Zarqālluh's Solar Theory", in Samsó 1994, Essay X.

Samsó, J. and E. Millás 1998. "The Computation of Planetary Longitudes in the *Zīj* of Ibn al-Bannā'", *Arabic Sciences and Philosophy*, 8:259–286.

Sánchez-Prieto, P. 1998. *Cómo editar los textos medievales. Criterios para su presentación gráfica*. Madrid.

Savage-Smith, E. 1985. *Islamicate Celestial Globes*. Washington, D.C.

Schabel, C. 1996. "John of Murs and Firmin of Beauval's Letter and Treatise on the Calendar Reform for Clement VI", *Cahiers de l'Institut du moyen-âge grec et latin*, 66:187–215.

Schroeter, J. F. 1923. *Spezieller Kanon der zentralen Sonnen- und Mondfinsternisse, welche innerhalb des Zeitraums von 600 bis 1800 n. Chr. in Europa sichtbar waren*. Kristiania.

Shore, L. A. 1989. "A Case Study in Medieval Nonliterary Translation: Scientific Texts from Latin to French", in Beer 1989, pp. 297–327.

Smart, W. M. 1965. *Text-Book on Spherical Astronomy*. Fifth edition. Cambridge.

Smith, A. M. 1992. "The Latin Version of Ibn Muᶜādh's Treatise 'On Twilight and the Rising of Clouds'", *Arabic Sciences and Philosophy*, 2:83–132.

Smith, A. M. 1996. *Ptolemy's Theory of Visual Perception: An English Translation of the Optics with Introduction and Commentary*. Transactions of the American Philosophical Society, 86.2. Philadelphia.

Smith, A. M. and B. R. Goldstein 1993. "The Medieval Hebrew and Italian Versions of Ibn Muᶜādh's Treatise 'On Twilight and the Rising of Clouds'", *Nuncius*, 8:611–643.

Stahlman, W. D. 1959. *The Astronomical Tables of Codex Vaticanus Graecus 1291*. Brown University, Ph. D. dissertation. University Microfilms, No. 62–5761.

Steele, J. M. 2000. *Observations and Predictions of Eclipse Times by Early Astronomers*. Dordrecht.

Suter, H. 1914. *Die astronomischen Tafeln des Muḥammad ibn Mūsā al-Khwārizmī*. Copenhagen.

Thorndike, L. 1948. "Astronomical Observations in Paris from 1312 to 1315", *Isis*, 38:200–205.

Thorndike, L. 1959. "Notes upon Some Medieval Astronomical, Astrological and Mathematical Manuscripts at Florence, Milan, Bologna and Venice", *Isis*, 50:33–50.

Thorndike, L. and P. Kibre 1963. *A catalogue of incipits of mediaeval scientific writings in Latin*. London.

Toomer, G. J. 1968. "A Survey of the Toledan Tables", *Osiris*, 15:5–174.

Toomer, G. J. 1984. *Ptolemy's Almagest*. New York.

Toomer, G. J. 1987. "The Solar Theory of Az-Zarqāl: An Epilogue", in King and Saliba 1987, pp. 513–519.

Vernet, J. 1952. *Contribución al estudio de la labor astronómica de Ibn al-Bannā'*. Tetuán.

Vernet, J. 1978. "Un texto árabe de la corte de Alfonso X el Sabio", *al-Andalus*, 43:405–421.

Vernet, J. (ed.) 1983. *Nuevos Estudios sobre la Astronomía Española en el Siglo de Alfonso X*. Barcelona.

Viladrich, M. 1987. "Una nueva evidencia de materiales árabes en la astronomía alfonsí", in Comes, Puig, and Samsó 1987, pp. 105–116.

Viladrich, M. and R. Martí 1983. "Sobre el *Libro dell Ataçir* de los *Libros del Saber de Astronomía* de Alfonso X el Sabio", in Vernet 1983, pp. 75–100.

Villuendas, M. V. 1978. "A Further Note on a Mechanical Treatise Contained in Codex Medicea Laurenziana Or. 152", *Journal for the History of Arabic Science*, 2:395–396.

Yeves, J. A. 1998. *Manuscritos españoles de la Biblioteca Lázaro Galdiano*, 2 vols. Madrid.

Zarco, J. 1924–1929. *Catálogo de los manuscritos castellanos de la Real Biblioteca de El Escorial*, 3 vols. Madrid.

NOTATION

α	true argument of anomaly
$\bar{\alpha}$	mean argument of anomaly
α	right ascension
α'	normed right ascension
β	celestial latitude
γ	ascensional difference
d	diurnal arc
$d(\lambda)$	length of daylight
δ	declination
Δ	increment
E	equation of time
ε	obliquity of the ecliptic
h	altitude
κ	true argument of center
$\bar{\kappa}$	mean argument of center
λ	true longitude
$\bar{\lambda}$	mean longitude
$n(\lambda)$	length of night
p	precession
π	lunar parallax
π_β	parallax in latitude
$\rho(\lambda)$	oblique ascension
$\sigma(\lambda)$	arc of setting
φ	terrestrial latitude
$\bar{\varphi}$	co-latitude
ω	argument of latitude

MANUSCRIPTS CITED

LIST OF PARAMETERS

INDEX

There are no entries for names, such as Alfonso X, that occur often. Also, entries in the glossary in chapter 3 have not been included in the index since they are already in alphabetical order.